TRANSITION METAL CHEMISTRY

TRANSITION METAL CHEMISTRY

A SERIES OF ADVANCES

Edited by

GORDON A. MELSON
Virginia Commonwealth University
Richmond, Virginia

BRIAN N. FIGGIS
The University of Western Australia
Nedlands, Western Australia

VOLUME 8

MARCEL DEKKER, INC.   New York and Basel

The Library of Congress Cataloged This Serial as Follows:

Transition metal chemistry. v. 1-
   New York, M. Dekker, 1965-
      v. illus. 24 cm.
   Editor: 1965-   R. L. Carlin.
   1. Transition metals. I. Carlin, Richard Lewis, 1935   ed.
QD172.T6T7   546   65-27431
Library of Congress   [4-1]
ISBN   0-8247-1656-6

MARCEL DEKKER, INC.
270 Madison Avenue, New York, New York 10016

Current printing (last digit):
10 9 8 7 6 5 4 3 2 1

PRINTED IN THE UNITED STATES OF AMERICA

Preface

Since the publication of *Transition Metal Chemistry* was suspended
with Volume 7, the subject has shown considerable changes in empha-
sis. Many workers, old and new alike, have turned from the tradi-
tional fields of physical inorganic chemistry toward metal organic
chemistry and biological inorganic chemistry. Each of these areas
is serviced by a review series of its own. However, the editors and
the publishers feel that there is a need to bring some features of
these fields together with the longer established subject matter
under a cohesive heading of Transition Metal Chemistry. It is with
that aim the series has been restarted with Volume 8.

The series will continue to include comprehensive reviews and
monographs at the research level. These reviews will be authored by
scientists active in their individual research areas and will seek
to cover transition metal chemistry widely and in-depth.

GORDON A. MELSON
BRIAN N. FIGGIS

Contributors

ALESSANDRO BENCINI Istituto per lo Studio della Stereochimica ed Energetica dei Composti di Coordinazione, Consiglio Nazionale delle Ricerche, Florence, Italy

DANTE GATTESCHI Faculty of Pharmacy, Institute of General and Inorganic Chemistry, University of Florence, Florence, Italy

FABRIZIO MANI Institute of General and Inorganic Chemistry, University of Florence, Florence, Italy

LUIGI SACCONI Institute of General and Inorganic Chemistry, University of Florence, Florence, Italy

ROBERT E. TAPSCOTT Department of Chemistry, University of New Mexico, Albuquerque, New Mexico

Contents

TRANSITION METAL CHEMISTRY

1

ESR Spectra of Metal Complexes of the First Transition Series in Low–Symmetry Environments

ALESSANDRO BENCINI

Consiglio Nazionale delle Ricerche
Florence, Italy

DANTE GATTESCHI

University of Florence
Florence, Italy

I. INTRODUCTION

Electron spin resonance (ESR) spectroscopy has become increasingly popular in recent years for the characterization of the electronic structure of transition metal complexes. Its field of application is now extremely broad, ranging from the study of metalloenzymes and metalloproteins to minerals, polymers, and metallozeolites. This wide variety of applications demands a thorough knowledge of the ESR spectra of simple model complexes in order to compare them with the complicated patterns observed, for instance, in the study of naturally occurring molecules.

In the initial development of the technique, simple complexes of relatively high symmetry were studied, and the results are now covered by many excellent textbooks [1-14] and review articles [15-20]. References to the current literature can be found in the *Chemical Society (London) Specialist Periodical Reports on Electron Spin Resonance,* and more concise reports are also published in *Analytical Chemistry* [21].

We wish to review here the ESR spectra of transiton metal complexes in low-symmetry environments, with the aim of showing how it

is possible to obtain a considerable insight into their electronic
properties by collecting accurate ESR data, and using sophisticated
spin Hamiltonian, ligand field, and molecular orbital calculations.
In order to keep the matter as tractable as possible, we shall limit
our interest to metals of the first transition series. We shall
consider in general only studies on solids, either single-crystal or
polycrystalline powders, since only in these cases can one meaning-
fully relate the spin Hamiltonian parameters to structure.

We shall consider only simple mononuclear complexes and shall
ignore the expanding field of spin-spin coupled systems [22-26]. In
general, we shall mention only "true" discrete complexes, mostly
with organic ligands, and shall ignore the subject of paramagnetic
impurities in ionic lattices. Also, we shall not mention the ESR
spectra of metalloenzymes and metalloproteins, which are covered by
some excellent reviews [27-33], and which would require a book by
themselves.

Of the possible oxidation states of a metal ion, only the most
important will be considered. Of course the expression "most important" is vague; by it we mean the oxidation states that are most
commonly encountered in classic transition metal chemistry, not
considering organometallic complexes in general.

Finally, we shall not touch upon electron spin relaxation
phenomena.

We shall treat in Sec. II the spin Hamiltonian formalism, showing all the subtleties that are required to extend it to the
interpretation of the low-symmetry transition metal complexes. The
restraints on the spin Hamiltonian parameters imposed by crystal
symmetry will be covered in Sec. III. Section IV will provide general solutions to the spin Hamiltonian, whereas in Sec. V the solutions for different S values will be given. Section VI will be
devoted to an outline of the computer analysis of single-crystal and
polycrystalline powder spectra. Sections VII and VIII will review
recent achievements in the molecular orbital and ligand field analyses of the spin Hamiltonian parameters. In Sec. IX a survey of the

experimental results of various metal ions in different oxidation
states will be provided.

II. THE SPIN HAMILTONIAN

The simplest approach to the interpretation of ESR spectra is that
of the spin Hamiltonian [34,35] formalism, which, as Griffith puts
it, "is a convenient resting place during the long trek from fundamental theory to the squiggles of an oscilloscope" [9]. Essentially,
the spin Hamiltonian is an operator equivalent that is designed to
act on spin coordinates only, giving the same results as the true
Hamiltonian, which may be difficult to handle, with respect to the
energies of the electronic and nuclear systems. The spin coordinates that are used are not necessarily the true spin coordinates,
but rather may be fictitious ones. Although essentially the same
as the true coordinates for systems with S = 1/2, they can be completely different in the case of S > 1/2 [1,9]. The choice of the
spin multiplicity to be used for interpreting the ESR spectra within
the spin Hamiltonian formalism is determined a posteriori by the
nature of the ESR spectra themselves.

Another advantage of the spin Hamiltonian formalism is that
many different interactions can easily be taken into consideration
by adding the appropriate terms to the Hamiltonian. A rather complete form of spin Hamiltonian, which can be used for the interpretation of the spectra of magnetically diluted complexes, neglecting
interactions involving the ligand nuclei, is

$$\underline{H} = \mu_B \underline{B} \cdot \underline{g} \cdot \underline{S} + \underline{I} \cdot \underline{A} \cdot \underline{S} + \underline{S} \cdot \underline{D} \cdot \underline{S} + \sum_{k=4}^{2S} \sum_{q=0}^{k} B_k^q O_k^q +$$

$$\underline{I} \cdot \underline{P} \cdot \underline{I} - \mu_N \underline{B} \cdot \underline{g}_N \cdot \underline{I} \quad \text{(k even)} \tag{1}$$

where μ_B and μ_N are the Bohr and nuclear magnetons, respectively; $\underline{B}$
is the magnetic flux density vector relative to the static magnetic
field; $\underline{S}$ and $\underline{I}$ are the electron and nuclear spin vector operators,
respectively; g, $\underline{A}$, $\underline{D}$, $\underline{P}$, and $\underline{g}_N$ are matrices; O_k^q are combinations
of spin operators equivalent to combinations of spherical harmonics,

and B_k^q are parameters. Although g, A, D, P, and g_N are often referred to as second-rank tensors, only D and P are true tensors; g, A, and g_N are not [1], since their behavior under a rotation of the reference frame cannot be represented in general by Eq. (2),

$$Y' = X \cdot Y \cdot X^{-1} \tag{2}$$

In Eq. (2), X is a rotation matrix, and Y and Y' stand for general tensors. In general, therefore, g, A, and g_N are not represented by symmetric matrices. This can present several difficulties in the use of the spin Hamiltonian, but these are easily overcome if one considers the $g^2 = \tilde{g}g$ and $A^2 = \tilde{A}A$ matrices, which are real symmetric and represent true second-rank tensors. The principal values and directions of these tensors are obtained by ESR experiment, yielding the moduli of g and A.

The terms that have been included in the spin Hamiltonian (1) represent the Zeeman interaction of electrons ($\mu_B B \cdot g \cdot S$); the electron spin-nuclear spin interaction ($I \cdot A \cdot S$), often referred to as hyperfine interaction; the electron spin-electron spin interaction ($S \cdot D \cdot S + \sum_{k=4}^{2S} \sum_{q=0}^{k} B_k^q O_k^q$), often referred to as fine interaction; the nuclear quadrupole interaction ($I \cdot P \cdot I$); and the nuclear Zeeman interaction ($\mu_N B \cdot g_N \cdot I$). Since in the spin Hamiltonian formalism there is no place for the electron orbital momentum, the electron spin-electron spin term will contain the contributions from electron spin-spin, spin-orbit, and orbit-orbit interactions. In the case of transition metal complexes, it is the spin-orbit interaction that essentially determines the fine structure term in the spin Hamiltonian. The most important O_k^q operators have been tabulated by Abragam and Bleaney [1] and are shown in Table 1. The same reference also gives the matrix elements of the O_k^q operators in the spin manifolds from $S = 1$ to $S = 8$.

The terms $\sum_{k=4}^{2S} \sum_{q=0}^{2S} B_k^q O_k^q$ have a different form compared to the others included in Eq. (1). They have their origin in a more general formalism by which in principle it is possible to express all

Table 1   The O_k^q Equivalent Operators Needed for 3d Ions

$$O_2^0 = 3\underset{\sim}{S}_z^2 - S(S + 1)$$

$$O_2^2 = (1/2)(\underset{\sim}{S}_+^2 + \underset{\sim}{S}_-^2)$$

$$O_4^0 = 35\underset{\sim}{S}_z^4 - 30S(S + 1)\underset{\sim}{S}_z^2 + 35\underset{\sim}{S}_z^2 - 6S(S + 1) + 3S^2(S + 1)^2$$

$$O_4^2 = (1/4)\{[7\underset{\sim}{S}_z^2 - S(S + 1) - 5](\underset{\sim}{S}_+^2 + \underset{\sim}{S}_-^2) + [7\underset{\sim}{S}_z^2 - S(S + 1) - 5]\}$$

$$O_4^3 = (1/4)[\underset{\sim}{S}_z(\underset{\sim}{S}_+^3 + \underset{\sim}{S}_-^3) + (\underset{\sim}{S}_+^3 + \underset{\sim}{S}_-^3)\underset{\sim}{S}_z]$$

$$O_4^4 = (1/2)(\underset{\sim}{S}_+^4 + \underset{\sim}{S}_-^4)$$

the terms of (1). As a matter of fact, the spin Hamiltonian can be expressed as a sum of terms having the dimensions $B^{\ell}S^sI^i$ with $s \leqslant 2s$ and $i \leqslant 2I$ in order to have nonzero representative matrix elements in the spin basis set and $\ell + s + i = 2n$ in order to be invariant under the Kramers operator. The terms with $\ell = i = 0$ are the fine structure terms, $s = 2$ corresponding to

$$B_2^0 O_2^0 + B_2^2 O_2^2 \tag{3}$$

The terms with $\ell = 0$, $s = 1$, $i = 1$, and $\ell = 0$, $s = 0$, $i = 2$ are equivalent to the terms $\underset{\sim}{S} \cdot \underset{\sim}{A} \cdot \underset{\sim}{I}$ and $\underset{\sim}{I} \cdot \underset{\sim}{P} \cdot \underset{\sim}{I}$ of Eq. (1). Higher-order terms involving the $\underset{\sim}{I}$ operators will not be discussed here. In general terms, with $i = 0$, $\ell = 1$ and $s \neq 1$ may be required. For S = 3/2 spin systems, terms in BS^3 have been used frequently [36,37]. It has to be noted that the latter terms are required for ions having appreciable "unquenched" orbital momentum. However, it has been shown [1] that for an orbitally nondegenerate state, the spin Hamiltonian (1) is correct to first order in perturbation theory up to S = 5/2 spin systems.

In general the spin Hamiltonian is expressed in the form of Eq. (1), the only $B_k^q O_k^q$ terms used being those with $k = 4$.

Since $\underset{\sim}{D}$ and $\underset{\sim}{P}$ are traceless tensors, it is customary to reformulate the spin Hamiltonian terms that contain them in such a way as to reduce by one the number of parameters. The relations required are

$$\underline{D}_{xx} + \underline{D}_{yy} + \underline{D}_{zz} = 0$$

$$\underline{P}_{xx} + \underline{P}_{yy} + \underline{P}_{zz} = 0 \tag{4}$$

In the case of orthorhombic symmetry, the spin Hamiltonian (1) takes the explicit form:

$$
\begin{aligned}
\underline{H} = \mu_B &\left(g_x \underline{B}_x \underline{S}_x + g_y \underline{B}_y \underline{S}_y + g_z \underline{B}_z \underline{S}_z \right) + \left(\underline{A}_x \underline{S}_x \underline{I}_x \right. \\
&+ \underline{A}_y \underline{S}_y \underline{I}_y + \underline{A}_z \underline{S}_z \underline{I}_z \right) + D \left[\underline{S}_z^2 - \frac{S(S+1)}{3} \right] + E \left(\underline{S}_x^2 - \underline{S}_y^2 \right) \\
&+ P_\parallel \left[\underline{I}_z^2 - \frac{I(I+1)}{3} + \eta \left(\underline{I}_x^2 - \underline{I}_y^2 \right) \right] + B_4^0 O_4^0 \\
&+ B_4^4 O_4^4 + B_4^2 O_4^2 + \text{nuclear Zeeman terms}
\end{aligned}
\tag{5}
$$

where $D = (3\underline{D}_z/2)$, $E = (\underline{D}_x - \underline{D}_y)/2$, $P_\parallel = (3\underline{P}_z/2)$, and $\eta = (\underline{P}_x - \underline{P}_y)/\underline{P}_z$.

Sometimes the $B_4^2 O_4^2$ term in Eq. (5) is assumed to be negligible and the quartic fine structure terms of (5) are written as

$$B_4 \left(O_4^0 + 5 O_4^4 \right) + B_4^0 O_4^0 \tag{6}$$

Strictly, this represents a cubic field plus a tetragonal distortion. A commonly used equivalent form of Eq. (6) is

$$
\begin{aligned}
&\frac{a}{6} \left[\underline{S}_x^4 + \underline{S}_y^4 + \underline{S}_z^4 - \frac{S(S+1)(3S^2 + 3S - 1)}{5} \right] \\
&+ \frac{F}{180} \left[35 \underline{S}_z^4 - 30 S(S+1) \underline{S}_z^2 + 25 \underline{S}_z^2 + 6 S(S+1) \right]
\end{aligned}
\tag{7}
$$

with

$$F = 180 B_4^0 \qquad a = 120 B_4 = 24 B_4^4 \tag{8}$$

When the $\underline{S} \cdot \underline{D} \cdot \underline{S}$ term is expressed according to the equivalent formalism as in Eq. (3), relations

$$D = 3 B_2^0 \qquad E = B_2^2 \tag{9}$$

hold.

In both the Hamiltonians (1) and (5), we have neglected terms relative to the interactions between the electron spin and the nuclear spins of ligands. Where these are present, the hyperfine term in the Hamiltonian (1) becomes

$$\sum_{i=1}^{N} \underline{S} \cdot \underline{A}_i \cdot \underline{I}_i \tag{10}$$

where N is the number of nuclei that interact with the unpaired electrons.

The form of the Hamiltonian (5) is fairly general. However, lower-symmetry environments often can be found for transition metal ions, and a more detailed analysis of the symmetry properties of the tensors that appear in the spin Hamiltonian (1) is appropriate. Abragam and Bleaney [1] showed, using Kramers doublet and time-reversal symmetry, that D_2 symmetry the principal axes of g and $\underline{A}$ tensors had to be coincident, whereas this is not the case in C_3 symmetry. These relations were generalized [38] to show that whenever the complex has a two (or higher)-fold axis and either a perpendicular binary axis C_2' or σ_v symmetry elements, g and $\underline{A}$ are necessarily collinear. Without the C_2' or σ_v symmetry elements, g and $\underline{A}$ have off-diagonal elements, resulting generally in noncollinear principal axes. The noncollinearity of g and $\underline{A}$ tensors in this case will cause $\underline{B}$ and $\underline{I}$ to be referred to different axis systems, and this will be observable if the nuclear Zeeman effect is measured together with the hyperfine terms. That can be achieved through electron nuclear double resonance (ENDOR) experiments.

In Table 2 the relations imposed by crystal symmetry on g and $\underline{A}$ are shown.

At the end of this section we want to mention, for the sake of completeness, an equivalent formulation of the spin Hamiltonian that can be obtained using tensor operators [36,39,40]. In this formalism the generalized spin Hamiltonian is

$$\underline{H} = \sum_{\ell,m} a_{\ell m} T_{\ell m}(B_{\ell_1}^{m_1} S_{\ell_2}^{m_2} I_{\ell_3}^{m_3}) \tag{11}$$

In Eq. (11) the summation over m goes from $-\ell$ to $+\ell$, where ℓ is an even integer. $T_{\ell m}(B_{\ell_1}^{m_1} S_{\ell_2}^{m_2} I_{\ell_3}^{m_3})$ is concise notation for the tensor product of the tensor operators $T_{\ell_1 m_1}(B)$, $T_{\ell_2 m_2}(S)$, and $T_{\ell_3 m_3}(I)$, which are the Racah tensor operators constructed with the components of B, S, and I, respectively.

The values of ℓ_1, ℓ_2, and ℓ_3 are restricted so that they sum up to give ℓ even, and the maximum possible values for ℓ_2 and ℓ_3 are $2S$ and $2I$, respectively. There are no general restrictions on ℓ_1 except those from parity considerations; usually, however, power terms greater than quadratic in B are neglected.

In this formalism the linear Zeeman term is obtained by putting $\ell_1 = 1$, $\ell_2 = 1$, and $\ell_3 = 0$ in Eq. (11):

$$H_Z = \sum_{q=-1}^{1} \sum_{k=-1}^{1} \mu_B g_{qk} T_{1q}(B) T_{1k}(S) \tag{12}$$

and the zero field term taking $\ell_1 = \ell_3 = 0$, $\ell_2 = 2S$ in (11) is

$$H_{zfs} = \sum_{\ell=0}^{2S} \sum_{m=-\ell}^{\ell} B_{\ell m} T_{\ell m}(S) \tag{13}$$

Buckmaster, Chatterjee, and Shing [41] have written a review article on the use of the tensor operator formalism in the analysis of ESR and ENDOR spectra and have reported several tables that are useful in computations.

III. ESR SPECTRA AND CRYSTAL SYMMETRY

In order to obtain the most complete experimental information from an ESR experiment, it is necessary to record single-crystal spectra. The appearance of single-crystal spectra depends on the site symmetry of the metal ion and the point symmetry of the space group of the lattice. Although often the ESR spectra of a low-symmetry complex may appear to be of higher symmetry than is allowed by the site symmetry, it must be understood that it can be true only to some level of approximation, and if the experimental data were obtained

Bencini and Gatteschi

Table 2 The Relations Imposed by Crystal Symmetry on $\underset{\sim}{g}$ and $\underset{\sim}{A}$

Crystal point group				
International symbol	Schonflies symbol	Crystal system	Restriction on the M_{ij} matrix element[a]	Are $\underset{\sim}{g}$ and $\underset{\sim}{A}$ collinear?
1 $\bar{1}$	C_1 S_2	$\Big\}$ Triclinic	No restriction	No
2 $\bar{2}$ = m 2/m	C_2 $C_s = C_{1h}$ C_{2h}	$\Big\}$ Monoclinic[b]	$\underset{\sim}{g}_{xz} = \underset{\sim}{g}_{yz} = \underset{\sim}{g}_{zx}$ $= \underset{\sim}{g}_{zy} = 0$	No
3 $\bar{3}$	C_3 S_6	$\Big\}$ Trigonal[b]	As for C_2 + $\underset{\sim}{g}_{xx}$ $= \underset{\sim}{g}_{yy}$, $\underset{\sim}{g}_{xy} = -\underset{\sim}{g}_{yx}$	No
4 $\bar{4}$ 4/m	C_4 S_4 C_{4h}	$\Big\}$ Tetragonal[b]	As for C_3	No
6 3/m 6/m	C_6 C_{3h} C_{6h}	$\Big\}$ Hexagonal[b]	As for C_3	No
222 mm2 mmm	D_2 C_{2v} D_{2h}	$\Big\}$ Orthorhombic	As for C_2 + $\underset{\sim}{g}_{xy}$ $= \underset{\sim}{g}_{yx} = 0$	Yes
32 3m $\bar{3}$m	D_3 C_{3v} D_{3d}	$\Big\}$ Trigonal[b]	As for $D_2 = \underset{\sim}{g}_{xx}$ $= \underset{\sim}{g}_{yy}$	Yes
422 4mm $\bar{4}$2m 4/mmm	D_4 C_{4v} D_{2d} D_{4h}	$\Big\}$ Tetragonal[b]	As for D_3	Yes
622 6mm $\bar{6}$m2 6/mmm	D_6 C_{6v} D_{3h} D_{6h}	$\Big\}$ Hexagonal[b]	As for D_3	Yes

Table 2 (Continued)

23	T			
m3	T_h			
432	O	Cubic	As for D_3 + g_{xx}	Yes
$\overline{4}$3m	T_d		$= g_{yy} = g_{zz}$	
m3m	O_h			

[a] M_{ij} is either g_{ij} or A_{ij} (i, j = x, y, z).

[b] The principal axis is z.

Source: Ref. 48.

with sufficient accuracy, the ESR spectra should conform with the
site symmetry. It is therefore important to know the site symmetry
of the metal ion. In general, there will be n crystallographically
equivalent sites for the metal ions in the lattice; however, metal
ions possessing identical site symmetry may be magnetically nonequivalent. "Magnetically nonequivalent" in this case means that whereas
the principal values of the tensors (g, A, . . .) are identical in
all the sites, the principal axes are physically distinct so that up
to n different spectra will contribute to the ESR spectra of the complex. In Table 3 are shown the maximum number of magnetically
nonequivalent sites that can be found in the different crystal systems

Table 3   Magnetically Nonequivalent Sites in Crystal Systems

	Number of sites for various orientations of the static magnetic field B						
Crystal system	B in general orientation	B along crystal axes a	b	c	B in crystal planes ab	ac	bc
Triclinic	1	1	1	1	1	1	1
Monoclinic	2	1	1	1	2	1	2
Orthorhombic	4	1	1	1	2	2	2
Trigonal	3	1	1	1	3	3	3
Tetragonal	4	2	2	1	2	4	4
Hexagonal	6	3	3	1	3	6	6
Cubic	12	3	3	3	6	6	6

Source: Ref. 48.

under the assumption of the metal ion being in general positions. In obtaining the number of distinct sites, it was taken into account that sites related by a center of inversion are indistinguishable in an ESR experiment, since for a tensor $\underset{\sim}{Y}$ the relation

$$\underset{\sim}{R} \cdot \underset{\sim}{Y} \cdot \underset{\sim}{R}^{-1} = (\underset{\sim}{R}\underset{\sim}{i}) \cdot \underset{\sim}{Y} \cdot (\underset{\sim}{R}\underset{\sim}{i})^{-1} \tag{14}$$

must hold, where $\underset{\sim}{R}$ and $\underset{\sim}{i}$ are matrices related to a rotation and an inversion, respectively. The values of Table 3 are valid if only one crystallographically independent molecule is present in the asymmetric unit.

It is apparent that if the site symmetry at the metal ion is higher than C_1, the number of magnetically nonequivalent sites will be reduced. In general, in order to have only one spectrum, it is sufficient that the site of the metal ion has the same symmetry of the crystal point group.

In general, one a priori does not have knowledge of the principal axes of the various tensors, and therefore laboratory axes are used and the tensors are obtained from the spectra. However, when a low-symmetry Hamiltonian is needed to interpret the spectra, it may not be a simple task to obtain g, A, D, and so on, directly, and computing facilities may be required (see Sec. VI).

The simplest case that one encounters is that of a spin system with $S = 1/2$ and no hyperfine interaction. Several methods are available for obtaining the g^2 tensor. The most common method uses three rotations in three orthogonal planes in order to obtain the g^2_{ij} terms. In general, least-squares fits of the g^2 values in each plane are obtained according to the relations [42]

$$\underset{\sim}{g}^2 = \underset{\sim}{g}^2_{ii} \cos^2 \theta + \underset{\sim}{g}^2_{jj} \sin^2 \theta + \underset{\sim}{g}^2_{ij} \sin 2\theta \tag{15}$$

where θ is the angle between the direction of the static magnetic field and the i laboratory axis. The i and j axes in Eq. (15) must conform to the cyclic rule, that is, x, y, z, x. It is important to understand that the sense of rotation must be known. This is particularly important in the case of triclinic crystals.

As a matter of fact, it is possible to choose arbitrarily the direction of rotation in two planes, leaving two different choices for the third plane [43]. If one is making the rotation on the same crystal, there is no problem once the sense of rotation of the crystal in all three planes has been carefully checked. If one is using different crystals and there is not some morphological feature that permits the exact orientation of the crystal, one can only hope that one of the two choices leads to an obviously absurd result.

The g_{ij}^2 values obtained according to the above procedure are subject to experimental errors. Billing and Hathaway have reported a system of eliminating these errors [44]. In order to obtain the principal $\underset{\sim}{g}$ values and directions the $\underset{\sim}{g}^2$ matrix must be diagonalized, using a method, for example, the Jacobi method, that yields the eigenvalues and eigenvectors of the matrix. The eigenvectors are the direction cosines of the principal directions in the laboratory axes frame.

Another method, which does not require the determination of complete rotations but only the maxima and minima of g^2 in each rotation, plus the g_{ii}^2 values, was suggested by Ayscough [4]. The relative formulas are collected in Eq. (16),

$$g_{ij}^2 = \left(\frac{1}{2}\right)\left[\Delta_{ij}^2 - \left(g_{ii}^2 - g_{jj}^2\right)^2\right]^{\frac{1}{2}} \qquad (i,\ j = x,\ y,\ z)$$

$$g_{ii}^2 = \left(\frac{1}{2}\right)(X_{ij} + X_{ik} - X_{jk}) \qquad (i,\ j = x,\ y,\ z) \qquad (16)$$

and those obtained from Eq. (16) by cyclic permutation of the i, j, k indices. In Eq. (16), X_{ij} indicates the sum of the maximum and minimum $\underset{\sim}{g}^2$ values in the ij plane, and Δ_{ij} is their difference. Usually g_{ij}^2 is taken as positive if $\underset{\sim}{g}^2$ takes the maximum value relative to the ij rotation in the first quadrant, and it is taken as negative if $\underset{\sim}{g}^2$ takes the minimum value.

Although it is common practice to make three rotations along three mutually orthogonal directions, this is not mandatory, and as a matter of fact, several reports of methods of using nonorthogonal rotations are available [45-49]. Of course, the number of rotations required can be reduced if the metal atom environment has some elements of symmetry.

The above discussion does not take into account the possibility of simplifying the experimental determination of the tensor by resorting to the crystal symmetry. Weil et al. [48], in a very detailed analysis of the effect of crystal point group symmetry on microscopic tensor properties, showed that except in the case of monoclinic and triclinic crystals, it is possible to obtain the six independent elements of a tensor $\underset{\sim}{Y}$ from the rotation in only one plane. This result is obtained by analyzing the spectrum of different sets of molecules present in the lattice. In Table 4 are shown the groups and the conditions whereby this can be obtained.

One experimental problem that appears in single-crystal measurements is that of the exact orientation of the crystal in the cavity. In general, the single crystal is oriented on a quartz (or perspex) rod either directly from the goniometer head used for the orientation of the crystal itself, or under a polarizing microscope using the physical features of the crystal. It is assumed that this error can be minimized, and it is not taken into consideration. On the other hand, an error can be made in the azimuthal angle in every orientation, so that the angular origin in the rotation is not the true origin one expects, but rather is rotated by a small angle from it.

It is apparent from Table 4 that the major problem in using only one rotation to determine the $\underset{\sim}{Y}$ tensor is to have the correct orientation of the crystal in the cavity. Often, indeed, the most highly developed faces of a crystal correspond to principal planes, so that they are not the planes required from Table 4 for obtaining the $\underset{\sim}{Y}$ tensor with a single rotation.

In principle, the same procedures can be applied to evaluate any symmetric second-rank tensor, for example, $\underset{\sim}{D}$, $\underset{\sim}{A}^2$, and so on. An example of the procedure for obtaining the fine and hyperfine structure tensors has been given by Lund and Vängård [49].

An important consequence of the existence of several sites in a certain space group is that it may be difficult to recognize to which

Table 4 The Crystallographic Proper Rotation Groups and the Conditions Under Which a Rotation in a Single Plane Permits the Determination of a $\underset{\sim}{Y}$ tensor[a]

Rotation group	Is it possible to determine Y by a rotation in a single plane?	Plane in which $\underset{\sim}{B}$ must lie: Miller indices	
C_1	No		
C_2	No		
C_3	Generally Yes[b]	{hkℓ}	ℓ ≠ 0
C_4	Yes	{hkℓ}	ℓ ≠ 0
C_6	Yes	{hkℓ}	ℓ ≠ 0
D_2	Yes	{hkℓ}	ℓ ≠ 0
D_3	Yes	{hkℓ}	ℓ ≠ 0
D_4	Yes	{hkℓ}	ℓ ≠ 0
D_6	Yes	{010} {100} {hkℓ}	ℓ ≠ 0
T	Generally Yes[b]	{hkℓ} {hk0}	ℓ ≠ 0, h, k ≠ 0
O	Yes	{hkℓ} {hk0}	ℓ ≠ 0, h, k ≠ 0

[a]Only general conditions are reported. Special conditions can occur, and other rotation axes can be used. For a general analysis of the question, see Ref. 48.

[b]Y can be obtained except in special conditions discussed in Ref. 48.

specific molecule the set of g^2, A^2, etc., tensors belongs. Often it is possible to propose some reasonable assignment on the basis of the idealized symmetry of the complex. For example, in the case of a pseudotetrahedral complex, one expects to find the principal axes of the tensors close to the bond directions; however, if ambiguity arises, the choice that is close to this limit will be preferred. In some cases, however, the idealized symmetry is of little help, and the ambiguity in the assignment of the tensors to individual molecules cannot be eliminated. As an example, consider the complexes Nip_3X, where p_3 is 1,1,1-tris(diphenylphosphinomethyl)ethane

and X = Cl, Br, I [50]. These crystallize in the orthorhombic
system space group C_{2v}^9, with the metal ion in a C_1 site [51]. The
ESR spectra of the nickel(I) complexes doped into the isomorphous
copper(I) complexes were recorded and the angular dependence of $\tilde{g}^2$
in the three principal planes was found as in Fig. 1 [52]. In the
C_{2v}^9 space group, four magnetically nonequivalent sites are expected,
but in the principal planes they are degenerate in pairs so that
only two signals are recorded for a general orientation of the crystal in the static magnetic field (see Table 3). Along the principal
axes, the four sites must all be equivalent, yielding only one signal. The coincidence of the spectra parallel to the crystal axes
can be used as a check of the orientation of the crystal in the ESR
spectrometer. From Fig. 1 it is apparent that it is quite easy to
follow the angular dependence of one molecule in one plane, but it
is difficult to know which line corresponds to this molecule in a
different plane. Since there are two choices in each plane, and
there are three planes, eight different possibilities exist. They
give rise to two sets of $\underline{g}$ values. Within each set all the four
crystallographically related sites show up. The principal values
and directions of the $\tilde{g}^2$ tensor are shown in Table 5. In the same
table the angles of the $\underline{g}$ directions with the four Ni-Cl bonds in
the unit cell are also shown. It is apparent that these are highly
misaligned when compared to the observed $\underline{g}$ directions, thus making
the assignment to a particular molecule an almost impossible task.

IV. GENERAL SOLUTIONS OF THE SPIN HAMILTONIAN

In the most general case it is impossible to solve analytically the
Hamiltonian (1). Rather it can be tackled only by computing techniques that solve it for a particular case of interest and find
eigenvalues and eigenvectors and determine the resonance conditions
and the transition probabilities. Often, however, several approximations are possible that make the solution of the Hamiltonian (1)
more tractable.

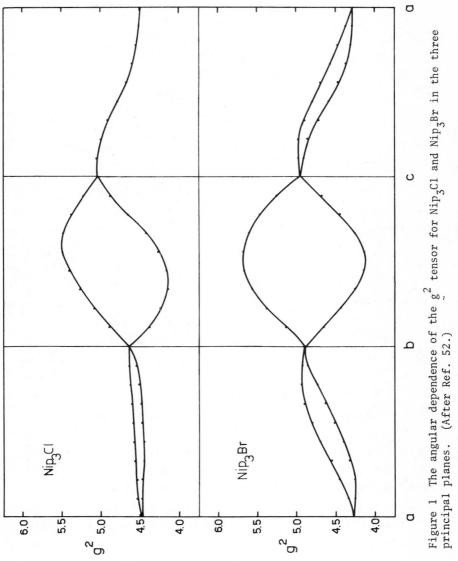

Figure 1 The angular dependence of the g^2 tensor for Nip_3Cl and Nip_3Br in the three principal planes. (After Ref. 52.)

Table 5 Principal g Values and Directions for Ni(p₃)Cl and Angles
with the Ni-Cl Bond Directions

g values	Directions cosines[a]			α (deg)[b]			
	a	b	c				
2.344	0.040	0.572	0.819	40.5	43.9	98.5	79.2
2.124	0.985	0.114	-0.128	57.7	116.3	69.3	51.4
2.040	0.170	-0.812	0.559	110.7	121.5	20.6	140.1

[a]The other three sets are obtained through the relations: ℓ, m, n;
$\bar{\ell}$, m, n; ℓ, $\bar{m}$, n; ℓ, m, $\bar{n}$.

[b] α is the angle of the principal g direction with the four Ni-Cl
bond directions in the unit cell.

One situation that is often encountered experimentally occurs
when the electronic Zeeman term is dominant. Here perturbation
techniques can be applied. Bleaney [53] derived expressions for the
case of axially symmetric g, $\underset{\sim}{D}$, and $\underset{\sim}{A}$ tensors, carrying the perturbation to second order; Azarbayejani [54], McClung [55], Golding [56],
and Sakaguchi et al. [57] generalized the treatment to orthorhombic
symmetry, whereas Lin [58], Kirmse et al. [59], Pilbrow and Winfield
[60], Golding and Tennant [61], and Rockenbauer and Simon [62,63],
extended it to the case where one or more principal axes of g, $\underset{\sim}{D}$,
and $\underset{\sim}{A}$ do not coincide. Finally, Golding and Tennant [64] gave a
general third-order solution to the spin Hamiltonian, in the case
of g, $\underset{\sim}{D}$, A, and $\underset{\sim}{P}$ tensors not coincident. Weil [65-67] generalized
the treatment to the case of more than one nuclei present, but res-
tricted the calculations to second-order perturbation, and did not
provide expressions for transition probabilities. In all the above
treatments it is assumed that the first-order hyperfine energy for
every nucleus is larger than any corresponding nuclear Zeeman or
quadrupole contribution.

One expression for the energies is that given by Weil [65]:

$$E(M, m_1, m_2, \ldots, m_N) = \mu_B \underline{g} B M + \frac{1}{2} Tr(\underset{\sim}{D})[S(S+1) - M^2]$$

$$+ \frac{1}{2} d_1 [3M^2 - S(S+1)] + (d_2 - d_1) \frac{[8M^2 + 1 - 4S(S+1)]M}{2\mu_B \underline{g} B}$$

$$+ \frac{1}{8\mu_B \underline{g} B}[\mathrm{Tr}(\underset{\sim}{D}^2) - 2d_2 + d_1^2 - 2d_{-1}\,\mathrm{Det}(\underset{\sim}{D})][2S(S+1) - 2M^2 - 1]M$$

$$+ \sum_i \left(K_i M m_i + \frac{1}{2\mu_B \underline{g} B}\{\frac{1}{2}[\mathrm{Tr}(\underset{\sim}{\tilde{A}}_i \cdot \underset{\sim}{A}_i) - k_i^2]M[I_i(I_i+1) - m_i^2]\right.$$

$$- \frac{\mathrm{Det}(\underset{\sim}{A}_i)}{K_i}[S(S+1) - M^2]m_i + (k_i^2 - K_i^2)M m_i^2 \tag{17}$$

$$+ 2(e_i - d_i)K_i[3M^2 - S(S+1)]m_i\} - \mu_N G_i \underline{B} m_i$$

$$+ \frac{\mathrm{Tr}(\underset{\sim}{P})}{2}[I_i(I_i+1) - m_i^2] + \frac{1}{2}P_{1i}[3m_i^2 - I_i(I_i+1)])$$

$$+ \frac{1}{\mu_B \underline{g} B} \sum_{i>j} (L_{ij} - K_i K_j)M m_i m_j + \frac{1}{M}\sum_i \{\mu_N^2 \frac{B^2}{2K_i}(g_i^2 - G_i^2)m_i$$

$$+ \frac{1}{2K_i}(p_{2i} - p_{1i}^2)[8m_i^2 - 4I_i(I_i+1) + 1]m_i + \frac{1}{8K_i}[\mathrm{Tr}(\underset{\sim}{P}_i^2) - 2p_{2i}$$

$$+ p_{1i}^2 - 2p_{-1i}\mathrm{Det}(\underset{\sim}{P}_i)][2I_i(I_i+1) - 2m_i^2 - 1]m_i$$

$$- \mu_N \frac{B}{K_i}(q_i - G_i p_{1i})[3m_i^2 - I_i(I_i+1)]\}$$

In Eq. (17), defining $\underset{\sim}{\eta} = \underset{\sim}{B}/|\underset{\sim}{B}|$, we have

$$\underline{g}^2 = \underset{\sim}{\tilde{\eta}}\cdot\underset{\sim}{\tilde{g}}\cdot\underline{g}\cdot\underset{\sim}{\eta}$$

$$\underline{g}^2 d_n = \underset{\sim}{\tilde{\eta}}\cdot\underset{\sim}{\tilde{g}}\cdot\underset{\sim}{D}^n\cdot\underline{g}\cdot\eta$$

$$\underline{g}^2 K_i^2 = \underset{\sim}{\tilde{\eta}}\cdot\underset{\sim}{\tilde{g}}\cdot\underset{\sim}{\tilde{A}}_i\cdot\underset{\sim}{A}_i\cdot\underline{g}\cdot\underset{\sim}{\eta}$$

$$\underline{g}^2 K_i^2 k_i^2 = \underset{\sim}{\tilde{\eta}}\cdot\underset{\sim}{\tilde{g}}\cdot\underset{\sim}{\tilde{A}}_i\cdot\underset{\sim}{A}_i\cdot\underset{\sim}{\tilde{A}}_i\cdot\underset{\sim}{A}_i\cdot\underline{g}\cdot\underset{\sim}{\eta}$$

$$\underline{g}^2 K_i^2 e_i = \underset{\sim}{\tilde{\eta}}\cdot\underset{\sim}{\tilde{g}}\cdot\frac{1}{2}(\underset{\sim}{D}\cdot\underset{\sim}{\tilde{A}}_i\cdot\underset{\sim}{A}_i + \underset{\sim}{\tilde{A}}_i\cdot\underset{\sim}{A}_i\cdot\underset{\sim}{D})\cdot\underline{g}\cdot\underset{\sim}{\eta} \tag{18}$$

$$\underline{g}_i^2 = \underset{\sim}{\tilde{\eta}}\cdot\underset{\sim}{\tilde{g}}_i\cdot\underline{g}_i\cdot\underset{\sim}{\eta}$$

$$\underline{g}K_i G_i = \underset{\sim}{\tilde{\eta}}\cdot\frac{1}{2}(\underset{\sim}{\tilde{g}}_i\cdot\underset{\sim}{\tilde{A}}_i\cdot\underline{g}_i + \underset{\sim}{\tilde{g}}_i\cdot\underset{\sim}{A}_i\cdot\underline{g}_i)\cdot\underset{\sim}{\eta}$$

$$\underline{g}^2 K_i^2 p_{ni} = \underset{\sim}{\tilde{\eta}}\cdot\underset{\sim}{\tilde{g}}\cdot\underset{\sim}{\tilde{A}}_i\cdot\underset{\sim}{P}_i^n\cdot\underset{\sim}{A}_i\cdot\underline{g}\cdot\underset{\sim}{\eta} \qquad n = \text{integer}$$

$$\underline{g}K_i q_i = \underset{\sim}{\tilde{\eta}}\cdot\frac{1}{2}(\underset{\sim}{\tilde{g}}\cdot\underset{\sim}{\tilde{A}}_i\cdot\underset{\sim}{P}_i\cdot\underline{g}_i + \underset{\sim}{\tilde{g}}_i\cdot\underset{\sim}{P}_i\cdot\underset{\sim}{A}_i\cdot\underline{g})\cdot\underset{\sim}{\eta}$$

$$\underline{g}^2 K_i K_j L_{ij} = \underset{\sim}{\tilde{\eta}}\cdot\underset{\sim}{\tilde{g}}\cdot\frac{1}{2}(\underset{\sim}{\tilde{A}}_i\cdot\underset{\sim}{A}_i\cdot\underset{\sim}{\tilde{A}}_j\cdot\underset{\sim}{A}_j + \underset{\sim}{\tilde{A}}_j\cdot\underset{\sim}{A}_j\cdot\underset{\sim}{\tilde{A}}_i\cdot\underset{\sim}{A}_i)\cdot\underline{g}\cdot\underset{\sim}{\eta}$$

In Eqs. (17) and (18), i and j refer to nuclei 1, 2, ..., N, $\underset{\sim}{g_i}$ is
the nuclear g tensor, and M and $\tilde{m}$ represent electronic and nuclear
spin quantum numbers, respectively.

In Eq. (17), a term varying as M^{-1} is present, which comes from
the above-mentioned assumptions on which (17) is based. The application of Eq. (17) to states with M = 0 is therefore not valid.

Although expression (17) appears cumbersome, it is important to
have a feeling of the effect of nonparallel axes of the g, A, and D
tensors. Golding and Tennant have discussed such effects at some
length [61]. In the single-crystal spectra the most relevant effect
is that of giving rise to hyperfine lines that would be forbidden in
the case of parallel axes. If D and A are not diagonal in the same
coordinate system as g, then $\Delta M_S = \pm 2$ transitions, which are strictly forbidden in the parallel-axes case, can appear when the field is
parallel to a principal axis of g.

More severe effects are to be expected in the case of measurements on polycrystalline samples. In Ref. 61, the polycrystalline
powder spectra for an S = 1/2 and I = 1/2 system with noncollinear g
and A tensors, and the spectra for S = 1, I = 0 with g and D not
collinear have been calculated. The powder spectra shown in Figs. 2
and 3 have been calculated using a gaussian line shape function.
Figure 2 represents a system with S = 1/2 and I = 1/2 having axial
symmetry with $g_\parallel$ = 2.0, $g_\perp$ = 2.1, $A_\parallel$ = 46.7 × 10^{-4} cm^{-1}, and $A_\perp$ =
24.5 × 10^{-4} cm^{-1}; Fig. 3 refers to an S = 1 spin system with $g_\parallel$ = 2.0,
$g_\perp$ = 2.1, D = 0.007 cm^{-1}. θ and φ are the zenithal and azimuthal
angles of the z axis of A and D in the g principal axes system. It
is interesting to note that a misalignment of 60° of g and D leads to
an apparent orthorhombic spectrum with no fine structure.

As mentioned above, the formulas reported are all in the high-
field limit. Although this is not in general a serious drawback for
the case where only hyperfine coupling is present, when the zero field
splitting tensor also contributes to the spectra, inconveniences can
occur. When D is of the same order of magnitude as the electronic
Zeeman term, the perturbation treatment must be abandoned, and the

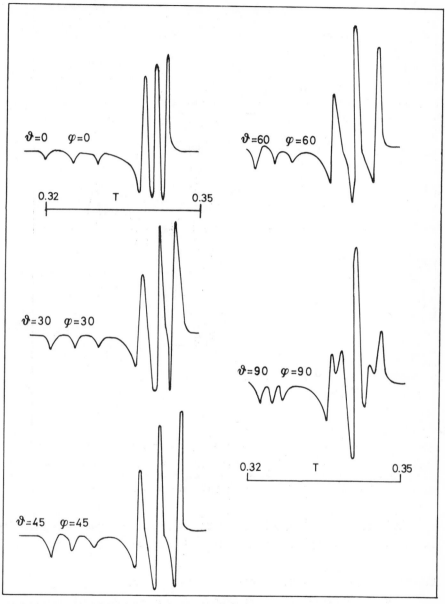

Figure 2   Simulated powder spectra for an S = 1/2, I = 1/2 system with noncollinear g and A tensors for various angles of rotation. θ and φ are the polar angles of A_z in the g principal axes frame. (After Ref. 61.)

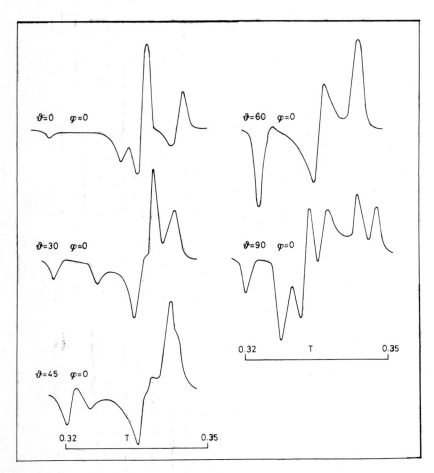

Figure 3   Simulated powder spectra for an S = 1 system with noncolinear g and D tensors for various angles of rotation.   θ and φ are the polar angles of $\underline{D}_z$ in the g principal axes frame.   (After Ref. 61.)

spin Hamiltonian must be diagonalized exactly. One case was reported in which this was applied to S = 3/2 systems [68].

V.   PARTICULAR SOLUTIONS OF THE SPIN HAMILTONIAN

The general solution of the spin Hamiltonian of the previous section is cumbersome, and suffers from some unescapable approximations. Therefore we shall provide a simpler formulation for the various

values of S, introducing into the spin Hamiltonian only the terms
relative to the electronic Zeeman and to zero field splitting. In
Tables 6 through 9, the spin Hamiltonian matrix and, when an exact
solution is possible, the energy levels and transitions for the
various spin systems in orthorhombic symmetry are reported. In this
symmetry the g and D tensors must be collinear.

A. S = 1 Spin Systems

The orthorhombic spin Hamiltonian for an S = 1 spin system is

$$\underline{H} = \mu_B\left(g_x B_x S_x + g_y B_y S_y + g_z B_z S_z\right) + D(S_z^2 - \tfrac{2}{3}) + E(S_x^2 - S_y^2) \quad (19)$$

When the zero field splitting is small compared to the energy of the
incident radiation, $h\nu$, two transitions are observed along each
principal direction of g. When, however, the zero field splitting
is larger than $h\nu$, one or even no signal may be detected. In Fig. 4
these two cases are depicted for $h\nu$ = 3D and for $h\nu$ = 0.5D.
    The matrix representation of Hamiltonian (19) in an S = 1 basis
for B having (ℓ, m, n) direction cosines in the orthorhombic

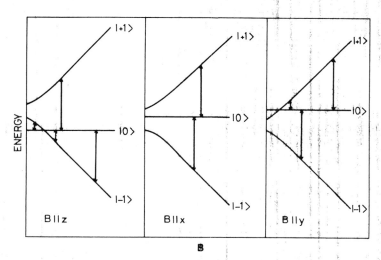

Figure 4 Energy levels as a function of static magnetic field for
an S = 1 system with E = (1/4)D. Heavy arrows correspond to the
allowed transitions for $h\nu$ = 3D; light arrows are for $h\nu$ = 0.5D.

reference frame [69] and the energy of the levels calculated [70,71] with $\underset{\sim}{B}$ along x, y, and z axes are shown in Table 6. Also reported are the analytical expressions for the resonant fields when the static magnetic field is parallel to one of the principal axes and the expressions for the resonance fields for $\Delta M_s = \pm 2$ transitions, both normal and double quantum [72,73], for polycrystalline powder spectra [74]. In Fig. 5 the Q-band spectrum of a polycrystalline sample of $Cu(chloroacetate)_2 \cdot \alpha$-picoline is shown. In this spectrum both the $\Delta M_s = \pm 2$ transition and the double quantum transition, involving the absorption of two microwave quanta, are found. The resonance fields corresponding to these two transitions are labeled, both in Fig. 5 and in Table 6, with B_{min} and B_{dq} for the $\Delta M_s = \pm 2$ and double quantum transitions, respectively.

Table 6   The Orthorhombic Spin Hamiltonian Matrix for S = 1 Spin Systems for a General Orientation of B and the Energy Levels and Transitions for B Along the Orthorhombic Axes$\underset{\sim}{}$

Spin Hamiltonian matrix[a]

	$|1\rangle$	$|0\rangle$	$|-1\rangle$
$\langle 1|$	$ng_z\mu_B B + \dfrac{D}{3}$	$(\tfrac{1}{2})^{1/2}(\ell g_x - img_y)\mu_B B$	E
$\langle 0|$	$(\tfrac{1}{2})^{1/2}(\ell g_x + img_y)\mu_B B$	$\dfrac{-2D}{3}$	$(\tfrac{1}{2})^{1/2}(\ell g_x - img_y)\mu_B B$
$\langle -1|$	E	$(\tfrac{1}{2})^{1/2}(\ell g_x + img_y)\mu_B B$	$-ng_z\mu_B B + \dfrac{D}{3}$

Energies[b,d]

$$W_{\pm 1,z} = \frac{D}{3} \pm (g_z^2 \mu_B^2 B^2 + E^2)^{1/2}$$

$$W_{0,z} = -\frac{2D}{3}$$

$$W_{\pm 1,i} = -\frac{D}{6} + \gamma_i \frac{E}{2} \pm \left[\frac{(D + E)^2}{4} + g_x^2 \mu_B^2 B^2 \right]^{1/2}$$

$$W_{0,i} = \frac{D}{3} - \gamma_i E$$

Resonant fields for $\Delta M_s = \pm 1$ lines[c,d]

$$Bx_{1,2} = \left(\frac{g_e}{g_x}\right)[(B_0 \mp D' \pm E')(B_0 \pm 2E')]^{1/2}$$

Table 6 (Continued)

$$By_{1,2} = \left(\frac{g_e}{g_y}\right)[(B_0 \mp D' \mp E')(B_0 \mp 2E')]^{1/2}$$

$$Bz_{1,2} = \left(\frac{g_e}{g_z}\right)[(B_0 \mp D')^2 - E'^2]^{1/2}$$

Resonant fields for $\Delta M_s = \pm 2$ lines in randomly oriented triplets[d]

$$B_{min} = \left(\frac{g_e}{g_{min}}\right)\left[\left(\frac{B_0}{2}\right)^2 - \left(\frac{D'^2}{3}\right) - E'^2\right]^{1/2}$$

$$B_{dq} = \left(\frac{g_e}{g_{av}}\right)\left[\left(B_0^2 - \left(\frac{D'^2}{3}\right)\right) - E'^2\right]^{1/2}$$

[a](ℓ, m, n) are the direction cosines of B in the orthorhombic system.

[b]The upper sign refers to the +1 spin component, the lower sign refers to -1; i = x, y.

[c]The upper and lower signs refer to transitions 1 and 2, respectively.

[d]The symbols are defined as follows:
$B_0 = (h\nu)/(\mu_B g_e)$; $D' = D/(\mu_B g_e)$; $E' = E/(\mu_B g_e)$; $g_e = 2.0023$; $g_{min} = (g_x g_y \sin^2 \alpha + g_z^2 \cos^2 \alpha)^{1/2}$; $\alpha^2 = \cos^{-1}\{[9 - 4(D/h\nu)^2]/[27 - 36(D/h\nu)^2]\}$; $g_{av} = [(g_x^2 + g_y^2 + g_z^2)/3]^{1/2}$; $\gamma_i = 1$ for i = x; $\gamma_i = -1$ for i = y.

Kottis and Lefebvre [75] have shown that B_{min}, which is the field at which the $\Delta M_s = \pm 2$ transition in a powder sample has the maximum transition probability, does not correspond to any of the principal directions.

In Fig. 6, plots of $D/h\nu$ against $B/h\nu$ are shown.

B. S = 3/2 Spin Systems

The orthorhombic spin Hamiltonian for an S = 3/2 spin system is

$$\underline{H} = \mu_B\left(g_x B_x \underline{S}_x + g_y B_y \underline{S}_y + g_z B_z \underline{S}_z\right) + D(\underline{S}_z^2 - \frac{5}{4}) + E(\underline{S}_x^2 - \underline{S}_y^2) \quad (20)$$

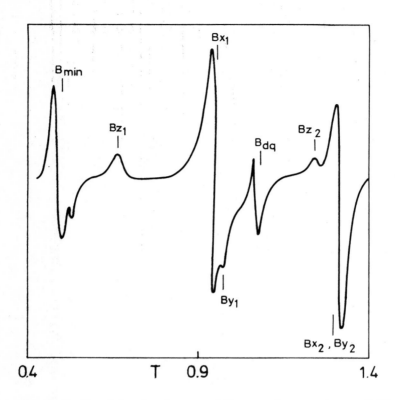

Figure 5   The Q-band polycrystalline powder spectrum of [Cu(chloroace-tato)$_2$α-picoline]$_2$.

In Table 7 the matrix representation of Hamiltonian (20) within an
S = 3/2 spin basis for $\underset{\sim}{B}$ along a general (ℓ, m, n) direction and the
energy levels calculated for $\underset{\sim}{B}$ along the principal axes are reported.
The appearance of the spectra is strongly dependent on the relative
order of magnitude of hν and the zero field splitting.   In the case
where the zero field splitting is small compared to hν, three $\Delta M_s = \pm 1$
transitions will be recorded for every principal direction of $\underset{\sim}{g}$.
When, on the contrary, the zero field splitting is larger than hν,
the two Kramers doublets constituting the S = 3/2 manifold will be
separated in energy by $2(D^2 + 3E^2)^{1/2}$ and only transitions within
each of them will be obtained.   In this case the spectra can be described by an effective S = 1/2 spin Hamiltonian.

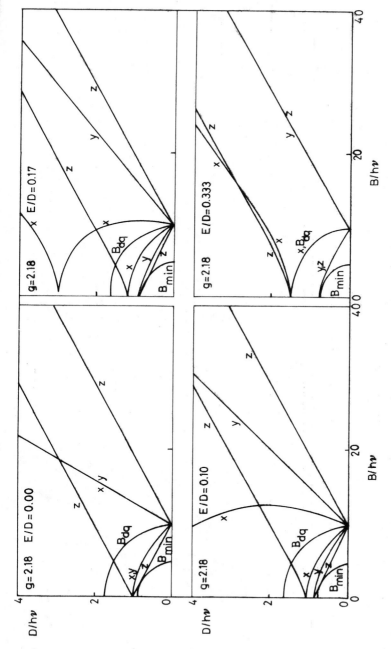

Figure 6   Plots of the magnetic field versus D/hν for various E/D ratios for S = 1 systems.   (After Ref. 74.)

27

Table 7 The Orthorhombic Spin Hamiltonian Matrix for S = 3/2 Spin Systems for a General Orientation of $\tilde{B}$ and the Energy Levels for $\tilde{B}$ along the Orthorhombic Axes

Spin Hamiltonian matrix[a]

	$\lvert 3/2\rangle$	$\lvert 1/2\rangle$	$\lvert -1/2\rangle$	$\lvert -3/2\rangle$
$\langle 3/2 \rvert$	$\frac{3}{2}ng_z\mu_B B + D$	$\left(\frac{3}{4}\right)^{1/2}(\ell g_x - img_y)\mu_B B$	$(3)^{1/2}E$	0
$\langle 1/2 \rvert$	$\left(\frac{3}{4}\right)^{1/2}(\ell g_x + img_y)\mu_B B$	$\frac{1}{2}ng_z\mu_B B - D$	$(\ell g_x - img_y)\mu_B B$	$(3)^{1/2}E$
$\langle -1/2 \rvert$	$(3)^{1/2}E$	$(\ell g_x + img_y)\mu_B B$ $-\frac{1}{2}ng_z\mu_B B - D$	$\left(\frac{3}{4}\right)^{1/2}(\ell g_x + img_y)\mu_B B$	$\left(\frac{3}{4}\right)^{1/2}(\ell g_x - img_y)\mu_B B$
$\langle -3/2 \rvert$	0	$(3)^{1/2}E$	$\left(\frac{3}{4}\right)^{1/2}(\ell g_x + img_y)\mu_B B$	$-\frac{3}{2}ng_z\mu_B B + D$

Energy levels[b]

$$W_{\pm 3/2,z} = \pm\left(\frac{g_z\mu_B B}{2}\right) + \left[(D \pm g_z\mu_B B)^2 + 3E^2\right]^{1/2}$$

$$W_{\pm 1/2,z} = \mp\left(\frac{g_z\mu_B B}{2}\right) + \left[(D \mp g_z\mu_B B)^2 + 3E^2\right]^{1/2}$$

$$W_{\pm 3/2,i} = \pm\left(\frac{g_i\mu_B B}{2}\right) \pm \left[(g_i^2\mu_B^2 B^2 + D^2 + 3E^2 \mp g_i\mu_B B(D - 3\gamma_i E)\right]^{1/2}$$

$$W_{\pm 1/2,i} = \mp\left(\frac{g_i\mu_B B}{2}\right) \pm \left[(g_i^2\mu_B^2 B^2 + D^2 + 3E^2 \pm g_i\mu_B B(D - 3\gamma_i E)\right]^{1/2}$$

[a] (ℓ, m, n) are the direction cosines of $\tilde{B}$ in the orthorhombic system.

[b] The upper and lower signs refer to the M_s and $-M_s$ spin states, respectively. $\gamma_i = 1$ for i = x and $\gamma_i = -1$ for i = y.

For an isotropic $\underset{\sim}{g}$ tensor, plots of the resonance fields against D have been reported by Pedersen and Toftlund [76] for $h\nu = 0.31$ cm^{-1} (X band).

In the case of separated Kramers doublets, Eq. (21) has been obtained relating the effective $\underline{g}'$ values of the S = 1/2 Hamiltonian along the principal axes to the principal $\underline{g}$ values of the true S = 3/2 spin Hamiltonian [77]:

$$\underline{g}'_i = \underline{g}_i \pm 2\underline{g}_i (\alpha_i D + 3\gamma_i E)(D')^{-1} \mp \underline{g}_i [\underline{g}_i \mu_B B (D')^{-1}]$$

$$[(\alpha_i D + 3\gamma_i E)(D')^{-1}]\{1 - [(\alpha_i D + 3\gamma_i E)(D')^{-1}]^2\} \tag{21}$$

where i = x, y, z. In Eq. (21) the upper and lower signs refer to $|\pm 3/2\rangle$ and $|\pm 1/2\rangle$ Kramers doublets, respectively, $D' = (D^2 + 3E^2)^{1/2}$, and α_i and γ_i are defined in Eq. (22);

$$\alpha_i = \begin{cases} 1/2 \text{ for } i = x, y \\ \\ 1 \text{ for } i = z \end{cases} \qquad \gamma_i = \begin{cases} 1/2 \text{ for } i = x, y \\ \\ 0 \text{ for } i = z \end{cases} \tag{22}$$

C. S = 2 Spin Systems

The orthorhombic spin Hamiltonian that has to be used for an S = 2 spin system must include fine structure terms up to fourth order in the spin operators:

$$\underline{H} = \mu_B (\underline{g}_x \underline{B}_x \underline{S}_x + \underline{g}_y \underline{B}_y \underline{S}_y + \underline{g}_z \underline{B}_z \underline{S}_z) + D(\underline{S}_z^2 - 2) + E(\underline{S}_x^2 - \underline{S}_y^2) +$$

$$\frac{a}{6}(\underline{S}_x^4 + \underline{S}_y^4 + \underline{S}_z^4 - \frac{102}{5}) + \frac{F}{180}(35\underline{S}_z^4 - 155\underline{S}_z^2) \tag{23}$$

Often, however, the F and a terms can be neglected as compared to D [78-81]. In Table 8 the spin Hamiltonian matrix of Hamiltonian (23) for a general (ℓ, m, n) direction of $\underset{\sim}{B}$ including only second-order terms is reported.

The form of the observed spectra depends strongly on D [82]. If D $\gg$ hν, only the "forbidden" transitions within the $|\pm 2\rangle$ and $|\pm 1\rangle$ doublets will be observed. The allowed $|-1\rangle \overset{\rightarrow}{\leftarrow} |0\rangle$ transition

Table 8 The Orthorhombic Spin Hamiltonian Matrix for S = 2 Spin Systems for a General Orientation of $\underset{\sim}{B}$[a]

	$\lvert 2\rangle$	$\lvert 1\rangle$	$\lvert 0\rangle$	$\lvert -1\rangle$	$\lvert -2\rangle$
$\langle 2\rvert$	$2ng_z\mu_B B + 2D - \frac{F}{3} + \frac{a}{10}$	$(\ell g_x - img_y)\mu_B B - 3.4a$	$(6)^{1/2}E - 3.4a$	$-3.4a$	$-2.9a$
$\langle 1\rvert$	$(\ell g_x + img_y)\mu_B B - 3.4a$	$ng_z\mu_B B - D - \frac{2F}{3} - 0.4a$	$\left(\frac{6}{4}\right)^{1/2}(\ell g_x - img_y)\mu_B B - 3.4a$	$3E - 3.4a$	$-3.4a$
$\langle 0\rvert$	$(6)^{1/2}E - 3.4a$	$\left(\frac{6}{4}\right)^{1/2}(\ell g_x + img_y)\mu_B B - 3.4a$	$-2D + 0.6a$	$(\ell g_x - img_y)\mu_B B - 3.4a$	$(6)^{1/2}E - 3.4a$
$\langle -1\rvert$	$-3.4a$	$3E - 3.4a$	$\left(\frac{6}{4}\right)^{1/2}(\ell g_x + img_y)\mu_B B - 3.4a$	$-ng_z\mu_B B - D - \frac{2F}{3} - 0.4a$	$(\ell g_x - img_y)\mu_B B - 3.4a$
$\langle -2\rvert$	$-2.9a$	$-3.4a$	$(6)^{1/2}E - 3.4a$	$(\ell g_x + img_y)\mu_B B - 3.4a$	$-2ng_z\mu_B B + 2D - \frac{F}{3} + \frac{a}{10}$

[a] (ℓ, m, n) are the direction cosines of $\underset{\sim}{B}$ in the orthorhombic system

will be observed depending on the strength of the magnetic field.
If $D < h\nu$, many $\Delta M_s = \pm 1$ transitions can be observed. In Fig. 7,
plots of the energy levels for the spin Hamiltonian (23) are shown
for B along the z and x axes, together with the observable transitions.

D. S = 5/2 Spin Systems

The orthorhombic spin Hamiltonian for an S = 5/2 spin system is

$$\underline{H} = \mu_B(g_x B_x \underline{S}_x + g_y B_y \underline{S}_y + g_z B_z \underline{S}_z) + D(\underline{S}_z^2 - \frac{35}{12}) + E(\underline{S}_x^2 - \underline{S}_y^2) +$$

$$B_4^0 O_4^0 + B_4^2 O_4^2 + B_4^4 O_4^4 \tag{24}$$

The spin Hamiltonian matrix of Hamiltonian (24) is reported in Table
9. The B_4^2 parameter is often neglected, and the spin Hamiltonian
(25) is used:

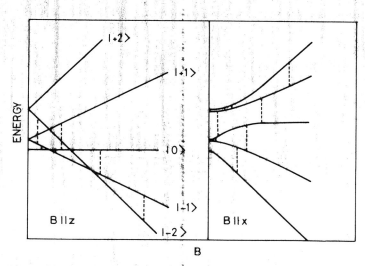

Figure 7   Energy levels as a function of the static magnetic field
for S = 2 systems with $|D| \gg |a|, |E|, |F|$.   Dashed lines indicate the
allowed transitions for $h\nu \doteq 2.6D$; solid lines are for $|D| \gg h\nu$.

Table 9 The Orthorhombic Spin Hamiltonian Matrix for S = 5/2 Spin Systems for a General Orientation of B[a]

	$\lvert 5/2\rangle$	$\lvert 3/2\rangle$	$\lvert 1/2\rangle$	$\lvert -1/2\rangle$	$\lvert -3/2\rangle$	$\lvert -5/2\rangle$
$\langle 5/2\rvert$	$\frac{5}{2}ng_z\mu_B B + \frac{10}{3}D + 60B_4^0$	$\left[\frac{5}{4}\right]^{1/2}(\ell g_x - img_y)\mu_B B$	$(10)^{1/2}E + 9(10)^{1/2}B_4^2$	0	$12(5)^{1/2}B_4^4$	0
$\langle 3/2\rvert$	$\left[\frac{5}{4}\right]^{1/2}(\ell g_x + img_y)\mu_B B$	$\frac{3}{2}ng_z\mu_B B - \frac{2}{3}D - 180B_4^0$	$(2)^{1/2}(\ell g_x - img_y)\mu_B B$	$(18)^{1/2}E - 15(2)^{1/2}B_4^2$	0	$12(5)^{1/2}B_4^4$
$\langle 1/2\rvert$	$(10)^{1/2}E + 9(10)^{1/2}B_4^2$	$(2)^{1/2}(\ell g_x + img_y)\mu_B B$	$\frac{1}{2}ng_z\mu_B B - \frac{8}{3}D + 120B_4^0$	$\frac{3}{2}(\ell g_x - img_y)\mu_B B$	$(18)^{1/2}E - 15(2)^{1/2}B_4^2$	0
$\langle -1/2\rvert$	0	$(18)^{1/2}E - 15(2)^{1/2}B_4^2$	$\frac{3}{2}(\ell g_x + img_y)\mu_B B$	$-\frac{1}{2}ng_z\mu_B B - \frac{8}{3}D + 120B_4^0$	$(2)^{1/2}(\ell g_x - img_y)\mu_B B$	$(10)^{1/2}E + 9(10)^{1/2}B_4^2$
$\langle -3/2\rvert$	$12(5)^{1/2}B_4^4$	0	$(18)^{1/2}E - 15(2)^{1/2}B_4^2$	$(2)^{1/2}(\ell g_x + img_y)\mu_B B$	$-\frac{3}{2}ng_z\mu_B B - \frac{2}{3}D - 180B_4^0$	$\left[\frac{5}{4}\right]^{1/2}(\ell g_x - img_y)\mu_B B$
$\langle -5/2\rvert$	0	$12(5)^{1/2}B_4^4$	0	$(10)^{1/2}E + 9(10)^{1/2}B_4^2$	$\left[\frac{5}{4}\right]^{1/2}(\ell g_x + img_y)\mu_B B$	$-\frac{5}{2}ng_z\mu_B B + \frac{10}{3}D + 60B_4^0$

$^a(\ell, m, n)$ are the direction cosines of B in the orthorhombic system.

$$\underline{H} = \mu_B(g_x B_x \underline{S}_x + g_y B_y \underline{S}_y + g_z B_z \underline{S}_z) + D(\underline{S}_z^2 - \frac{35}{12}) + E(\underline{S}_x^2 - \underline{S}_y^2) +$$

$$\frac{a}{6}(\underline{S}_x^4 + \underline{S}_y^4 + \underline{S}_z^4 - \frac{707}{16}) + \frac{F}{36}(7\underline{S}_z^2 + \frac{665}{14}\underline{S}_z^2 + \frac{567}{16}) \tag{25}$$

with a and F defined by Eq. (8).

In general, the values of the parameters associated with the quartic operators, a and F, are smaller than the parameters arising from the quadratic operators, D and E, and often in the literature they are not included in the Hamiltonian. Since S = 5/2 ions have in general isotropic g values close to the spin only value, the main factor affecting the appearance of the spectra is the zero field splitting tensor, and we try to show here how different spectra must be expected for different values of the D, E, a, and F parameters. In the most general environment, the six S = 5/2 levels will be split into three Kramer doublets according to the scheme shown in Fig. 8. In the limit of cubic symmetry, D = E = 0, and the only splitting can be determined by a. Under these conditions only one isotropic line will be observed in the ESR spectra if a = 0, with

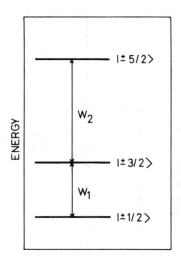

Figure 8 The splitting of the energy levels of an S = 5/2 system in the most general environment. $W_1 = -(3/2)[a + (3/2)F] + 4D - \{[a + (2/3)F + 2D]^2 + (5/4)a^2\}1/2$; $W_2 = 2\{[a + (2/3)F + 2D]^2 + (5/4)a^2\}1/2$.

$\underline{g} \simeq 2.0023$, whereas five lines can be observed if a $\neq$ 0 (Fig. 9).
If D and E are different from zero but small, five ESR lines are
seen. It must be mentioned that often together with these lines intense ΔM_s = ± 2 forbidden transitions are observed.

Another interesting limit case is that in which D and/or E is
large compared to the Zeeman splitting, so that the three Kramers

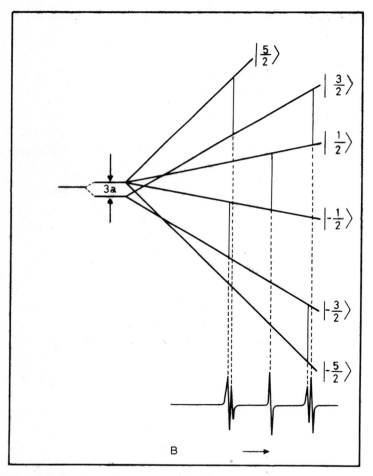

Figure 9 A typical spectrum for an S = 5/2 system in cubic symmetry
with a $\neq$ 0 for B parallel to a principal axis and hν $\gg$ a. (After
Ref. 2.)

Kramers doublets can be considered as effective $S = 1/2$ spin systems. In the case where D is large but E is close to zero, the lowest Kramers doublet has effective $\underline{g}_\parallel = 2$ and $\underline{g}_\perp = 6$. When $D = 0$ and $E \neq 0$, the middle doublet has an isotropic $\underline{g}$ value of 4.29 [83-85]. This condition is equivalent to $D \neq 0$ and $E/D = 1/3$ [86]. As a matter of fact, D and E are defined in terms of the D_{ii} components:

$$D = \frac{3D_{zz}}{2} = \frac{-3(D_{xx} + D_{yy})}{2}$$

$$E = \frac{D_{xx} - D_{yy}}{2} \tag{26}$$

Figure 10 shows how the D_{ii} components are expected to vary when $\lambda = E/D$ is varied in the range -1 to +1. In the negative range D_{yy} is kept fixed, whereas in the positive range D_{xx} is kept fixed. It is apparent that for $\lambda = -1$, 0, and +1, the zero splitting tensor is axial, with the unique axis x, y, and z, respectively. The values for which the anisotropy has its maximum are $\lambda = \pm1/3$. When $\lambda \longrightarrow \infty$ (i.e., D goes to zero, whereas $E \neq 0$, it is easy to show that

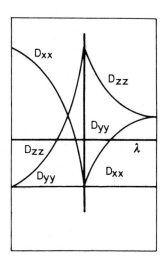

Figure 10 Values of the D_{ii} components as a function of $\lambda = E/D$. Right: D_{xx} fixed, λ varying from zero to 1/3. Left: D_{yy} fixed, λ varying from zero to -1/3.

$$D_{zz} = 0 \qquad D_{xx} = -D_{yy} \tag{27}$$

that is, the anisotropy is maximum.

Recently Golding et al. [87] have explored the conditions that give rise to an isotropic effective g tensor in terms of the spin Hamiltonian (25). They found that there are six cases under which one of the Kramers doublets yields an isotropic $\underset{\sim}{g}_{eff}$; four of them yield $\underset{\sim}{g}_{eff} = 30/7$ and two cases yield $\underset{\sim}{g}_{eff} = -10/3$. The six cases are reported in Table 10 together with the expressions of the $\underset{\sim}{g}$ principal values.

Dowsing and Gibson [88] published useful D versus B plots for various values of λ, which can be used to obtain the resonance fields of the various transitions. They are shown in Fig. 11. Although they may be used to interpret the ESR spectra of powders, they must be considered with some suspicion, since the intensities of the spin-forbidden $\Delta M_s = \pm 2$ transitions can be very high, compared with the intensities of spin-allowed bands. An alternative to this model involves the diagonalization of the spin Hamiltonian matrix and the computation of the transition probabilities after a proper spherical averaging of the intensities has been performed (see Sec. VI).

A more correct approach was suggested by Aasa [89], who determined the probabilities of the transitions that have their turning points not coincident with the principal axes.

VI. COMPUTER ANALYSIS OF ESR DATA

The computer techniques useful in the analysis of ESR data can be divided into two main branches: the fitting procedures and the simulation techniques. By fitting procedures we mean routines that, using as input data the observed transition fields, yield the best spin Hamiltonian parameters that reproduce them. It often happens, however, that the transition fields cannot be guessed directly from the spectra. Such are the cases of polycrystalline powder spectra or of overlapping lines in single-crystal spectra [90]. Under those conditions the problem of extracting useful information from the spectra

Table 10   The Possible Cases in Which an S = 5/2 Spin System Gives Rise to an Isotropic $\underset{\sim}{g}$ tensor and the General Expressions for the $\underline{g}$ Principal Values

$\underline{g}$ principal values[a]

$$\underline{g}_x = 2[3\beta^2 + (20)^{1/2}]\alpha\gamma + (32)^{1/2}\beta\gamma$$

$$\underline{g}_y = 2[3\beta^2 + (20)^{1/2}]\alpha\gamma - (32)^{1/2}\beta\gamma$$

$$\underline{g}_z = 2(5\alpha^2 + \beta^2 - 3\gamma^2)$$

The six cases in which $\underline{g}_x = \underline{g}_y = \underline{g}_z = \underline{g}$

Case	α	β	γ	$\underline{g}$
1	$\left(\frac{2}{7}\right)^{1/2}$	$\left(\frac{5}{7}\right)^{1/2}$	0	4.2857
2	$\left(\frac{2}{7}\right)^{1/2}$	$-\left(\frac{5}{7}\right)^{1/2}$	0	4.2857
3	$\left(\frac{9}{14}\right)^{1/2}$	0	$\left(\frac{5}{14}\right)^{1/2}$	4.2857
4	$\left(\frac{9}{14}\right)^{1/2}$	0	$-\left(\frac{5}{14}\right)^{1/2}$	4.2857
5	$\left(\frac{1}{6}\right)^{1/2}$	0	$-\left(\frac{5}{6}\right)^{1/2}$	-3.3333
6	$\left(\frac{1}{6}\right)^{1/2}$	0	$-\left(\frac{5}{6}\right)^{1/2}$	-3.3333

[a]The Kramers doublets have the general expression

$$|\psi^+\rangle = \alpha|5/2\rangle + \beta|1/2\rangle + \gamma|-3/2\rangle$$

$$|\psi^-\rangle = \alpha|-5/2\rangle + \beta|-1/2\rangle + \gamma|3/2\rangle$$

Source: Ref. 87.

can be solved only by the use of simulation routines, that is, by algorithms that can reproduce the observed shape of the spectra as a function of transition fields and intensities (or, more generally, of spin Hamiltonian parameters).

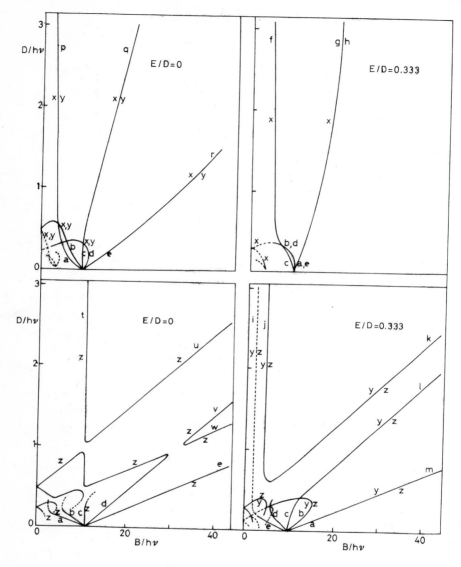

Figure 11 Plots of the magnetic field versus $D/h\nu$ for various E/D ratios for $S = 5/2$ systems. The transitions are (high-field labels) $a = -3/2 \rightarrow -5/2$; $b = -1/2 \rightarrow -3/2$; $c = 1/2 \rightarrow -1/2$; $d = 3/2 \rightarrow 1/2$; $e = 5/2 \rightarrow 3/2$; $f = 5/2(-3/2) \rightarrow -5(3/2)$; $g = -3/2(1/2) \rightarrow -5/2(-1/2)$; $h = 5/2(1/2) \rightarrow 3/2(-1/2)$; $i = 5/2(1/2) \rightarrow -5/2(-1/2)$; $j = 3/2(-1/2) \rightarrow -3/2(1/2)$; $k = -5/2(-1/2) \rightarrow 1/2(-3/2)$; $\ell = 1/2(-3/2) \rightarrow -5/2(-1/2)$; $m = -3/2(1/2) \rightarrow -5/2(-1/2)$; $p = 5/2(1/2)(-3/2) \rightarrow 5/2(1/2)(3/2)$; $q = 5/2(-3/2) \rightarrow 3/2(-1/2)(-5/2)$; $r = 5/2 \rightarrow 3/2$; $s = 3/2 \rightarrow -3/2$; $t = 1/2 \rightarrow -1/2$; $u = -3/2 \rightarrow -1/2$; $v = -1/2 \rightarrow -3/2$; $w = -5/2 \rightarrow -3/2$. (After Ref. 88.)

A. Fitting Procedures

In Sec. III we outlined the methods of obtaining the spin Hamiltonian tensors from experiment. The analysis is straightforward for $S = 1/2$ spin systems (either true or fictitious) and becomes more complicated when hyperfine coupling is included.

For $S \geqslant 1$ spin systems, the angular dependence of the transition fields is complicated by the presence of zero field splitting terms, which are often comparable in magnitude with the Zeeman term so that the perturbation formulas reported in Secs. IV and V cannot be used for fitting. In such cases exact diagonalization of the appropriate spin Hamiltonian matrix is needed.

Various methods based on exact numerical diagonalization of the spin Hamiltonian matrix have been proposed for fitting the observed resonance fields. Generally all methods require the minimization of a function expressing the poorness of fitting which is of the form

$$F = \sum_{i=1}^{n} [\Delta E(B_i) - h\nu]^2 \tag{28}$$

In Eq. (28), B_i $(i = 1, \ldots, n)$ are the n observed resonance fields, and $\Delta E(B_i)$ represents the energy difference at the observed field value B_i calculated with a set of spin Hamiltonian parameters. The various methods differ in the method by which the minimum value of F is calculated.

All methods of fitting must work satisfactorily for two or more magnetic field directions. This allows one to increase the number of resonance fields beyond the number of parameters and improve the consistency of the fit between orthogonal directions, which is a way to be confident that the minimum value of F is actually an absolute minimum.

In the following we give a brief survey of the principal fitting methods based on exact diagonalization of the spin Hamiltonian matrix.

Nonlinear least-squares fitting procedures have been described that are usually more effective and faster when compared to other methods [91-94]. These procedures, however, require the computation

of the derivatives of the eigenvalues of the spin Hamiltonian (or of
the transition fields) with respect to the parameters which, in the
case of a nonlinear dependence of the energy levels on the parameters,
can be difficult to achieve.

Jain and Upreti [95] describe a method for the analysis of spectra of ions with $S > 1/2$, based on Uhrin's iterative method [96],
that is applicable to the analysis of orthorhombic or higher-symmetry
crystal spectra. The authors formulate the method for the case of a
spin Hamiltonian of the form

$$\underline{H} = \mu_B \underline{B} \cdot \underline{g} \cdot \underline{S} + \sum_{k=2}^{2S} \sum_{q=0}^{k} B_k^q O_k^q \tag{29}$$

An extension of the method to include hyperfine coupling is also discussed, and a fitting procedure when the number of parameters is
greater than that of the observables is suggested.

The method of systematic variation of parameters by Buckmaster
et al. [97] has been modified by Misra and Sharp [98] in order to
vary all the parameters at the same time and to be able to consider
data obtained from more than one crystal orientation simultaneously.
The method fits data with the Hamiltonian (29). A similar method
has been contemporaneously proposed by Smith et al. [99], which allows one to fit transitions with $\Delta M = \pm 2$. It must be noted, however, that the methods based on a systematic variation of parameters
usually require a good initial guess of the starting values of the
parameters in order to converge. The authors suggest as starting
values those obtained by second-order perturbation formulas.

B. Simulation Techniques

Although in principle only single-crystal measurements can give the
experimental information required to determine the spin Hamiltonian
parameters, often it is experimentally difficult to record these
spectra. Powder spectra are more easily obtained, so much work is
concerned with extracting parameters from ESR polycrystalline powder
spectra. For the general case this requires computer simulation of

the spectra.   In App. A we report the relevant formulas needed for the simulation of both single-crystal and polycrystalline spectra. An exhaustive review article dealing with magnetic resonance spectra of polycrystalline solids has been published by Taylor et al. [100].

An efficient method for computing the general formula (79) that gives the derivative, with respect to the field, of the absorption of radiation by a polycrystalline sample is to approximate it by (30):

$$\frac{d}{dB}W(B) \propto \sum_{i=1}^{n} \sum_{j=1}^{m} \sum_{a} <M_a> \left| \frac{\partial B_a}{\partial h\nu} \right| \frac{d}{dB} f_a (B - B_a) \sin \theta_i \frac{\pi}{nm} \tag{30}$$

in which the integration paths are subdivided by the relations

$$\theta_i = \left(\frac{\pi}{n}\right) i \qquad \phi_j = \left(\frac{\pi}{m}\right) j \tag{31}$$

and the other symbols are defined in App. A.   Usually the number of subintervals in θ and ϕ required in order to have a reasonable line shape is about 300 times the ratio between field range and line width, depending on the anisotropy in the g tensor [101].   In order to reduce this lengthy calculation, symmetry can be used to reduce the integration limits in Eq. (79).   In Table 11, suggested integration limits for various centrosymmetric point groups are given.

Further reduction in computer time is obtained through interpolation procedures that allow one to calculate B_a, $<M_a>$, and $(\partial B_a / \partial h\nu)$ for a limited number of orientations of B and to have approximate values for all the other orientations [102,103].   The greatest inconvenience related to all the interpolation procedures is that, in the case where more than one transition contributes to the spectrum (a > 1), an algorithm is needed that can correlate transitions at different orientations.   Van Veen [101] has reported an interpolation procedure that effects this correlation almost everywhere in (θ,ϕ) space, so that direct calculations of B_a, $<M_a>$, and $(\partial B_a / \partial h\nu)$ can be limited to a narrow set of field orientations.

The line shape function $f_a (B - B_a)$ occurring in Eq. (30) is defined generally in App. A.   This is a rather complicated expression determined by the line-shape function of each line-broadening

Table 11   Integration Limits for Various Point Groups

Point group[a]	Integration limits in θ From	To	Integration limits in ϕ From	To
S_2 (C_1)	0	$\pi/2$	0	
C_{2h} (C_2, C_s)	0	$\pi/2$	0	
D_{2h} (D_2, C_{2v})	0	$\pi/2$	0	$\pi/2$
C_{4h} (C_4, S_4)	0	$\pi/2$	0	$\pi/2$
D_{4h} (D_4, C_{4v}, D_{2d})	0	$\pi/2$	0	$\pi/4$
S_6 (C_3)	0	π	0	$\pi/3$
	0	$\pi/2$	0	$2\pi/3$
D_{3d} (D_3, C_{3v})	0	π	0	$\pi/6$
	0	$\pi/2$	0	$\pi/3$
C_{6h} (C_6, C_{3h})	0	$\pi/2$	0	$\pi/3$
D_{6h} $(D_{3h}, C_{6v}, C_{6h}, D_6)$	0	$\pi/2$	0	$\pi/6$
T_h (T)	As for S_6 or D_{2h}			
O_h (T_d, O)	As for D_{3h} or D_{4h}			
$D_{\infty h}$ $(C_{\infty v})$	0	$\pi/2$		

[a]In parentheses are reported noncentrosymmetric point groups. $\theta = 0$ corresponds to the principal axis. $\theta = \pi/2$, $\phi = 0$ correspond to a twofold axis perpendicular to the unique axis; when this twofold axis is not present, the origin in the integration over ϕ is not important. *Source:* Ref. 101.

mechanism that contributes to the total feature. The line shape can be defined as homogeneous or nonhomogeneous, depending on whether the mechanism responsible for the broadening acts in the same way for each molecule in the system or not. A general line shape is computed as the convolution of the line-shape functions for the nonhomogeneously, $G(B - B_a)$, and homogeneously, $L(B - B_a)$, broadened line shape:

$$f_a(B - B_a) = \int G_a(B - B_a)L_a(B - t) \, dt \tag{32}$$

Often nonhomogeneous broadening is due to mechanisms such as imperfection in the single crystal, strain, unresolved hyperfine structure, and so on, so the L function is the convolution of the component ones. Usually a nonhomogeneously broadened line shape is approximated by a Gaussian function and a homogeneously broadened line shape by a Lorentzian function of $x = B - B_a$. For the derivative spectrum the line-shape functions are

$$G'(x) = \frac{3\Gamma x}{\pi \sqrt{3} \left[\left(\frac{3\Gamma^2}{4} \right) + x^2 \right]^2} \tag{33}$$

$$L'(x) = \frac{6\Gamma x}{\sqrt{3\pi (3\Gamma^2 + 4x^2)^2}} \tag{34}$$

$G'(x)$ and $L'(x)$ are the first derivatives with respect to x of a Gaussian and a Lorentzian function, respectively. In Eqs. (33) and (34), Γ is the peak-to-peak amplitude in the derivative ESR spectrum.

The peak-to-peak amplitude obtained by a convolution of two or more Gaussian functions is given by Eq. (35);

$$\Gamma^2 = \sum_{i=1}^{n} \Gamma_i^2 \tag{35}$$

where Γ_i is the peak-to-peak amplitude for a Gaussian component and n is the number of components. The convolution product of a Gaussian and a Lorentzian function is known as a Voigt function. Values and approximate expressions for the Voigt function have been reported [104,105].

When the experimental lines cannot be considered to a good approximation as either simple Gaussian or Lorentzian, a phenomenological line width can be used in the form [101]:

$$\Gamma_a = \Gamma_0 \left| \frac{\partial B_a}{\partial h\nu} \right| + \sum_p \Gamma_p \left| \frac{\partial B_a}{\partial h\nu} \right| \tag{36}$$

where p refers to a spin Hamiltonian parameter and Γ_p is its standard deviation. Quite general expressions for the line shapes for $S = 5/2$ spin systems have been published by Coffman [106].

When perturbation theory cannot be applied in order to calculate the resonance fields B_a, exact numerical diagonalization of the

complete spin Hamiltonian matrix is needed in order to obtain the
energy of the states and their eigenvectors. In App. B the relevant
formulas for computing the resonance fields and their required
derivatives are reported.

VII. MOLECULAR ORBITAL APPROACH

The first derivation of the spin Hamiltonian in terms of the molecular orbital (MO) model was made by Maki and McGarvey [107] in 1958.
In this early treatment, several approximations were inherent, which
become less satisfactory for low-symmetry molecules [108,109]. A
similar approach was used by Kivelson and Neiman [110] for trigonal
copper(II) complexes.

Recently, Keijzers and de Boer [111-112] extended the model to
cope with these inconveniences by comparing directly the spin
Hamiltonian parameters with the results of extended Hückel calculations and suggested general formulations for the g and A tensors in
the case of one unpaired electron. Using a perturbation treatment,
they showed that the states that can contribute to the calculation
of g and A are (1) those with an electron excited into the MO of the
unpaired electron, (2) with the unpaired electron excited into an
initially empty MO, and (3) with an electron excited from a doubly
occupied MO into an empty one. The resulting expressions are [113]

$$g_{\alpha\beta} = g_e \delta_{\alpha\beta} + g_e \sum_{m \neq n} \sum_{k,k'} \frac{<\chi_n^k | \xi_k(\underline{r}_k) L_\alpha^k | \chi_m^k> <\chi_m^{k'} | L_\beta^{k'} | \chi_n^{k'}>}{\varepsilon_n - \varepsilon_m}$$

$$A_{\alpha\beta}^k = P \left\{ <\psi_n \left| \frac{F_{\alpha\beta}^k}{\underline{r}_k^3} \right| \psi_n> + \sum_{m \neq n} \sum_k \left| \frac{2<\psi_n | \xi(\underline{r}_k) L_\alpha^k | \psi_m> <\psi_m | L_\beta^k/\underline{r}_k^3 | \psi_n>}{\varepsilon_n - \varepsilon_m} \right. \right.$$

$$\left. \left. + \sum_{\gamma,\delta} \frac{i \exp(\gamma\delta\alpha) <\psi_n | \xi_k(\underline{r}_k) L_\gamma^k | \psi_m> <\psi_m | F_{\delta\beta}^k/\underline{r}_k^3 | \psi_n>}{\varepsilon_n - \varepsilon_m} \right| \right\}$$

(37)

$$P = g_e g_N \mu_B \mu_N$$

where n indicates the ground state, the summation over m runs over
all excited states, and the summations k and k' run over all atoms.

α and β are Cartesian components. L_α^k is the α component of the angular momentum vector operator centered on the nucleus k; ξ_k $(\underline{r}_k)$ is the one-electron spin-orbit coupling operator for the atom k; χ_m^k is the part of the m-th MO that is centered on the atom k; and $F_{\alpha\beta}^k$ is a component of the symmetrical traceless tensor operator related to the dipolar electron-nucleus interaction. $\underline{A}_{\alpha\beta}^k$ is the anisotropic part of the hyperfine tensor relative to the nucleus k, to which the isotropic part

$$\underline{A}_{iso}^k = \frac{8\pi P}{3}[\psi_n(0)]^2 \tag{38}$$

must be added. This term represents the contribution of s electrons to the isotropic hyperfine splitting of nucleus k, either direct or through spin polarization.

The authors used these formulas within an extended Hückel formalism, neglecting several multiple-center integrals because of their $\underline{r}^{-3}$ dependence. The complexes they considered were copper bis(diethyldithiocarbamate) and copper bis(di-t-butyldiselenocarbamate) [24-114], and they were able to obtain good agreement between experimental and calculated values. In particular, they showed that for these complexes the ligand contribution to the $\underline{g}$ and $\underline{A}$ values is relevant, so that the use of simple formulas to obtain MO coefficients is in this case meaningless. This point had been raised previously by other authors [115-117]. Keijzers and de Boer had to include the 4d functions of selenium in the calculation in order to have good agreement with the principal $\underline{g}$ values and directions.

The molecular orbital approach has been used for the calculation of the ligand hyperfine structure, which cannot be accomplished within a ligand field formalism. In particular, in the case of octahedral complexes, the anisotropic transferred hyperfine structure, that is, the contribution to the ligand hyperfine coupling originating from spin density transfer from the metal ion to the ligands, has been related to the difference $f_\sigma - f_\pi$, where f_σ and f_π are the fractions of unpaired spin on the ligands in σ and π orbitals, respectively [118]. The important point to be made is that,

in order to have successful calculations, the polarization of the
core orbitals by the unpaired electron must be taken into account.
Usually this is made by using different orbitals for different spins.
The main limitations for this kind of calculation is that of having
a good molecular orbital scheme, and this is not in general a simple
matter for transition metal complexes. Literature reports on some
octahedral hexahalogeno complexes of different metal ions [119-121]
have shown reasonable agreement with the experimental data.

In the past few years the multiple scattering Xα (MSXα) [122-
124] method has attained some popularity for the interpretation of
the electronic properties of transition metal complexes [125-127].
The method is essentially an ab initio one-particle approach in
which the molecular orbitals do not depend on atomic orbitals as
basis functions as in the LCAO (linear combination of atomic orbitals) approach. In the muffin tin approximation to the potential,
the atomic sites are surrounded by nonoverlapping spheres and the
potential is spherically averaged within these spheres. The spheres
touch each other along the bond directions. All the spheres are
surrounded by an outer sphere, which touches all the ligand spheres.

Larsson [128,129] used the method to calculate the spin and
bonding properties and optical spectra of TiF_6^{3-}, CrF_6^{3-}, $CrCl_6^{3-}$,
MnF_6^{3-}, MnF_6^{4-}, FeF_6^{3-}, NiF_6^{2-}, NiF_6^{4-}, and $NiCl_6^{4-}$. Since the method
does not use a LCAO approach, the orbitals obtained through the
calculation are characterized through integration of the probability
distribution for each orbital within each sphere. In this respect
the method is similar to the Mulliken population analysis [130].
Surprisingly enough, in the case of $NiCl_6^{4-}$ the analysis shows that
the bonding $\underline{e}_g$ orbital with spin up is 54% within the nickel sphere.
Undoubtedly this produces a large covalency with the composition of
bonding and antibonding orbitals reversed as compared to general
expectations. The calculated orbitals were used to evaluate the
transferred hyperfine interactions defined through the integrals
representing the "occupancy" of the atomic orbitals:

$$q_R(\phi) = \int_V \overline{\phi}(\underline{r}) \phi(\underline{r}) \, dv \qquad\qquad (39)$$

where the integration volume V is the ligand sphere with radius R and ϕ is a suitable molecular orbital. In octahedral complexes the relation between f_σ and f_π and the above-defined quantities is

$$f_\sigma = 2[\sum_i q_L(\underline{e}_g^i\uparrow) - \sum_j q_L(\underline{e}_g^j\downarrow)]$$

$$f_\pi = \frac{3}{2}[\sum_i q_L(\underline{t}_{2g}^i\uparrow) - \sum_j q_L(\underline{t}_{2g}^j\downarrow)] \qquad (40)$$

where $\underline{e}_g^i$ and $\underline{t}_{2g}^i$ are bonding and antibonding orbitals, and $\uparrow$ and $\downarrow$ refer to the orbitals with spin up and down, respectively. By the same method the contact contribution to the ligand hyperfine coupling can be calculated according to the relation

$$\underline{A}_{iso} = \frac{P}{2S}\frac{8\pi}{3}[\rho_\uparrow(R_a) - \rho_\downarrow(R_a)] \qquad (41)$$

where the ρ's are defined as

$$\rho_\uparrow(\underline{r}) = \sum_i \overline{\phi}_i\uparrow(\underline{r})\phi_i\uparrow(\underline{r})$$

$$\rho_\downarrow(\underline{r}) = \sum_i \overline{\phi}_i\downarrow(\underline{r})\phi_i\downarrow(\underline{r}) \qquad (42)$$

and evaluated at nucleus a. The summations in Eq. (42) run over all the orbitals with spin up and down, respectively.

The calculated spin densities were compared satisfactorily with experimental nuclear magnetic resonance and ESR data.

VIII. LIGAND FIELD APPROACH

A. General Considerations

The application of the ligand field model to the interpretation of the spectral and magnetic properties of transition metal complexes has been "up and down" over the years. After the initial success of classical crystal field theory, which began to be used by inorganic chemists about 30 years after its introduction by Bethe [131] and van Vleck [132], the inadequacy of its physical basis was early recognized , and ESR experiments were crucial in this respect

[133-135]. Ligand field theory overcame most of the inconveniences
of crystal field theory, but at the same time molecular orbital
methods, using Wolfsberg-Helmholtz and similar approaches, became
equally popular, since it was hoped that the latter approach could
provide the ultimate answers. However, the formidable difficulties
inherent in a molecular orbital treatment of open shells induced
many people to return to the simpler ligand field model.

Recently, the angular overlap model has been developed [136,
137] in order to express the electronic energies with parameters
that are more MO oriented than those of the classical ligand field
theory. In particular, it was shown that both the angular overlap
and the classic ligand field models are similar in approach, that
is, the electronic energies are calculated by a perturbation treatment of the metal orbital only. Under some assumptions the two
models are mathematically equivalent. Several excellent reviews of
the theory and applications are now available [138-141], and we
shall not go through it here.

The main perturbation formulations of the g, D, and A tensors
in the ligand field formalism were reviewed by McGarvey [15]. Although often valid, in some cases it has now been recognized that
perturbation calculations are inaccurate and that more sophisticated
approaches must be attempted.

One of the main drawbacks of the perturbation models is that
the principal axes of the spin Hamiltonian tensors are not known and
must be guessed. Although this can be done when the symmetry of the
metal ion is orthorhombic or higher, it is impossible to know where
the axes are to be placed when the symmetry is lower. In the survey
of experimental data, many examples will be shown where the principal directions of the g tensor are rotated by large angles from the
positions they would be expected to have on the basis of an idealized higher symmetry.

Recently, Gerloch et al. developed a ligand field treatment
for the interpretation of the magnetic susceptibility of unsymmetric
transition metal complexes [142], where they also suggested a convenient form for the calculation of the g^2 tensor in its general

nondiagonal form from which the principal values and directions can be obtained by diagonalization.

The central assumption of this model is in the definition of the g value by the first-order Zeeman splitting of a degenerate manifold as $\mu_B g B$ between adjacent levels. Therefore the model can be successfully applied when the ground state, obtained by ligand field spin-orbit coupling calculations, is a multiplet separated from the excited levels by an energy gap that is larger than the Zeeman interaction. If there is only one unpaired electron, the ground multiplet will be a Kramers doublet, and the calculated g values will be the actual g values. If there is more than one unpaired electron, the ground multiplet may or not correspond to the spin multiplicity of the metal ion.

High-spin cobalt(II) often has a ground Kramers doublet well separated by other multiplets, so that the ESR spectra are interpreted using an effective spin Hamiltonian with S = 1/2. The values of g that are calculated with this model are exactly the g values within the ground doublet, therefore the effective g values can be compared directly with experiment. Horrocks and Burlone [143] used a similar approach to calculate the g values of tetrahedral cobalt-(II) complexes.

Recently, Bencini and Gatteschi [144] extended the formalism to calculate the metal hyperfine splitting. The relevant formulas for the calculation of g^2 and A^2 are

$$g^2_{\alpha\beta} = n \sum_{i,k} <i|\mu_\alpha|k><k|\mu_\beta|i> \tag{43}$$

$$A^2_{\alpha\beta} = M^{-2} \sum_{i,k} <i|\tilde{S}_\alpha|k><k|\tilde{S}_\beta|i> \tag{44}$$

where α and β are Cartesian components; μ_α is the α component of the magnetic moment operator $\mu = L + g_e S$; $\tilde{S}_\alpha I_\alpha$ is the α component of the hyperfine coupling Hamiltonian written in the form $H_{hf} = P \sum_\alpha \tilde{S}_\alpha I_\alpha$; n is the number of degenerate levels at zero field; and $M^2 = \sum_{m=-S}^{+S} m^2$.

The summation in Eqs. (43) and (44) is extended over all the $|i>$ and $|k>$ states that are degenerate at zero magnetic field.

Although in principle any ligand field model is well suited for this kind of calculations, the angular overlap model has been most widely used. There are several reasons for this. The first is that the relevant matrix elements are well defined in the case when no symmetry element is present. The second is that the AOM can easily take into account nonlinear bonding [145] of a donor atom. By this we mean that the ligand-metal bonding axis does not possess cylindrical symmetry. For instance, if one considers a bond as shown in Fig. 12, where a pyridine ring is lying in the yz plane, the π interaction parallel to x will be different from the π interaction parallel to y. The AOM can take it into account by setting $\underline{e}_{\pi y} = 0$, and $\underline{e}_{\pi x} \neq 0$.

An example of the efficacy of the AOM in the interpretation of ESR spectra has been provided by the $M(\text{pyridine-N-oxide})_6(ClO_4)_2$ complexes, where M = Mn, Fe, Co, Ni, Cu, or Zn. All these complexes

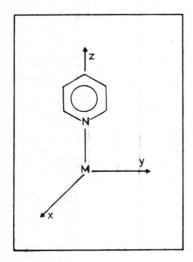

Figure 12   An example of a nonlinearly ligating ligand. The pyridine ring lies in the yz plane and can have a π interaction only in the x direction.

are isomorphous, belonging to the R_3 space group with the metal atom in a site S_6 [146]. However, the O-M-O angles are as close as possible to the octahedral angles, being 89.3° and 90.7° [147]. Therefore, using idealized symmetry, one would anticipate that the ESR spectra would be very close to the limit of O_h symmetry. This is not the case, however, and both manganese(II) [148] and nickel-(II) complexes [149] show very large zero field splittings, and the cobalt(II) complex has $g_\parallel$ = 2.26, $g_\perp$ = 4.77 [150]. The origin of this low symmetry has been suggested to be due to the nonlinear ligating behavior of the oxygen donors [151,152]. As a matter of fact, the M-O-N angle γ (Fig. 13) is close to 120°, so that the oxygen atom can be considered to be sp^2 hybridized, giving rise to very anisotropic π interaction with the metal M. As a consequence, the actual symmetry is distinctly lower than O_h, as a matter of fact, S_6, as seen in the structure. Mackey et al. [152] used the AOM to calculate the magnetic and ESR properties of cobalt(II) and nickel-(II) pyridine N-oxide complexes and was able to reproduce them with

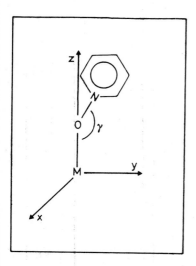

Figure 13 The nonlinear bonding of pyridine-N-oxide in the complexes $M(pyO)_6(ClO_4)_2$, where M = Mn, Fe, Co, Ni, Cu, or Zn.

physically reasonable values of the $\underset{\sim}{e}_\sigma$ and $\underset{\sim}{e}_\pi$ parameters. Wood [153] studied electron density maps and found substantial support for this view.

B. Effects of Covalency

It is well known that by using the ligand field approach the calculated values of χ, g, and so on, are larger than the experimentally obtained values. In order to obtain agreement with experiment, it is necessary to reduce the values of several integrals from the values they have in the free ion [154]. The integrals that are relevant for the calculation of the spin Hamiltonian parameters are $<du|\underset{\sim}{L}|dv>$ and $<du|\xi(\underline{r})L\cdot S|dv>$. In a molecular orbital treatment these integrals are not computed with pure d metal functions, but in a LCAO approximation, with linear combinations of ligand and metal orbitals. If the assumptions of the ligand field model are valid, the coefficient of the d orbital in the molecular orbital will be close to 1. Stevens [133], Owen and Thornley [118], and Tinkham [155-156] showed how, in cubic symmetry, the integrals referring to the $\underset{\sim}{L}$ operator can be expressed as

$$<t_2|\underset{\sim}{L}|t_2> = \underline{k}(t_2t_2)<dt_2|\underset{\sim}{L}|dt_2>$$
$$<t_2|\underset{\sim}{L}|e> = \underline{k}(t_2e)<dt_2|\underset{\sim}{L}|de> \tag{45}$$

where $|t_2>$ and $|e>$ stand for the true MO functions and $|dt_2>$ and $|de>$ are the corresponding pure d functions. Both $\underline{k}(t_2t_2)$ and $\underline{k}(t_2t_2)$ are smaller than 1. It is also commonly assumed that $\underline{k}(t_2t_2) = \underline{k}(t_2e) = \underline{k}$, so that only one parameter is used. This has not much theoretical basis, the best justification being that with this assumption the treatment becomes as simple as possible. A similar simplification is made for the electron-repulsion energy terms. In spherical symmetry only two parameters are needed, B and C, whereas in a cubic environment nine would be required, but only two, B' and C', are actually used.

The orbital reduction factor $\underline{k}$ is generally considered to be unique in symmetries lower than cubic, but in tensorial form,

k_{ij} (i, j = x, y, z) [157]. An early review on the role of k in magnetic susceptibility measurements has been published [158].

The other relevant integrals $<du|\xi(\underline{r})L \cdot S|dv>$ will be reduced accordingly, since the same kind of orbital momentum integrals are involved. However, there is a difference in principle, since $\xi(\underline{r})$ has a $\underline{r}^{-3}$ dependence, which makes the integrals $<d|\xi_L(\underline{r}_L)|\phi_L>$, $<d|\xi(\underline{r})|\phi_L>$, $<d|\xi_L(\underline{r}_L)|d>$ and $<\phi_L|\xi(\underline{r})|\phi_L>$ negligible. Consequently, the spin-orbit reduction factor, for which the same limitations mentioned above for k are valid, should be indicated as k' and should be different from k [155]. Finally, it must be mentioned that another mechanism is also operative to reduce the value of the spin-orbit coupling integrals. This is sometimes referred to as central field covalency [159,160], as opposed to the above-described mechanism, which is referred to as symmetry-restricted covalency. In the former mechanism the effective charge on the metal nucleus in the complex will be lower than the charge on the free ion, thus producing an expanded radial function and a different value for ξ, the spin-orbit coupling constant.

The effect of different reductions on the calculated g values has been considered at some length by Hathaway and Billing [161-162]. Smith has made detailed calculations of g values for tetragonal copper(II) complexes [116].

Covalency will also affect the hyperfine splitting. Usually the metal hyperfine constant is given by the sum of three contributions, the contact, the dipolar, and the pseudo-dipolar term. The last depends on the magnetic anisotropy on the central metal ion, therefore it is reduced exactly as the g values. The dipolar term, which is determined by the spin operators, is reduced according to the square of the coefficient of the d metal orbital in the molecular orbital. Also, the contact term is different from that in the purely ionic case. If the contact interaction is due solely to core polarization [163], the observable K is the expectation value of the polarization operator K averaged over the electron wave function [164]. In the simple case of a molecular orbital that can be described as

$$|\psi> = \alpha|d> - \beta|\psi_L> \tag{46}$$

with $\alpha^2 + \beta^2 - 2\alpha\beta<d|\psi_L> = 1$, the experimental contact term is given by

$$\underline{K} = <\psi|\underline{K}|\psi>$$

$$= \alpha^2<d|\underline{K}|d> + \beta^2<\psi_L|\underline{K}|\psi_L> - \alpha\beta(<d|\underline{K}|\psi_L> + <\psi_L|\underline{K}|d>) \tag{47}$$

The terms in parentheses should be smaller than the other two, so that

$$\underline{K} \simeq \alpha^2 K_d + \beta^2 K_L \tag{48}$$

showing that the observed contact term is slightly different from the calculated value for the free ion.

Finally, it must be mentioned that since the charge on the metal in the complex is different from that in the free ion, $P' = \underline{g}_e g_N \mu_B \mu_N <\underline{r}^{-3}>$ will be reduced as compared to the free ion value.

IX.   SURVEY OF EXPERIMENTAL RESULTS

In this section results of ESR studies are given and discussed. The subsections are organized according to the metal atom, so that iron(III) and iron(II), for instance, will be covered under the same heading. At the beginning of each subsection, short information on the isotopes possessing nuclear spin is provided. Although this survey is not comprehensive, the maximum possible information is gathered in the tables. In general, the literature has been scanned to the end of 1978. Results that can be found in the review articles of McGarvey [15] and Goodman and Raynor [16] often are not quoted.

A.   Titanium

Titanium has five naturally occurring isotopes, ^{48}Ti (abundance 73.45%) being the most abundant. Of these only ^{47}Ti (abundance 7.75%) and ^{49}Ti (abundance 5.51%) possess nonzero nuclear spin quantum numbers, $I = 5/2$ and $I = 7/2$, respectively, and have nearly equal gyromagnetic ratios.

1. Titanium(III)

Titanium(III) is a d^1 ion, and its ESR spectra can be interpreted using an $S = 1/2$ spin Hamiltonian. Because of the existence of ^{47}Ti and ^{49}Ti with nearly equal gyromagnetic ratios, the ESR spectra of Ti(III) compounds are characterized by strong central lines and satellite lines caused by the overlapping of two spectra of six and eight lines, respectively.

A few molecular complexes of titanium(III) have been characterized by ESR spectroscopy, and they have been reviewed by Goodman and Raynor [16].

The most common stereochemistry of titanium(III) is octahedral. In this environment the free ion ^2D is split into a $^2T_{2g}$ ground level and ^2E_g excited one. Low-symmetry effects split the $^2T_{2g}$ level, leaving in most cases low-lying excited states that produce short spin-lattice relaxation times and make the ESR spectra undetectable at room temperature.

Some selected literature data on the spin Hamiltonian parameters of titanium(III) are reported in Table 12.

The best-characterized titanium(III) compounds are trigonally distorted octahedral [165,166]. In D_3 symmetry the $^2T_{2g}$ state splits into $^2A_1$ and ^2E, and spin-orbit coupling produces three Kramer doublets. Gladney and Swalen have performed complete ligand field spin-orbit coupling calculations using the ligand field parameters Δ, v, and v' representing, respectively, the cubic splitting, the diagonal splitting of the ground manifold, and the off-diagonal interaction of the ground manifold with the upper one. For $v > 0$, the lowest-lying doublet (independent of v'/Δ) is derived from ^2E and has the g values [165,167]

$$g_\parallel = - \frac{4\sqrt{2}t'}{1 + t/3 - t'/\sqrt{2}} \left(1 - \frac{\tau\sqrt{2}}{t'} + \frac{3\tau^2}{2t'^2} \right)^{1/2}$$

$$g_\perp = 0 \tag{49}$$

with $t = v/\Delta$, $t' = v'/\Delta$, and $\tau = \xi/\Delta$.

Table 12 Spin Hamiltonian Parameters of Some Titanium(III) Complexes

Compound	Chromophore[a]	Spectra[b]	g_1^c	g_2^c	g_3^c	A_1^c	A_2^c	A_3^c	Reference
CsTi(SO$_4$)$_2$·12H$_2$O	TiO$_6$	C	1.14		1.25				169
K(Al,Ti)(SO$_4$)$_2$·12H$_2$O	TiO$_6$	C	1.828	1.898	1.979				170
(Al,Ti)(acac)$_3$	TiO$_6$	C	1.921		2.000				175
[C(NH$_2$)$_3$](Al,Ti)(SO$_4$)$_2$·6H$_2$O	TiO$_6$	C	1.295		1.136				166
TiCl$_3$·6H$_2$O	TiO$_6$	P	1.91		1.84				165
(CH$_3$NH$_3$)(Al,Ti)(SO$_4$)$_2$·12H$_2$O	TiO$_6$	C	1.61		1.37				176
KTi(C$_2$O$_4$)$_2$·2H$_2$O	TiO$_6$	P	1.96		1.86				165
TiCl$_3$(aq)	TiO$_6$	G	1.896		1.994				174
trans-[Ti(H$_2$O)$_4$F$_2$]$^+$	TiO$_4$F$_2$	G	1.932		1.968		16.2		179
[(π-C$_5$H$_5$)$_2$Ti(PPh$_2$)$_2$]$^-$	Ticp$_2$P$_2$	G	1.979	1.988	1.998		8.8		180
[(π-C$_5$H$_5$)$_2$Ti(GePh$_3$)$_2$]$^-$	Ticp$_2$Ge$_2$	G	1.977	1.993	2.001		7.2		180
(π-C$_5$H$_5$)$_2$TiH$_2$Na	Ticp$_2$H$_2$	S	1.994		1.994		6.7		181
(π-C$_5$H$_5$)$_2$TiH$_2$AlCl$_3$	Ticp$_2$H$_2$	S	1.992		1.992		5.2		181

[a] Metal and nearest-neighbor atoms.

[b] C = single crystal; P = polycrystalline powder; G = frozen solution; S = fluid solution.

[c] (1, 2, 3) refer to (x, y, z) in single-crystal spectra. When two values are reported, the first is the perpendicular component; when one value is reported, it refers to the average value. The A values are in cm^{-1} × 10^{-4}.

For the case in which $v < 0$, but not $v'/\Delta \gg v$, the ground Kramers doublet is derived from $^2\underline{A}_1$. First-order perturbation theory [168] gives for the $\underline{g}$ values

$$\underline{g}_\| = 3 \frac{(\xi/2) - v}{S - 1}$$

$$\underline{g}_\perp = \frac{v + (3\xi/2)}{S + 1} \tag{50}$$

with $S = \{[(\xi/2) - v]^2 + 2\xi^2\}^{1/2}$, yielding $\underline{g}_\| > \underline{g}_\perp$ for every v/ξ value. The $\underline{g}_\| > \underline{g}_\perp$ pattern has been found for the hexaaquoion $[Ti(H_2O)_6]^{3+}$ in various host lattices [169-173] and in solution [174], and for $(Al,Ti)(acetylacetonate)_3$ [175].

A reverse $\underline{g}$ pattern, $\underline{g}_\| < \underline{g}_\perp$, has been observed in $TiCl_3 \cdot 6H_2O$ and $KTi(C_2O_4)_2 \cdot 2H_2O$ [165], but no X-ray data are available for these compounds, and no structural correlation can be made. More recently, Rumin et al. [176] have measured the ESR spectra of the $[Ti(H_2O)_6]^{3+}$ in the $[CH_3NH_3]Al(SO_4)_2 \cdot 12H_2O$ host lattice (space group P_{a3}, site D_3) and have found $\underline{g}_\perp = 1.61 > \underline{g}_\| = 1.37$. This behavior has been interpreted by assuming Jahn-Teller coupling between the $\underline{E}_g$ normal modes of lattice vibration and the $\underline{T}_{2g}$ orbital ground state. Ham [177] has shown that the effect of this coupling is to reduce the trigonal field and spin-orbit coupling operators from their values in the absence of a Jahn-Teller interaction by a quantity that to first order is $\exp\{-3E_{JT}/2\hbar\omega\}$, where E_{JT} is the minimum in the potential energy. As a consequence, the orbital contribution in Eq. (50) is partially quenched according to the value of the exponential. McFarlane derived formulas for $\underline{g}$ in the presence of Jahn-Teller coupling [178] that gave good agreement with experimental data.

A number of π-cyclopentadienyltitanium(III) complexes have been characterized by ESR spectroscopy [180-183], and data for some of them are reported in Table 12. Wasson et al. performed extended Hückel molecular orbital calculations for compounds of general formula $Cl_2TiX_2M(Y_2)$ (X = H, Cl; M = Li, Na, Mg, Al, Zn; Y = Br, Cl) in order to interpret the superhyperfine splitting observed both with the bridging X atom and with the trans-annular M atom. He found

that the main contributions to the interaction of the unpaired
electron with the M nucleus is due to through-space metal-3d-trans-
annular-atom-s orbital coupling [184].

B. Vanadium

The most abundant isotope of vanadium is ^{51}V (abundance 99.8%). It
has a nuclear spin I = 7/2 and a gyromagnetic ratio $\gamma = \mu_N/I =$
1.4683.

1. Vanadium(IV)

Vanadium(IV) is a d^1 electron system. In its most common
stereochemistries (derived from octahedral, tetrahedral, and square
pyramidal), it has a nondegenerate ground state with excited states
more than 10,000 cm^{-1} higher in energy, so its relaxation time is
sufficiently long to give an observable signal at room temperature.
In the majority of its complexes, vanadium(IV) is present as the
oxovanadium(IV) (vanadyl) ion with stereochemistries derived from
octahedral or square pyramidal. A few complexes containing V^{4+} have
been characterized by ESR spectroscopy. In Table 13 the spin
Hamiltonian parameters for some well-characterized complexes are reported.

Complexes of Oxovanadium(IV) Ion. The vanadyl ion in hexacoordinate complexes imposes a strong tetragonal distortion to the octahedral environment. The free ion ^2D term is split by a tetragonal
distortion according to the scheme shown in Fig. 14 for C_{4v} symmetry
with z along C_4 and x along a bond direction. Ballhausen and Gray
[185] have performed a molecular orbital analysis of the spectro-
scopic and magnetic properties of $VO(SO_4) \cdot 5H_2O$. The complex contains
the chromophore $VO(H_2O)_5^{2+}$ in nearly C_{4v} symmetry with the V-O bond
length for VO^{2+} = 167 pm and for water ligands = 230 pm (average distance). The authors, neglecting the π bonding of water because of
the large distance, calculate the energy levels in Fig. 14 with the
ground state being 90% d(xy) metal in character. This energy-level
scheme has been discussed by several authors [186-190] and seems now
to be accepted [191-193].

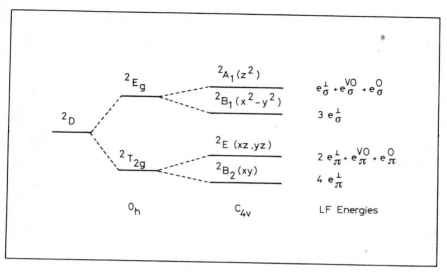

Figure 14   The energy-level scheme for $VO(H_2O)_5^{2+}$ assuming C_{4v} symmetry and the ligand field energies. The superscript refers to the four basal oxygen atoms, the VO refers to the vanadyl oxygen, and the O refers to the axial water oxygen opposite the vanadyl oxygen. (Reprinted with permission from C. J. Ballhausen and H. B. Gray, *Inorg. Chem.*, *I*:11. Copyright 1962 American Chemical Society.)

For a B_2 ground state, a simple molecular orbital approach [15] gives the spin Hamiltonian parameters

$$g_{\parallel} = g_e - \frac{8\beta_1^2\beta_2^2\xi}{\Delta E|d(xy) \rightarrow d(x^2 - y^2)|}$$

$$g_{\perp} = g_e - \frac{2\beta_1^2\beta_3^2\xi}{\Delta E|d(xy) \rightarrow d(xz), d(yz)|} \tag{51}$$

$$A_{\parallel} = -P'\left[K + \frac{4\beta_1^2}{7} + (g_e - g_{\parallel}) + \frac{3}{7}(g_e - g_{\perp})\right]$$

$$A_{\perp} = -P'\left[K - \frac{2\beta_1^2}{7} + \frac{11}{14}(g_e = g_{\perp})\right]$$

where β_1, β_2, and β_3 are the molecular orbital coefficients of the $d(xy)$, $d(x^2 - y^2)$ and $d(xz)$, $d(yz)$ metal orbitals, respectively.

Table 13 Spin Hamiltonian Parameters for Some Vanadium(IV) Complexes

Compound	Chromophore[a]	Spectra[b]	$g_1{}^c$	$g_2{}^c$	$g_3{}^c$	$A_1(N)^{c,d}$	$A_2(N)^{c,d}$	$A_3(N)^{c,d}$	Reference
Octahedral oxovanadium(IV) chromophores									
$VOSO_4 \cdot 5H_2O$	$VO(O)_5$	P	1.98	1.99	1.94	~70		180	185
$(VO,Zn)SO_4 \cdot 7H_2O$	$VO(O)_5$	P	1.979		1.914	73		180	191
$(VO,Mg)(NH_4)_2(SO_4)_2 \cdot 6H_2O$	$VO(O)_5$	P	1.973		1.925	68		177	191
$(VO,K)SO_4 \cdot 7H_2O$	$VO(O)_5$	P	1.975		1.912	76		188	191
$(VO,Cd)SO_4 \cdot \frac{8}{3}H_2O$	$VO(O)_5$	P	1.975		1.932	66		182	191
$Rb(Al,VO)(SO_4)_2 \cdot 12H_2O$	$VO(O)_5$	P	1.979		1.932	66		183	228
$Cs(Al,VO)(SO_4)_2 \cdot 12H_2O$	$VO(O)_5$	P	1.987		1.932	64		175	228
$(Al,VO)(urea)_6(ClO_4)_3$	$VO(O)_5$	C	1.987		1.936	64		179	226
$(Al,VO)(urea)_6Cl_3 \cdot 3H_2O$	$VO(O)_5$	C	1.987		1.934	64		182	226
$(Al,VO)(urea)_6I_3$	$VO(O)_5$	C	1.977		1.936	68		177	226
$K(Al,VO)(SO_4)_2 \cdot 12H_2O$	$VO(O)_5$	C	1.978		1.940	67		176	196
$NH_4(Al,VO)(SO_4)_2 \cdot 12H_2O$	$VO(O)_5$	G							196
$[BVW_{11}O_{40}]^{7-}$	$VO(O)_5$	G	1.967		1.918	59.4		167.3	199
$[H_2VW_{11}O_{40}]^{8-}$	$VO(O)_5$	G	1.966		1.922	57.6		164.9	199
$[ZnVW_{11}O_{40}]^{8-}$	$VO(O)_5$	G	1.967		1.921	58.5		164.8	199
$[PVMo_{11}O_{40}]^{5-}$	$VO(O)_5$	G	1.974		1.939	53.4		151.2	199
$[SiVMo_{11}O_{40}]^{6-}$	$VO(O)_5$	G	1.975		1.936	54.2		149.5	199

$[Mo_{10}V_2PO_{40}]^{6-}$	$VO(O)_5$	G	1.882		1.928		67.8		173	200
$[PVW_{11}O_{40}]^{5-}$ into										201
$K_5BW_{12}O_{40} \cdot nH_2O$	$VO(O)_5$	C	1.968		1.925		59.6		169.3	
$[VW_5O_{19}]^{5-}$ into										202
$(MeNH_3)_2Na_2V_2^VV_4O_{19} \cdot 6H_2O$	$VO(O)_5$	C	1.969		1.949		61		166.9	

Five-coordinate square pyramidal oxovanadium(IV) complexes

trans-$(Pd,VO)(benzac)_2$	$VO(O)_4$	C	1.9829	1.9807	1.9432	61		62	164	204
cis-$(Zn,VO)(benzac)_2$	$VO(O)_4$	C	1.9805	1.9758	1.9408	55		60	165	204
cis-$VO(benzac)_2$	$VO(O)_4$	C	1.982		1.943		70		170	231
$VO(acac)_2$	$VO(O)_4$	G	1.984	1.981	1.942	63.0		64.2	173.4	207
$VO(tropolonate)_2$	$VO(O)_4$	G	1.986	1.979	1.9500	41.6		52.7	153.8	207
$VO(maltolate)_2$	$VO(O)_4$	G	1.986	1.979	1.9480	48.3		57.4	161.1	207
$[VO(pfp)_2]^{2+}$	$VO(O)_4$	G	1.981		1.955		42		143	197
$VO(acen)$	$VO(O)_4$	G	1.984		1.954		57		166.0	230
$VO(salen)$	$VO(O)_4$	G	1.985		1.949		57		165.9	230
$VO(acpn)$	$VO(O)_4$	G	1.984		1.954		57		166.9	230
$VO(bzen)$	$VO(O)_4$	G	1.986		1.952		57		166.8	230
$VO(bzpn)$	$VO(O)_4$	G	1.984		1.957		59		166.3	230
$VO(tfen)$	$VO(O)_4$	G	1.983		1.955		59		167.0	230
$VO(quinO)_2$	$VO(O_2N_2)$	G	1.973	1.979	1.931	52.8		66.4	156.6	206

Table 13 (Continued)

Compound	Chromophore[a]	Spectra[b]	g_1[c]	g_2[c]	g_3[c]	$A_1(N)$[c,d]	$A_2(N)$[c,d]	$A_3(N)$[c,d]	Reference
VO^{2+} in (NH$_4$)$_2$SbCl$_5$	VO(Cl)$_4$	C	1.9793		1.9478	62.8		168.8	232
VO(mnt)$_2$[MeP(Ph$_3$)$_3$]$_2$	VO(S)$_4$	G	1.986		1.973	43		135	227
(Ni,VO)[S$_2$P(C$_6$H$_{11}$)$_2$]$_2$	VO(S)$_4$	P	1.98		1.959	63 32(^{31}P)		17 28(^{31}P)	211
VO[S$_2$P(C$_6$H$_5$)$_2$]$_2$	VO(S)$_4$	G	1.981		1.960	55 32(^{31}P)		154 31(^{31}P)	229
VO[S$_2$P(OEt)$_2$]$_2$	VO(S)$_4$	G	1.987		1.967	54 47(^{31}P)		155 48(^{31}P)	229
VO[S$_2$P(CH$_3$)$_2$]$_2$	VO(S)$_4$	G	1.963		1.981	52 30(^{31}P)		153 32(^{31}P)	229
VOCl$_2$·2 tmu	VO(Cl$_2$O$_2$)	G	1.983		1.941	64.1		174.9	208
(Ni,VO) (Mesalim)	VO(N$_2$O$_2$)	C	1.9842	1.9796	1.9543	59.7	48.4	156.3	203

Other complexes

Complex	Metal and nearest-neighbor atoms[a]	b[b]	g[c]			A[d]				Ref
$[(C_6H_5)_4As]_2(Mo,V)(mnt)_3$	VS_6	C	1.974		2.000		−91		9	225
$(Ti,V)(cp)_2S_5$	$V(cp)_2S_5$	C	1.9964	1.9997	1.9689	62.1		103.9	21.6	216
$(Ti,V)(Mecp)_2Cl_2$	$V(Mecp)_2Cl_2$	C	1.9802	1.9695	2.0013	74.5		115.4	19.2	217
$(Ti,V)(cp)_2Cl_2$[e]	$V(cp)_2Cl_2$	C	1.965	1.946	1.967	74.9		118.9	26.7	218
$V(cp)_2(SCN)_2$[e]	$V(cp)_2(SCN)_2$	G	1.988	1.988	2.000	71.2		115.3	19.2	219
$V(cp_2)(OCN)_2$[e]	$V(cp_2)(OCN)_2$	G	1.985	1.957	2.000	73.4		119.4	18.6	219
$V(cp)_2(CN)_2$	$V(cp)_2(CN)_2$	G	1.994	1.986	2.000	58.7		99.1	11.0	219
$V(NMe_2)_4$	$V(NMe_2)_4$	G	1.985	1.985	1.955		27.5		123	220
$V(NEt_2)_4$	$V(NEt_2)_4$	G	1.986	1.986	1.949		26		123	223
$(Ti,V)(OBut)_4$	$VO(OBut)_4$	P	1.984	1.994	1.940		36		125	221

[a]Metal and nearest-neighbor atoms.

[b]C = single crystal; P = polycrystalline powder; G = frozen solution.

[c](1, 2, 3) refer to (x, y, z) in single-crystal spectra. When two values are reported, the first is the perpendicular component.

[d]Values in $cm^{-1} \times 10^{-4}$. N is the nucleus giving rise to the hyperfine coupling; when omitted, $N = {}^{51}V$.

[e]The reference axes system is defined as in Ref. 216.

ξ is the spin-orbit coupling constant of vanadium, and $P' = g_e g_N \mu_B \mu_N \langle r^{-3} \rangle$. Several authors have chosen a value of ξ between 135 and 250 cm^{-1} [185,194-197], assuming different charges on the vanadium atom as a consequence of the predominantly covalent V-O bond. The values of P' have been calculated by McGarvey [163] for various charges on the metal atom.

Formula (51) predicts a $\underline{g}_\perp > \underline{g}_\parallel$ pattern with small deviations from the $\underline{g}_e$ = 2.0023 free-electron value because of the small value of the spin-orbit coupling constant and $\underline{A}_\parallel > \underline{A}_\perp$. The values reported in Table 13 agree well with these predictions.

Viswanath [191] studied the single-crystal spectra of VO^{2+} doped into $ZnSO_4 \cdot 7H_2O$ and found that it is necessary to assume $\underline{A}_\parallel$ and $\underline{A}_\perp$ negative in order to calculate a P' value close to the expected values [163].

A series of molybdo- and tungsteno-vanadates has been investigated through ESR spectroscopy [198-202]. Single-crystal studies of $PVW_{11}O_{40}^{5-}$ doped into $K_5BW_{12}O_{40}$ [201] and $VW_5O_{19}^{4-}$ doped into $(CH_3NH_3)Na_2(V_2W_4O_{19}) \cdot 6H_2O$ [202] in which vanadium is present as a nearly C_{4v} $VO(O_5)$ chromophore have been interpreted using formula (51). The slight differences in the spin Hamiltonian parameters have been explained by a greater degree of electron delocalization in the molybdovanadate complexes as compared to the others [199].

Square pyramidal complexes of the vanadyl ion are generally found with chelating ligands. The symmetry of the vanadium ion is then lowered to C_2 at best. The main effect of this low symmetry on the ESR spectra is to increase the in-plane anisotropy and rotate the in-plane $\underline{g}$ directions from the bonds. These effects were observed in the single-crystal spectra of VO^{2+} doped into Ni(N-methylsalicilaldiminato)$_2$ [203] and into Zn(benzoylacetonate)$_2 \cdot C_2H_5OH$ [204] and explained by Hitchman et al. [205] through the mixing of the excited $d(z^2)$ orbital into the ground state.

Assuming the y axis coincident with the bisector of the chelate angle, the composition of the ground state has been considered as a mixture of $d(x^2 - y^2)$ and $d(z^2)$ orbitals, thus giving essentially the same $\underline{g}$ and $\underline{A}$ pattern as for $VO(O_5)$ chromophores. The relevant perturbation formulas for the calculation of the components of g and $\underline{A}$ are given by McGarvey [15]. The ground-state composition has been shown to be consistent with the ESR spectra of several five-coordinate oxovanadium(IV) complexes [197,206-210].

Stoklosa et al. [211] have discussed the ESR spectrum of the complex $VO[S_2P(C_6H_{11})_2]_2$, which contains the square pyramidal VOS_4 chromophore. They have interpreted the ^{31}P superhyperfine splitting as being due to a trans-annular interaction of the $d(x^2 - y^2)$ orbital of vanadium with the 3s orbital of the phosphorus atoms and have determined that the 3s orbital contributes about 1% to the ground state. Analogous interpretations have been given for similar complexes [212-215].

Other Complexes. A few complexes of vanadium(IV) that do not contain the oxovanadium(IV) ion have been characterized by ESR spectroscopy [16,216-223].

The single-crystal ESR spectra of tetrahedral complexes of the type $V(cp)_2S_5$ and $V(Mecp)_2Cl_2$ have been studied [216,217]. The complexes contain the $V(L_2)X_2$ (L = cp, Mecp; X = S, Cl) chromophores with site symmetry C_1 and C_2, respectively. Assuming an idealized C_{2v} symmetry and using the formulas reported [15], Petersen and Dahl have calculated that the unpaired electron is in an orbital that is mainly d metal in character. Using the (x, y, z) reference system shown in Fig. 15, the ground-state orbital for the S and Cl derivatives is a molecular orbital mainly composed of $d(z^2)$ with a small amount of $d(x^2 - y^2)$. The ratios $d(z^2)/d(x^2 - y^2)$ are 0.127 and 0.200 respectively. These findings have been also reproduced by a Fenske-Hall type of molecular orbital calculation [224].

The ESR spectra of the pseudo-octahedral complex $[(C_6H_5)_4As]_2$-$V(mnt)_3$ (mnt = maleonitriledithiolate) diluted into the Mo analog have been interpreted [225] using idealized D_3 symmetry. The $\underline{g}_{\parallel} \simeq$ 2.0023 $> \underline{g}_{\perp}$ and $|\underline{A}_{\perp}| > |\underline{A}_{\parallel}|$ patterns are indeed in accord with a mainly $d(z^2)$ molecular orbital as found in titanium(III) trigonal octahedral complexes. The authors showed that $\underline{A}_{\parallel}$ and $\underline{A}_{\perp}$ must have opposite signs in order to agree with the $<\underline{A}>$ value in solution.

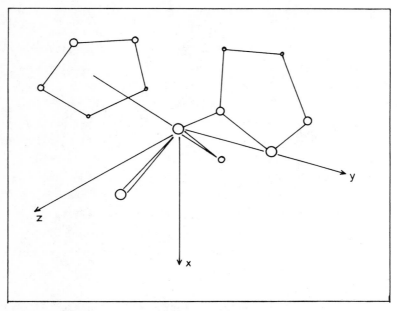

Figure 15 Reference system used for the interpretation of the ESR
spectra of VL_2X_2-type complexes. (Reprinted with permission from
J. L. Petersen and L. F. Dahl, *J. Am. Chem. Soc.*, 97:6422. Copyright 1975 American Chemical Society.)

2. *Vanadium(III)*

Vanadium(III) is a d^2 electronic system, and its ESR spectra
must be interpreted using an S = 1 spin Hamiltonian.

In octahedral environments, vanadium(III) has an orbital triplet ground state, and spin-orbit coupling is operative to give large
zero field splittings and short spin-lattice relaxation times, which
make ESR spectra undetectable. For hexaureavanadium(III) bromide
trihydrate, D has been evaluated by magnetic susceptibility measurements to be 5.9 ± 0.4 cm^{-1} (E $\approx$ 0) [233], and for V^{3+} doped into
Al_2O_3 it has been found to be 8.29 ± 0.02 [234] by ESR measurements.

$V(H_2O)_6^{3+}$ has been investigated by ESR spectroscopy at 4.2 K,
doped into $CsAl(SO_4)_2 \cdot 12H_2O$ [235] and $[C(NH_2)_3]Al(SO_4)_2 \cdot 6H_2O$ single
crystals [166], the observed transition being $|+1> \leftrightarrow |-1>$. In both
cases the chromophore is assumed to be trigonal octahedral. Table
14 lists the observed spin Hamiltonian parameters.

Table 14 Spin Hamiltonian Parameters for Some Vanadium(III) Complexes

Compound	Chromophore[a]	Spectra[b]	$g_\perp$	$g_\parallel$	D^c	E/D	$A_\perp^d$	$A_\parallel^d$	Reference
$[C(NH_2)_3][(Al,V)(SO_4)_2 \cdot 6H_2O]$	VO_6	C		1.86	3.8			89	166
$Cs(Al,V)(SO_4)_2$ $12H_2O$	VO_6	C		1.975	10			101.9	235
V^{3+} into Al_2O_3	VO_6	C	1.72	1.96	8.29	0.0043			234

[a] Metal and nearest-neighbor atoms.
[b] C = single crystal.
[c] Values in cm^{-1}.
[d] Values in $cm^{-1} \times 10^{-4}$.

67

3. *Vanadium (II)*

Vanadium(II) is a d^3 ion, and the ESR spectra can be fitted by an S = 3/2 spin Hamiltonian.

Because of the small number of complexes reported, no extensive studies have been made on molecular complexes; some of them have been reviewed by Goodman and Raynor [16]. Spin Hamiltonian parameters for V^{2+} doped into various hosts of nearly octahedral symmetry are reported by McGarvey [15].

Octahedral coordination is most common for vanadium(II). In O_h symmetry the free-ion ^4F state is split into $^4A_{2g}$, $^4T_{2g}$, and $^4T_{1g}$ states, with $^4A_{2g}$ lying lowest and the excited states at energy greater than 10,000 cm^{-1}, so that the spin lattice relaxation time is sufficiently long to observe a signal at room temperature. Spin-orbit coupling mixes the $^4T_{2g}$ functions into the ground state, so that to first approximation, the g values are expected to be quite close to

$$g = 2.0023 - \frac{8\xi}{\Delta E(^4A_{2g} \to ^4T_{2g})} \tag{52}$$

and the low value of the spin-orbit coupling constant for vanadium-(II) (free-ion value ξ = 170 cm^{-1}) produces a small deviation of g from 2.0023.

As a matter of fact, the reported spectra have been interpreted with g values slightly lower than 2.0023, the mean deviation being about 0.02, and with D values smaller than 1 cm^{-1}.

Spin Hamiltonian parameters for some vanadium(II) complexes are reported in Table 15.

C. Chromium

The only stable isotope of chromium having nonzero nuclear spin is ^{53}Cr (abundance 9.5%); it has I = 3/2 and a gyromagnetic ratio γ = μ_N/I = -0.3157.

Table 15 Spin Hamiltonian Parameters for Some Vanadium(II) Complexes

Compound	Chromophore[a]	Spectra[b]	g_1^c	g_2^c	g_3^c	$A_1^{c,d}$	$A_2^{c,d}$	$A_3^{c,d}$	D^d	E/D	Reference
Cs(Ca,V)Cl$_3$	VCl$_6$	C		1.958			79.9				236
Cs$_s$(Cd,V)Cl$_3$	VCl$_6$	Site 1	1.9638		1.9635	78		78	∓0.0043		237
		Site 2 ^{C}	1.9559		1.9559	80		80	∓0.0444		237
V^{2+} in KZnF$_3$	VF$_6$	C		1.972			-86.2				238
Cs(Mg,V)Cl$_3$	VCl$_6$	C	1.976		1.974	74		76	0.0858		239
Cs(Mg,V)Br$_3$	VBr$_6$	C	1.999		1.996	70		70	0.128		239
Cs(Mg,V)I$_3$	VI$_6$	C	2.038		20.35	65		65	0.208		239

[a] Metal and nearest-neighbor atoms.

[b] C = single crystal

[c] (1, 2, 3) refer to (x, y, z) in single-crystal spectra. When two values are reported, the first is the perpendicular component; when one value is reported, it refers to the isotropic value.

[d] Values in cm^{-1} × 10^{-4}.

1. Chromium(III)

Chromium(III) is a d^3 ion, and its ESR spectra can be interpreted using a spin Hamiltonian with S = 3/2. The most common coordination number of chromium is 6. In hexacoordinate complexes, the ion possesses an orbital singlet ground state with all the excited states well separated in energy (more than 10,000 cm^{-1}); and the spin-lattice relaxation time is sufficiently long to give ESR spectra at room temperature.

Spin Hamiltonian parameters for several well-characterized chromium(III) complexes are reported in Table 16.

The ESR spectra can in general be interpreted with a quasiisotropic $\tilde{g}$ tensor with principal values in the range 1.98 ± 0.02 and zero field splitting less than 2 cm^{-1}. In most cases $\underline{D}$ values smaller than 1 cm^{-1} have been reported. In some cases the hyperfine coupling with ^{53}Cr has been observed. The reported $\underline{A}$ values are consistent with a near isotropic tensor with principal values in the range 0.0014-0.0018 cm^{-1}. The small variations from the free-electron value of $\underline{g}$, due to the small value of the spin-orbit coupling constant of chromium(III) (free-ion value ξ = 273 cm^{-1}), make the $\tilde{g}$ tensor of little utility in the characterization of the chromium(III) complexes. Usually it is the zero field splitting that is more sensitive to the bonding and structural properties of the complexes.

Several formulas have been reported for the zero field splitting parameters D and E in tetragonal and trigonal environments. The second-order formulas expressing D and E have been derived with the assumption that the main interaction determining the zero field splitting is spin-orbit coupling [15].

McGarvey [240] reported the relevant formulas for the calculation of D and E in D_{2h} and D_3 symmetries, including interactions with doublet states. He justified the high D value observed in trans-$CrCl_2(en)_2^{2+}$ on the basis of the crystal field mixing of the ground state with excited states of the same symmetry. The same effect must be operative in D_3 and lower symmetries. This may be responsible for the unusually high D value that was observed in solution spectra [241].

Table 16 Spin Hamiltonian Parameters of Some Chromium(III) Complexes

Compound	Chromophore[a]	Spectra[b]	g_1^c	g_2^c	g_3^c	D^d	E/D	$A_\perp^e$	$A_\parallel^e$	Reference
Octahedral chromophores										
$Cr(en)_3Cl_3 \cdot 3H_2O$	CrN_6	C		1.9900		0.036	0.			240
$Cr(en)_3Cl_3$	CrN_6	C		1.9871		0.0413	0.			240
$Cr(en)_3Cl_3 \cdot NaCl \cdot 6H_2O$	CrN_6	C		1.9874		0.00495	0.			240
$(Co,Cr)(NH_3)_6InCl_6$	CrN_6	P	1.9854		1.9855	0.0458	0.0096			255
$(Co,Cr)(NH_3)_6SbCl_6$	CrN_6	P	1.9858		1.9860	0.0948	0.0026			255
$(Co,Cr)(NH_3)_6BiCl_6$	CrN_6	P	1.9857		1.9859	0.1084	0.0020			255
$(Co,Cr)(NH_3)_6CdCl_5$	CrN_6	P	1.974		1.984	0.0949	0.			267
$Co(NH_3)_6(In,Cr)Cl_6$	$CrCl_6$	P		1.9856		0.	0.		16.2	255
$(Al,Cr)Cl_3$	$CrCl_6$	C	1.9889		1.9860	0.116	0.0172			262
		α	1.984	1.9809	1.982	∓ 0.3268	∓ 0.1921			
$K_3(Al,Cr)(ox)_3 \cdot 3H_2O$	CrO_6	C								256
		β	1.980	1.980	1.975	± 0.3914	∓ 0.0682			
		297 K		1.975	1.975	0.0681	0.			
$Rb(Al,Cr)(SO_4)_2 \cdot 12H_2O$	CrO_6	C								251, 252
		4.2 K		1.975	1.975	0.0349	0.	17.36	17.27	
		297 K		1.975	1.975	-0.0762	0.			
$Cs(Al,Cr)(SO_4)_2 \cdot 12H_2O$	CrO_6	C								251, 252
		4.2 K		1.975		-0.0715	0.	17.20	17.46	

Table 16 (Continued)

Compound	Chromophore[a]	Spectra[b]		$g_1{}^c$	$g_2{}^c$	$g_3{}^c$	D^d	E/D	$A_\perp{}^e$	$A_\parallel{}^e$	Reference
$C(NH_2)_3(Al,Cr)(SO_4)_2 \cdot 6H_2O$	CrO_6	C	297 K		1.975		-0.0577	0.			251, 252
			4.2 K		1.975	1.975	-0.0889	0.	17.10	17.10	
$(Al,Cr)Cl_3 \cdot 6H_2O$	CrO_6	C	297 K		1.975		-0.0326	0.			251, 252
			4.2 K		1.975	1.975	-0.0341	0.	17.30	17.52	
$NaMg(Al,Cr)(ox)_3 \cdot 9H_2O$	CrO_6	C		1.9822	1.9812	1.9780	0.7786	0.039	16.1	15.7	257
$NaMg(Al,Cr)(ox)_3 \cdot 8H_2O$	CrO_6	C		1.980	1.980	1.975	0.7783	0.			257
$(NH_4)_3(Al,Cr)(ox)_3 \cdot 8H_2O$	CrO_6	C			1.976	1.978	0.700	0.036			266
$K_2(Sb,Cr)F_6$	CrF_6	C		1.9744	1.9744	1.9717	0.2750	0.0304			260
$Na(Al,Cr)(ox)_3 \cdot 5H_2O$	CrO_6	C			1.973	1.978	0.773	0.0624			266
$(Al,Cr)(urea)_6(ClO_4)_3$	CrO_6	C			1.975	1.977	0.168	0.9643			258
$[(Co,Cr)(bipy)_2ox]Cl \cdot 4H_2O$	CrN_4O_2	P		1.98	1.99	2.00	0.490	0.13			268
$[(Co,Cr)(phen)_2ox]Cl \cdot 4H_2O$	CrN_4O_2	P		1.989	1.983	1.988	0.398	0.311			268
$trans\text{-}Cr(en)_2Cl_2 \cdot HCl \cdot 2H_2O$	CrN_4Cl_2	C			1.9765		0.504	0.071			240
$trans[Cr(en)_2(H_2O)_2]^{3+}$	CrN_4O_2	G			1.99		0.455	0.059			263
$[(Co,Cr)(NH_3)_5(H_2O)](ClO_4)_3$	CrN_5O	P		1.988	1.990	1.990	0.0986	0.022			248
$[(Co,Cr)(NH_3)_5Cl]Cl_2$	CrN_5Cl	P		1.9860	1.9856	1.9850	0.08812	0.0499			250

Complex	Metal and nearest-neighbor atoms	b	g	g	g	D	E	A	A	Ref
[(Co,Cr)(NH$_3$)$_5$Br]Br$_2$	CrN$_5$Br	P	1.990	1.993	1.986	0.2043	0.0040			248
[(Co,Cr)(NH$_3$)$_5$I]$^{2+}$	CrN$_5$I	G		1.98		0.469	<0.002			247
(NH$_4$)$_2$(In,Cr)Cl$_5$H$_2$O	CrCl$_5$O	C		1.9828	1.9871	0.0595	0.0853	14.7	15.4	242
(In,Cr)(ethylxantate)$_3$	CrS$_6$	C		1.9987	1.9947	0.169	<0.005			264
(Co,Cr)(dtp)$_3$	CrS$_6$	C	1.990	1.990	1.991	-0.115	-0.413			264
[(Rh,Cr)(NH$_3$)$_4$Cl$_2$]Cl	CrN$_4$Cl$_2$	P		1.9869	1.9862	0.1580	0.095			265
[(Co,Cr)(NH$_3$)$_4$ClH$_2$O]Cl$_2$	CrN$_4$ClO	P		1.9867	1.9860	0.1952	0.062			265
[Cr(NH$_3$)$_4$Br$_2$]$^+$	CrN$_4$Br$_2$	G				>0.4	<0.2			76
[(Rh,Cr)(py)$_4$Br$_2$]Br·6H$_2$O	CrN$_4$Br$_2$	P				0.710	0.01			76
[(Co,Cr)(NH$_3$)$_4$ClH$_2$O]Cl$_2$	cis-CrN$_4$ClO	P	1.9860	1.9860	1.9850	0.08800	0.0043			248
Square pyramidal complexes										
(NH$_4$)$_2$(Sb,Cr)Cl$_5$	CrCl$_5$	C	1.9835	1.9839	1.9853	0.03706	0.0820		16.5	232
Other complexes										
Cr[N(SiMe$_3$)$_3$]$_3$	CrN$_3$	P		1.98		-1.85	0.003			261

[a] Metal and nearest-neighbor atoms.

[b] C = single crystal; P = polycrystalline powder; G = frozen solution.

[c] (1, 2, 3) refer to (x, y, z) in single-crystal spectra. When two values are reported, the first is the perpendicular component; when one value is reported, it is the isotropic value.

[d] Values in cm^{-1}.

[e] Values in cm^{-1} × 10^{-4}.

Garrett et al. [242,243] derived the molecular orbital formulas for $\underline{g}$ and D in C_{4v} symmetry. They found that the D value depends essentially on the radial expansion of the d orbitals, electron delocalization onto the ligands, spin-orbit coupling interaction onto the ligands, and charge transfer state mixing with the ground state. With this model they reproduced the data for the chromophore $CrCl_5(H_2O)^{2+}$. Macfarlane performed complete ligand field calculations [244-246] on a d^3 configuration that gave satisfactory results for the zero field splitting in ruby, a CrO_6 chromophore.

Pedersen and Toftlund [76] performed complete ligand field calculations in D_4 symmetry and illustrated the dependence of the zero field splitting on ligand field and spin-orbit coupling parameters. The authors applied the calculations to a series of CrN_4XY chromophores [76] and to a series of $Cr(NH_3)_5X^{n+}$ complexes [247], obtaining in general fair agreement with the experimental results. For the $Cr(NH_3)_5X^{n+}$ complexes, some differences in the spin Hamiltonian parameters can be found among the values reported by different authors [243,247-249] as a result of different experimental conditions. An accurate determination of the spin Hamiltonian parameters for $[Cr(NH_3)_5Cl]Cl_2$ doped into the cobalt(III) analog has been made by Andriessen [250], who suggested that the powder spectra of this compound can be used for the calibration of the magnetic field over the range 0.1-0.6 tesla.

The chromium(III) hexaaquoion doped into various host lattices shows a marked temperature dependence of D. Manoogian and Leclerc [251] studied a wide series of trigonal lattices containing the $Cr(H_2O)_6^{3+}$ ion and obtained three different categories of D versus temperature curves: that is, a positive D value with a large decrease in magnitude between 297 and 4.2 K, a negative D value with a large increase in magnitude between 297 and 4.2 K, and a negative D value with only a small variation in D as a function of temperature. These variations of D with temperature have been interpreted in terms of contributions from both the static and dynamic coupling of the paramagnetic ion to lattice vibrations [252]. A similar model has been applied by Owens [253] to the interpretation of the temperature dependence of D for $Cr(H_2O)_6^{3+}$ in $NH_4Al(SO_4)_2 \cdot 12H_2O$.

Lattice effects causing a dependence of D and E on temperature have also been observed in chromium amine complexes [249,254,255]. Stout and Garrett observed increasing D values for $Cr(NH_3)_6^{3+}$ chromophores on going from In to Bi hexaamine host lattices. They attributed this increasing trigonal distortion to lattice interactions at the cation sites [255].

Sites α and β in $K_3Al(oxalate)_3 \cdot xH_2O$ doped with Cr^{3+} [256] correspond to x = 3 and 2, respectively, the less hydrated form being the one less distorted from D_3 symmetry. This is analogous to what was observed for Cr^{3+} doped into $NaMgAl(oxalate)_3 \cdot nH_2O$ (n = 9, 8) [257].

The reported D and E parameters for $Cr(urea)_6(ClO_4)_3$ doped into the Al analog [258,295] can be transformed to standard form by interchanging the y and z axes, yielding D = -0.0327 and E = -0.0003 cm^{-1}. It should be noted, however, that the parameters have been obtained by assuming the direction of the z axis, which by no means is fixed by symmetry.

On the basis of molecular orbital and ligand field calculations, the magnetic properties of Cr^{3+} diluted into $(NH_4)_2SbCl_5$ have been interpreted as due to a five-coordinate square pyramidal $[CrCl_5]^{2-}$ species [232]. In K_2SbF_5, however, the dominant species seems to be six-coordinate with D_3 symmetry [260].

An unusual three-coordinate chromophore $Cr[N(SiMe_3)_3]_3$ has been described with a very high zero field splitting, which has been confirmed by ligand field calculations [261].

D. Manganese

The only stable isotope of manganese is ^{55}Mn (abundance 100%). It has a nuclear spin quantum number I = 5/2 and a gyromagnetic ratio $\gamma = \mu_N/I = 1.378$.

1. *High-Spin Manganese(II)*

Manganese(II) is a d^5 ion. In the high-spin form it has five unpaired electrons, and the ESR spectra can be interpreted using an S = 5/2 spin Hamiltonian. The spin-lattice relaxation time is generally sufficiently long to give observable signals at room temperature. The spectra are characterized by a quasiisotropic g tensor with principal values close to the spin-only value. The hyperfine coupling of the unpaired electrons with ^{53}Mn is generally observable in solution or diluted solid spectra. The isotropic hyperfine coupling constant has been related to the percentage of ionic character [269], since theoretical considerations suggest that covalent interactions decrease its value (see Secs. VII and VIII). However, the largest reported values of A have been found in complexes with uncharged ligands containing oxygen donors [270,271], suggesting that other factors beyond ionicity-covalency must be operative in determining the value of A.

Although the papers published reporting ESR spectra of manganese(II) are quite numerous, single-crystal determinations are rather rare, so that the reported data are often not very precise. One must add that many works of the early 1970s are erroneous, in part because the authors did not recognize forbidden transitions that appeared with high intensity in the powder spectra.

Dowsing et al. [272] reported the features of the spectra of some typical manganese(II) complexes at X-band frequency. They are shown in Fig. 16. The values of the relative spin Hamiltonian parameters are shown in Table 17. Regular octahedral and tetrahedral complexes have a strong line at g_{eff} = 2, since in manganese(II) the value of a [Eq. (25)] is generally small. When the distortion from cubic symmetry increases and D becomes significant (D > 0.15 cm^{-1}), a strong line at g_{eff} = 6 is present together with other bands. Because of large zero field splitting and large orthorhombic components, an isotropic line at g_{eff} = 4.29 results, which was mentioned and justified in Sec. V.D.

Recently, Golding and Singhasuwich [273] provided expressions that correlate the spin Hamiltonian parameters to ligand field

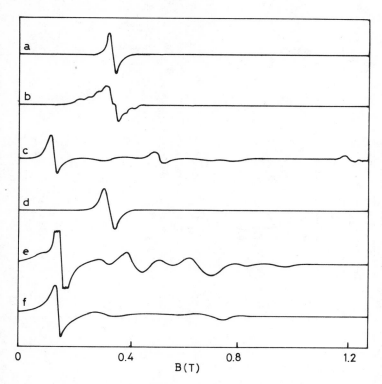

Figure 16.   Polycrystalline powder ESR spectra of some typical high-spin manganese(II) complexes.   (a) Cubic complexes (a ≈ 0).   (b) Mn(γ-pic)$_4$(NCS)$_2$.   (c) Mn(γ-pic)$_4$Br$_2$.   (d) Polymeric Mn(γ-pic)$_2$Cl$_2$. One single line at $\underline{g}_{eff}$ = 2 is observed as a result of magnetic interaction with neighboring Mn(II) ions.   (e) Mn(Ph$_3$AsO)$_2$Br$_2$.   (f) Mn(dipyam)$_2$I$_2$.   Reprinted with permission from *Nature*, 219:1037. Copyright © 1968 Macmillan Journals United.)

parameters.   In this operator equivalent approach, they considered the admixture of $^4\underline{T}_1(\underline{t}_2^4\underline{e})$, $^4\underline{T}_2(\underline{t}_2^4\underline{e})$, and $^2\underline{T}_2(\underline{t}_2^5)$ into the ground state $^6\underline{A}_1(\underline{t}_2^3\underline{e}^2)$ via spin-orbit coupling.   The resulting expressions are rather cumbersome, and they include the spin-orbit coupling constant ξ, the energy separations between the ground state and the three excited states, and the crystal field parameters.   As an example, the method is applied to tetragonal distortions, and the agreement with experimental data is considered satisfactory.   However, the routine interpretation of spectra can hardly be made on

Table 17 Spin Hamiltonian Parameters for Some High-Spin Manganese(II) Complexes

Compound	Chromophore[a]	Spectra[b]	g_1[c]	g_2[c]	g_3[c]	D[d]	E/D	$A_\perp$[e]	$A_\parallel$[e]	a[d]	F[d]	Reference
Octahedral chromophores												
Mn^{2+} into $MgK_2(SO_4)_2 \cdot 6H_2O$	MnO_6	C		2.001		-0.0343	-0.180	-86	-87	0.0011		290
$(Zn,Mn)(pyO)_6(ClO_4)_2$	MnO_6	C		2.002		0.0383	0	-84	-81	0.0008[g]		148
$(Zn,Mn)(4MepyO)_6(ClO_4)_2$	MnO_6	P		2.002		0.0063	0.06					274
$(Cd,Mn)(4MepyO)_6(ClO_4)_2$	MnO_6	P		2.002		0.0061	0.06					274
$(Hg,Mn)(4MepyO)_6(ClO_4)_2$	MnO_6	P		2.002		0.0059	0.06					274
$(Mg,Mn)(apy)_6(ClO_4)_2$	MnO_6	P		2.008		-0.0045	0	-87	-87	0.0007[g]		291
$(Mg,Mn)(1-Me-2-pyridone)_6(NO_3)_2$	MnO_6	P		1.999		0.0010	0	-89				292, 293
$Na[(Zn,Mn)(acac)_3] \cdot nH_2O$	MnO_6	P		2.002		-0.063	0					284
$Na[(Cd,Mn)(acac)_3] \cdot nH_2O$	MnO_6f	P		2.002		0.113	0.03					284
$(Mg,Mn)(acac)_2(H_2O)_2$	MnO_4O_2	C		1.98		0.0649	0.293	-83	-87			148
$(Mg,Mn)(OMPA)_3(ClO_4)_2$	MnO_6	P		2.011		0.0147	0	-101	-100	0.0008[g]		270, 271
$(Zn,Mn)(NIPA)_3(BF_4)_2$	MnO_6	P		2.002		0.010			-90			277
$Mn(picolinate)_2(H_2O)_2$	$MnN_2O_2O_2$	C	2.0003		2.0007	0.06432	0.2778			-0.0014	0.0026	281
$Mn(\gamma-picoline)_4(NCS)_2$	MnN_4N_2	P		2.002		0.035						278
$Mn(\gamma-picoline)_4Cl_2$	MnN_4Cl_2	P		2.002		0.16	0.1					278
$Mn(\gamma-picoline)_4Br_2$	MnN_4Br_2	P		2.002		0.54	0.01					278
$Mn(\gamma-picoline)_4I_2$	MnN_4I_2	P		2.002		0.87	0.01					278
$(Co,Mn)(\beta-picoline)_4(H_2O)_2I_2$	MnN_4O_2	P		2.002		0.057	0.033					294
$Mn(pyrazole)_4I_2$	MnN_4I_2	P		2.002		0.98	0					295
$Mn(dipyam)_2I_2$	MnN_4I_2	P		2.002		0.975	0.300					283
$Mn(biet)_2I_2$	MnN_4I_2	P		2.002		0.68	0.03					283

Square pyramidal complexes

Compound	Coordination[a]	Sample[b,c]	g	D[d]	E/D	Ref
(Zn,Mn)(Ph₃AsO)₄(ClO₄)₂	MnO₄O	P	2.002	0.1775	0.100	286
(Zn,Mn)(Ph₃AsO)₄I₂	MnO₄I	P	2.002	0.18	0.100	286
(Zn,Mn)(Ph₃AsO)₄(NCS)₂	MnO₄N	P	2.002	0.19	0.100	286
(Zn,Mn)(Ph₃PO)₄(ClO₄)₂	MnO₄O	P	2.002	0.0875	0.067	286
(Zn,Mn)(Ph₃PO)₄(NCS)₂	MnO₄N	P	2.002	0.0375	0.067	286

Trigonal bipyramidal complexes

Compound	Coordination[a]	Sample[b,c]	g	D[d]	E/D	Ref
(Ni,Mn)(HL)₂Cl₃	MnN₂Cl₃	P	2.002	0.31	0.03	287
(Ni,Mn)(HL)₂Br₃	MnN₂Br₃	P	2.002	0.60	0	287

Tetrahedral complexes

Compound	Coordination[a]	Sample[b,c]	g	D[d]	E/D	Ref
(Zn,Mn)(Ph₃PO)₂Cl₂	MnO₂Cl₂	C	298 K: 2.02, 2.03; 16 K: 2.03, 2.03, 2.02	0.172; 0.191	0.273; 0.263	288
Mn(3-Me-isoquin)₂Br₂	MnN₂Br₂	P	2.002	0.365	0.267	285
Mn(HMPA)Br₂	MnO₂Br₂	P	2.002	0.42	0.25	277
Mn(NIPA)₂Br₂	MnO₂Br₂	P	2.002	0.44	0.20	277
[N(Me)₄]₂(Zn,Mn)Cl₄	MnCl₄	C	2.002	0.0122	0	289
(Zn,Mn)(HL)Cl₃	MnNCl₃	P	2.002	0.19	0.04	287
Mn(Ph₃AsO)₂Br₂	MnO₂Br₂	P	2.002	0.425	0.267	285

Additional parameters for [N(Me)₄]₂(Zn,Mn)Cl₄: −75, −77, −0.0014[g]

[a] Metal and nearest-neighbor atoms.

[b] C = single crystal; P = polycrystalline powder.

[c] When one value is reported, it is the isotropic value.

[d] Values in cm⁻¹.

[e] Values in cm⁻¹ × 10⁻⁴.

[f] Trigonal prismatic.

[g] The value reported is a − F.

this basis, and the qualitative considerations made by Griffith [83] are still the tool more often used to help rationalize the spin Hamiltonian parameters.

In his approach Griffith uses intermediate symmetry considerations; that is, he assumes that although on a strict basis information regarding the matrix elements should be obtained in the symmetry group H of the molecule, in certain cases extra information can be deduced by working in a suitably chosen larger group G. He concludes that large rhombic components are to be expected for meridional isomers of octahedral MA_3B_3 complexes, MA_6 complexes of D_{2h} symmetry (i.e., complexes where the ligands in the plane form a rectangle rather than a square), cis-MA_4B_2 complexes, and tetrahedral MA_2B_2 complexes. These predictions have been verified experimentally several times and have been used in a large number of cases to assign the stereochemistry of manganese(II) complexes.

The zero field splitting of regular octahedral complexes is found to be very small, <0.01 cm^{-1}, the noticeable exception being the hexakis-substituted pyridine-N-oxide complexes [274-276].

The metal hyperfine coupling constant for oxygen donors is in the range 80-90 $\times$ 10^{-4} cm^{-1}. The largest values were observed with some phosphoramides: 100 $\times$ 10^{-4} cm^{-1} with OMPA (octamethylpyrophosphoramide) [270,271], 90 $\times$ 10^{-4} cm^{-1} for NIPA (nonamethylimidodiphosphoramide) [277].

The effect of changing the axial ligands in a series of tetragonal octahedral complexes is shown by the Mn(γ-picoline)$_4X_2$ complexes [278]. As the ligand field strength of the axial ligand is decreased, the tetragonality of the complex is expected to increase, giving a larger D. The zero field splitting is also affected by the size of the host metal ion. D decreases for [Mn(4-methylpyridine-oxide)$_6$](ClO_4)$_2$ doped into zinc, cadmium, and mercury analogs, respectively [274]. Since the size of the host metal ion increases on passing from zinc to mercury, this behavior can be explained by considering that the axial ligand field is decreased by the larger metal ligand distance [279,280].

Accurate single-crystal studies have been reported for a couple of manganese(II) bis-chelate complexes having water molecules in the axial positions, $Mn(acac)_2(H_2O)_2$ and $Mn(picolinate)_2(H_2O)_2$ [281,148]. Both have very similar D values and are also similar in their rhombicity as measured from the E/D parameter. This feature is easily understood in the case of the picolinate, since in the plane there are two inequivalent donors, but it is less easily understood in the case of acetylacetonate, since the four oxygen donors form an almost regular square plane. It is probable that the quasiaromatic nature of the chelate ring produces a large anisotropy in the x and y directions (x is bisecting the chelate angle). Another interesting feature of the zero field splitting tensor seen in the two complexes is that the principal axes bisect the metal ligand bonds in the acetylacetonate, whereas they are closer to the bond directions in the picolinate complex.

The complexes $[Mn(dipyam)_2]X_2$ and $[Mn(o-phen)_2]X_2$ (dipyam = di-2-dipyridylamine; o-phen = 1,10-phenanthroline; X = Cl, Br, I, NCS) were considered to be cis-distorted octahedral on the basis of their large E/D values [282,283].

Birdy and Goodgame [284] studied the ESR spectra of manganese-(II) in several host lattices of general formula M'[M''(acetylacetonate)$_3$]·nH$_2$O (M' = Na, K; M'' = Co, Ni, Zn, Cd). In the cadmium lattice the metal ion has trigonal prismatic geometry rather than octahedral, and the D value (0.113 cm^{-1}) is large for MnO_6 chromophores, as confirmed by the spectra in the octahedral cobalt, nickel, and zinc lattices, where D = 0.060 cm^{-1}.

No unquestionable examples of ESR spectra of square pyramidal five-coordinate complex have been reported. Goodgame et al. reported the ESR spectra of $[Mn(Ph_3AO)_4X]X$ complexes (A = P, As; X = ClO_4, I, NCS) [285,286]. They conclude that at least one anion is bound to the metal ion, but it is very difficult on the basis of the ESR spectra to distinguish between an elongated trans-octahedral complex and a square pyramidal complex.

The complexes $[Mn(HL)_2X_3]$ (X = Cl, Br; HL = 1,4-diazabicyclo-[2.2.2]octane) doped into the corresponding nickel lattice are trigonal bipyramidal [287]. In both cases D is fairly large, and the rhombicity is small.

The tetrahedral $[MnCl_4]^{2-}$ ion has been studied in the tetramethyl-ammonium tetrachlorozincate lattice [289]. The spectra were found to be temperature dependent, corresponding to a phase transition of the host lattice.

Among the tetrahedral complexes of MA_2X_2 stoichiometry, the best characterized is $Mn(Ph_3PO)_2Cl_2$, for which a single-crystal study was reported [288]. As expected on the basis of the Griffith's model [83], E/D is close to 1/3. The zero field splitting parameters were found to be temperature dependent, the dependence thought to be due to an increase of the metal halogen bond length with increasing temperature.

2. Low-Spin Manganese(II)

Examples of low-spin manganese(II) complexes are far less numerous than the high-spin ones. The ESR spectra can be interpreted using an S = 1/2 spin Hamiltonian and can generally be observed at room temperature. The spin Hamiltonian parameters of several low-spin manganese(II) complexes are reported in Table 18. Among the

Table 18   Spin Hamiltonian Parameters for Some Low-Spin Manganese(II) Complexes

Compound	Spectra[a]	$g_\perp$	$g_\parallel$
$(Ru,Mn)(cp)_2$	C	1.069	3.548
$(Os,Mn)(cp)_2$	C	1.126	3.534
$(Fe,Mn)(cp)_2$	P	1.223	3.519
$(Fe,Mn)(Mecp)_2$	G	1.847	3.063
$(Fe,Mn)(Mecp)_2$	G	1.891	3.060
$(Mg,Mn)(Mecp)_2$	G	1.887	2.998

[a] C = single crystal; P = polycrystalline powder; G = frozen solution.
Source: Data from Ref. 296.

well-characterized ESR spectra are those of manganocene, $Mn(cp)_2$, (cp = cyclopentadienyl ion), which has been studied in a large variety of diamagnetic host systems [296,297]. Both high-spin and low-spin species can be formed, depending essentially on the dimensions of the host lattice [296]. In systems favoring large metal-to-ring distances [e.g., $Mg(cp)_2$], the high-spin state is formed, whereas the low-spin state is obtained in $Fe(cp)_2$, $Ru(cp)_2$, and $Os(cp)_2$. Systems exhibiting spin equilibrium have also been found [298-300]. The order of the energy levels for metallocenes [301] is shown in Fig. 17; therefore, for a low-spin d^5 species, a E_{2g} ground state is expected. The $\underline{g}$ values of low-spin $Mn(cp)_2$ confirm a large orbital contribution [296], with $\underline{g}_\parallel$ = 3.548 and $\underline{g}_\perp$ = 1.069. However, the spin Hamiltonian parameters are found to be extremely lattice dependent. All these results have been interpreted considering the

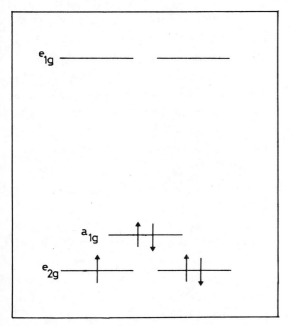

Figure 17. Order of the energy levels for a low-spin d^5 configuration, D_{5d} symmetry notation. (After Ref. 301.)

presence of a dynamic Jahn-Teller effect, whose main influence is that of reducing all the integrals of the orbital momentum by vibronic quenching. A very detailed analysis of d^5 and d^7 metallocenes has been reported [296].

E. Iron

The only stable isotope of iron having a nonzero nuclear spin is ^{57}Fe (abundance 2.19%; $I = 1/2$; $\gamma = \mu_N/I = 0.0451$). Because of the low abundance of this isotope and its small gyromagnetic ratio, no hyperfine structure is observed in the spectra of paramagnetic iron compounds.

1. *High-Spin Iron(III)*

Iron (III) is a d^5 ion, which in the high-spin form has ground state S = 5/2. As such it is characterized by almost isotropic g values and lines sufficiently narrow that spectra are readily obtained even at room temperature. The appearance of the spectrum is determined mainly by the value of the zero field splitting, including both second- and fourth-order terms. In general, a has been neglected compared to the second-order terms D and E. However, it should be mentioned that in some cubic lattices, such as MgO [302], a has been found to be 0.02038 cm^{-1}, so that the D and E parameters obtained from the spectra should be distinctly larger than this value if they are to be considered with any confidence. For the qualitative features of the spectra, refer to Fig. 16.

The origin of the spin Hamiltonian parameters has been discussed at length in the literature [1,133,303-307]. Relevant data of well-characterized complexes are shown in Table 19. In general, hexakis monodentate ligand complexes show small but definite zero field splittings. It is now accepted that, at least in molecular iron(III) complexes, the distortion of the ligand field determines the zero field splitting tensor. The values of D shown in Table 19 range from 0.07 to 0.17 cm^{-1}, and E/D from 0 to 0.07, with the exception of Fe(pyridine-N-oxide)$_6$(ClO$_4$)$_3$, which has a very large D and rhombic spectra, with E/D = 0.233 [308]. In terms of effective g values, this corresponds to $\underline{g}_{eff} = 4.3$, and thus it was suggested that the octahedron is severely distorted. However, it has been shown that the low-symmetry spectral properties of the $[M(\text{pyridine-N-oxide})_6]^{2+}$ complexes (M = Mn, Co, Ni) is due to the nonlinear ligation of the oxygen atom of the pyridine-N-oxide ligand [151,152] (see also Sec. VIII), and we are inclined to suspect that a similar effect is also operative for iron(III).

When one or more monodentate ligands is substituted by different ligands, one observes in general an increase in the value of D and E. A large rhombicity was deduced from single-crystal spectra of the $[FeCl_5H_2O]^{2-}$ chromophores [309,310]. This was related to the π-bonding interaction with the water ligand [309]. High values of the zero field splitting were calculated for the trans-FeO_4X_2 chromophores, but with small orthorhombic components [311].

In the case of MA_3X_3 complexes [312], meridional isomers have been tentatively identified by their larger D and E/D values as compared to facial isomers, in accord with the suggestions of Griffith [85] (see Sec. IX.C).

The $[Fe(\text{oxalate})_3]^{3-}$ ion has been studied in several lattices [313], and the zero field splitting has been found to be extremely sensitive to the environment. Also, the dithio derivative has been studied [314]. In $[M(Ph_3P)_2]_3Fe(\text{dithioox})_3$ (M = Cu, Ag), the structure is as shown in Fig. 18. In the copper derivative, D is similar to that found in the $K_3Al(\text{oxalate})_3 \cdot 3H_2O$ lattice, whereas the silver derivative has very small D. The two dithiooxalate complexes are isomorphous and the iron atom has trigonal symmetry, as confirmed also by the ESR spectra. The observed large difference in the D value is attributed to the different strain exerted by the silver and copper ions, whose radii are different (126 pm for Ag^+ and 96 pm for Cu^+).

Five-coordinate trigonal bipyramidal geometry was assigned to the $Fe(Ph_3AO)_2X_3$ complexes (A = P, As; X = NCS, NO_3) [311] on the basis that the E/D ratio is closer to the limit expected for axial symmetry than to that of a complex of orthorhombic symmetry, therefore suggesting the presence of a trigonal axis.

Table 19 Spin Hamiltonian Parameters for Some High-Spin Iron(III) Complexes

Compound	Chromophore[a]	Spectra[b]	$g_\perp$[c]	$g_\parallel$[c]	D[d]	E/D	a[d]	F[d]	Reference
Octahedral complexes									
$Fe(Me_2SO)_6(NO_3)_3$	FeO_6	P			0.166	0.033			327
$Fe(Me_2SO)_6(ClO_4)_3$	FeO_6	P			0.130	0.067			327
$Fe(urea)_6Cl_3$	FeO_6	P			0.130	0.067			327
$Fe(urea)_6(NO_3)_3$	FeO_6	P			0.070	0.067			327
$Fe(pyO)_6(ClO_4)_3$	FeO_6	P			0.36	0.233			308
$Fe(H_2O)_6(NO_3)_3 \cdot 3H_2O$	FeO_6	P			0.080	0.03			328
$(In,Fe)Cl_5 \cdot H_2O$	$FeOCl_5$	C	2.0091	2.0109	-0.1519	-0.186	0.00215	0.000054	309, 310
$(In,Fe)(phenazine)_6(ClO_4)_3$	FeO_6	P			0.07	0.00			327
$(In,Fe)(Me_2NCHO)_6(ClO_4)_3$	FeO_6	P			0.088	0.01			327
$(In,Fe)(pyrazole)_3Cl_3$	$fac\text{-}FeN_3Cl_3$	P			0.24	0.133			312
$(In,Fe)(Mepyrazole)_3Cl_3$	$mer\text{-}FeN_3Cl_3$	P			0.90	0.31			312
$(In,Fe)(py)_3Cl_3$	$fac\text{-}FeN_3Cl_3$	P			0.164	0.10			312
$(In,Fe)(py)_3Br_3$	$fac\text{-}FeN_3Br_3$	P			0.665	0.033			312
$[Fe(Ph_3PO)_4Br_2]FeBr_4$	$trans\text{-}FeO_4Br_2$	P			1.20	0.0			311
$[Fe(Ph_3PO)_4Cl_2]FeCl_4$	$trans\text{-}FeO_4Cl_2$	P			0.63	0.10			311
$[Fe(Ph_3AsO)_4Br_2]FeBr_4$	$trans\text{-}FeO_4Br_2$	P			1.50	0.00			311
$[Fe(Ph_3AsO)_4Cl_2]FeCl_4$	$trans\text{-}FeO_4Cl_2$	P			0.55	0.10			311
$K_3(Al,Fe)(ox)_3 \cdot 3H_2O$	FeO_6	C α		2.012	0.2990	0.0826			313
		β		2.000	0.2382	0.0901			
$NaMg(Al,Fe)(ox)_3 \cdot 9H_2O$	FeO_6	C α	2.003[e]	2.003[e]	0.0466	0.0160			313
		β	2.002	2.002	0.0663	0			

Compound	Nearest-neighbor atoms[a]	State[b]	g					Ref.
[Cu(Ph₃P)₂]₃Fe(dithiox)₃	FeO_6	P		0.2268	0.			314
[Ag(Ph₃P)₂]₃Fe(dithiox)₃	FeO_6	P, C		0.087	0.			314
FeEDTA	FeN_2O_4	P	2.00	0.83	0.31			89
Square pyramidal complexes								
FeCl(acac)₂	FeO_4Cl	P	$g_{eff} = 4.22$					315
(NH₄)₂(Sb,Fe)F₅	FeF_5	C	1.9956	0.069	0.0015	0.0032	0.0006	316
Trigonal bipyramidal complexes								
Fe(Ph₃PO)₂(NCS)₃	FeO_2N_3	P	2.002	0.37	0.10			311
Fe(Ph₃PO)₂(NO₃)₂	FeO_2O_3	P	2.002	0.55	0.04			311
Fe(Ph₃AsO)₂(NO₃)₂	FeO_2O_3	P	2.002	0.62	0.067			311
Seven coordinate complexes								
Fe(EDTAH)H₂O	FeN_2O_4O	G	2.00	0.7	0.327			329
FeB(NCS)₂(BF₄)	FeN_5N_2	G	2.00	0.5	0.03			329
Planar complexes								
Fe(Ph₃PO)₄(ClO₄)₃	FeO_4	P	2.002	0.84	0.005			311
Fe(Ph₃AsO)₄(ClO₄)₃	FeO_4	P	2.002	1.05	0.000			311

[a] Metal and nearest-neighbor atoms.

[b] C = single crystal; P = polycrystalline powder; G = frozen solution.

[c] When one value is reported, it is the isotropic value.

[d] Values in cm^{-1}.

[e] The first value is g_x, the second is g_y.

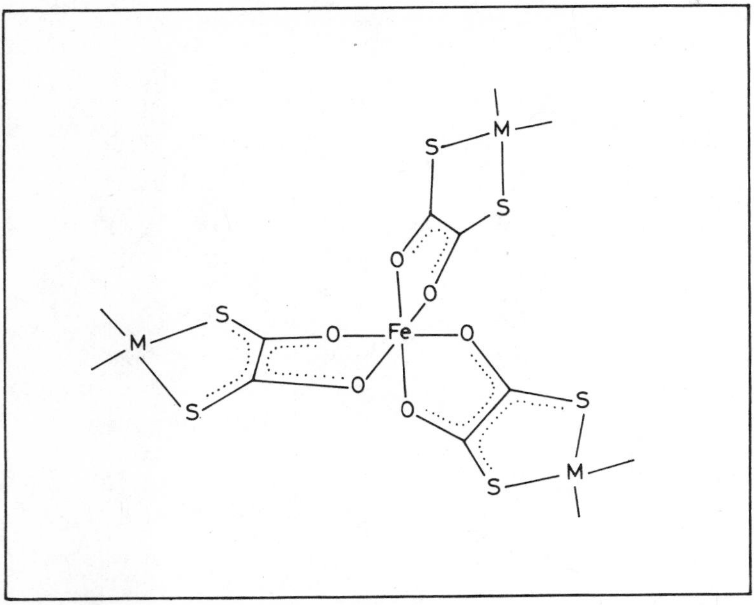

Figure 18   The FeO_6 chromophore in $[M(Ph_3P)_2]_3Fe(dithiox)_3$ (M = Cu, Ag).   (After Ref. 314.)

The square pyramidal five-coordinate geometry is well established in $FeCl(acac)_2$ [315] and $(NH_4)_2FeF_5$ [316] doped into the
antimony analog.   The former shows an isotropic $\underline{g}_{eff}$ = 4.22 spectrum
indicative of a large orthorhombic distortion and zero field splitting, whereas the latter has a very small zero field splitting similar to what one would expect for a regular octahedron.

Square planar geometry was assigned to $Fe(Ph_3AO)_4(ClO_4)_3$ (A = P,
As), and the observed D value was rather large [311].

Iron(III) has a unique biological role, and this has been
responsible for much research activity on model compounds.   Particularly interesting are the tetraphenylporphine (TPP) derivatives,
which have been used as models to explain the ESR spectra of cytochrome P-450.   The latter shows absorptions at $\underline{g}$ = 8, 3.7, and 1.7
[317,318], and a somewhat similar pattern was observed in the ESR
spectra of TPPFeCl coprecipitated with free $TPPH_2$ [319].   A careful
analysis of these spectra [320] revealed the presence of transitions

within the first excited Kramers doublet beyond those within the ground doublet. This observation allowed the authors to determine the zero field splitting parameters from the analysis of g values in the two Kramers doublets at the same temperature. They found D in the range 3.5-4.1 cm^{-1}, and E/D = 0.085, which compare quite well with the values reported for cytochrome P-450. Good agreement with the lattice spectra was also reported for other synthetic analogs [321]. Larger zero field splitting were measured for other ferric porphyrin complexes [322].

Another relevant field for the application of ESR spectroscopy has been the analysis of the iron(III) sites in minerals [323-326].

2. Low-Spin Iron(III)

In sufficiently strong ligand fields, doublet states become of lowest energy so that the spectra can be interpreted using a spin Hamiltonian with S = 1. The large majority of low-spin complexes are pseudooctahedral. In O_h symmetry the ground level is $^2T_{2g}$, which has its origin in the t_{2g}^5 configuration, which alternatively can be considered as a single hole in the t_{2g} orbitals. In general, the levels are split by low-symmetry components and spin-orbit coupling, so that several excited levels are close enough to the ground doublet to give fast spin-lattice relaxation. As a consequence, the ESR spectra can be detected in general only at low temperatures.

The two Kramers doublets can be described as [330]

$$|+> = A|1^+> + (B/2^{1/2})(|2^-> - |-2^->) + C|-1^+>$$
$$|-> = -A|-1^-> + (B/2^{1/2})(|2^+> - |-2^+>) - C|1^-> \qquad (53)$$

with $A^2 + B^2 + C^2 = 1$. It is easy to show that in orthorhombic symmetry the g values are [133,331]

$$g_x = 2[-2AC + B^2 + (2k^2B^2)^{1/2}(A - C)]$$
$$g_y = 2[-2AC - B^2 - (2k^2B^2)^{1/2}(A + C)] \qquad (54)$$
$$g_z = 2[A^2 - B^2 + C^2 + k(A^2 - C^2)]$$

where $\underline{k}$ is the orbital reduction factor (see Sec. VIII.B). In the limit of axial symmetry the $\underline{t}_{2g}$ levels will split into an orbitally degenerate and a nondegenerate level. The $\underline{g}$ values expected for a strong distortion are $g_\| = 2$, $g_\perp = 2$ for the unpaired electron in the orbitally nondegenerate level, and $\underline{g}_\| = 4$, $\underline{g}_\perp = 0$ for the unpaired electron in the orbitally degenerate state. It must be emphasized that relations (53) and (54) are only approximate, since contributions from the excited states are neglected. That these actually make relevant contributions is shown by the following considerations. If Eq. (54) were the exact expression, the relation

$$\underline{g}_x^2 + \underline{g}_y^2 + \underline{g}_z^2 \leqslant 16 \tag{55}$$

should hold for $0 < \underline{k} < 1$.

As is apparent from Table 20, there are many examples where $\underline{g}_x^2 + \underline{g}_y^2 + \underline{g}_z^2 > 16$. It has been common practice to cope with this situation by using $\underline{k}$ values larger than 1. If, however, the contributions of excited $^4T_2(\underline{t}_2^4 e^1)$ levels are taken into account, the $\underline{k}$ value in expressions (54) becomes [330]

$$\underline{k} = \underline{k}_{tt} + 6B\underline{k}_{te}(-0.2E_1^{-1} + 2.20E_2^{-2}) \tag{56}$$

where $\underline{k}_{tt}$ and $\underline{k}_{te}$ are defined by Eq. (45), E_1 and E_2 are the energy separations between the ground state and the $^2T_2[\underline{t}_2^4 e^1, d(x^2 - y^2)]$ and $^2T_2[\underline{t}_2^4 e^1, d(z^2)]$ states, respectively. The $\underline{k}$ value in relation (54) is given therefore by the sum of two terms. Although $\underline{k}_{tt}$ and $\underline{k}_{te}$ are smaller than 1, $\underline{k}$ may be larger than this value.

The experimental values are compared to the values calculated using Eq. (54), and the A, B, C, and $\underline{k}$ parameters are determined. From these values information about the order of the levels originating from $^2T_2$ can be gathered by Eq. (57):

$$\frac{\varepsilon}{\xi} = - \frac{[(2A^2)^{1/2} + B]C}{(18)^{1/2}(C^2 - A^2)}$$

$$\frac{E_1}{\xi} = -\left(\frac{A^2}{2B^2}\right)^{1/2} - \frac{2\Delta}{3\xi} \tag{57}$$

$$\frac{\Delta}{\xi} = -\left(\frac{A^2}{2B^2}\right)^{1/2} - \frac{1}{2} + \left(\frac{3A}{C}\right)\left(\frac{\varepsilon}{\xi}\right)$$

where ξ is the spin-orbit coupling constant, $6\varepsilon = (E_{xz} - E_{yz})$, and $\Delta = E_{xy} - [(E_{xz} + E_{yz})/2]$.

If ξ can be evaluated independently, for example, from magnetic susceptibility or Mössbauer measurements, then the complete energy pattern of the ground $^2T_2$ levels can be obtained.

Several single-crystal studies of iron(III) low-spin complexes have been reported. One of the most recent is that of $[Fe(CN)_5NH_3]^{2-}$ doped into $Na_2Fe(CN)_5NO \cdot 2H_2O$ [332]. The principal $\underline{g}$ values are listed in Table 20. Whereas $\underline{g}_z$ is parallel to the Fe-NH$_3$ bond, $\underline{g}_y$ and $\underline{g}_x$ are found to bisect the in-plane bond angles, so that a modified version of Eq. (54) was used in order to allow for a $d(x^2 - y^2)$ rather than for a $d(xy)$ orbital. In the host lattice the iron atom sits in a symmetry plane, and this symmetry is also maintained in the $[Fe(CN)_5NH_3]^{2-}$ ion. It may be somewhat surprising to observe such a large $\underline{g}$ value anisotropy, especially when compared to the doubly substituted $Fe(CN)_4N_2$ complexes of axial symmetry [333]. The calculated splitting of the $d(xz)$ and $d(yz)$ orbitals is 500 cm^{-1} and may result from angular distortions. The $d(x^2 - y^2)$ orbital is found to lie lowest, seemingly in contrast with simple ligand field strength considerations. However, in the angular overlap formalism, the energy difference between $d(xz)$ or $d(yz)$ and $d(x^2 - y^2)$ orbitals is

$$E(xz) - E(x^2 - y^2) = -2e_{\underline{\pi}}^{CN} \tag{58}$$

for an orthoaxial chromophore, assuming that $e_{\underline{\pi}}^{NH_3} = 0$. Since the experimental data show that $E(xz) - E(x^2 - y^2)$ is positive, it must be concluded that $e_{\underline{\pi}}^{CN}$ is negative. In the angular overlap formalism, this means a π-bonding effect of the ligand on the metal orbitals, in line with the accepted view of the strong metal π back-donation to the CN$^-$ ion.

Another case where the single-crystal $\underline{g}$ values yielded orbital energy order in contrast to ligand field expectations is that of

Table 20 Spin Hamiltonian Parameters for Some Low-Spin Iron(III) Complexes

Compound	Chromophore[a]	Spectra[b]	g_1[c]	g_2[c]	g_3[c]	Reference
$K_3(Co,Fe)(CN)_6$	FeC_6	C	2.10	0.91	2.35	331
$[Fe(CN)_5NH_3]^{2-}$ into						
$Na_2Fe(CN)_5NO \cdot 2H_2O$	FeC_5N	C	2.177	0.845	2.995	332
$[Fe(CN)_4bipy]^-$	FeC_4N_2	P	2.50		1.65	333
$[Fe(CN)_4phen]^-$	FeC_4N_2	P	2.7		1.6	333
$Fe(phen)_3(ClO_4)_3$	FeN_6	C	2.615	2.727	1.459	334
$Fe(bipy)_3(ClO_4)_3$	FeN_6	G	2.61		1.61	335
$Fe(4,4'-dmb)_3(ClO_4)_3$	FeN_6	P	2.64		1.38	349
$Fe(5,5'-dmb)_3(ClO_4)_3$	FeN_6	P	2.60		1.60	349
$Fe(CN)_2(bipy)_2ClO_4$	FeN_4C_2	P	2.50		2.00	335
$Fe(CN)_2(bipy)_2ClO_4$	FeN_4C_2	G	2.47	2.74	1.54	345
$Fe(CN)_2(phen)_2ClO_4$	FeN_4C_2	P	2.70		1.86	345
$Fe(CN)_2(phen)_2ClO_4$	FeN_4C_2	G	2.63		1.42	333
$[Ph_4P]_3Fe(mnt)_3$	FeS_6	P	2.114	2.225	1.986	339

Compound	Nearest-neighbor[a]	Method[b]					Ref.
Fe(S$_2$COEt)$_3$	FeS$_6$	G	2.143		2.193	1.992	339
Fe(dtb)$_3$	FeS$_6$	G	2.094		2.155	2.008	339
Fe(MeCSCHCOMe)$_3$	FeS$_3$O$_3$	G	2.182		2.341	1.930	339
KBa[Fe(S$_2$C$_2$O$_2$)$_3$]·6H$_2$O	FeS$_6$	P		2.16		1.99	339
Fe(dtc)$_3$	FeS$_6$	P	2.076		2.111	2.015	343
Fe(ttd)(dtt)$_2$	FeS$_6$	G	2.097		2.156	2.018	340
Fe(sacsac)$_3$	FeS$_6$	G	2.090		2.161	2.004	339, 341, 342
[Fe(NCS)$_2$(cyclam)]$^+$	FeN$_4$N$_2$	G	2.23		3.20	1.03	338
[FeBr$_2$(cyclam)]$^+$	FeN$_4$Br$_2$	G	2.34		3.25	1.08	338
Fe(C$_5$H$_5$)$_2$	Fecp$_2$	G		1.26		4.35	347
Fe(MeC$_5$H$_4$)$_2$	Fecp$_2$	G		1.47		4.17	347
Fe(C$_5$H$_4$CHO)$_2$	Fecp$_2$	G		1.77		3.63	347
Fe(DMFc$^+$)(PF$_6$)	Fecp$_2$	P		1.92		4.002	348
Fe(DMFc$^+$)(TCA$^-$)(TCAA)	Fecp$_2$	P		1.22		4.44	348

[a] Metal and nearest-neighbor atoms.

[b] C = single crystal; P = polycrystalline powder; G = frozen solution.

[c] (1, 2, 3) refer to (x, y, z) in single crystal spectra. When two values are reported, the first is the perpendicular component.

Fe(1,10-phenathroline)$_3$(ClO$_4$)$_3$ [334]. The $\underline{g}$ values, which are
quasiaxial, require a $^2A_{1g}$ ground level, which implies that the
orbitally nondegenerate component of the split $\underline{t}_{2g}$ orbitals lies at
higher energy. Since the chromophore can be described as a trigonally compressed octahedron, one would expect the reverse order.
However, the $\underline{t}_{2g}$ orbitals are of π symmetry in octahedral geometry,
and again the ESR data indicate π bonding effects of the phenanthroline ligand on the metal d orbitals. This is in line with the current view of substantial π back-bonding of iron(III) toward
phenanthroline.

A similar pattern of levels was found for Fe(bipy)$_3$(ClO$_4$)$_3$
[335] using magnetically perturbed Mössbauer spectra [336], which
also proved to be very useful in the analysis of iron protein spectra
[337].

Very highly anisotropic $\underline{g}$ values have been reported for the
macrocyclic complex FeX$_2$cyclam [338]. The usual ligand field
treatment shows that the d(xy) orbital is lower in energy than the
d(xz) and d(yz) orbitals, in accord with the smaller ligand field
strength of the axial as compared to the equatorial ligands.

Many studies have been devoted to tris(bischelate) complexes
with sulfur donor atoms [339-343]. In general, the $\underline{g}$ values observed for these complexes are closer to the spin-only value, and
the ground state obtained by relation (54) is essentially composed
of the d(xz) $\pm$ d(yz) hole, implying a very small distortion from O_h
symmetry. However, this view did not seem to be reasonable to
De Simone [343], and he tried to obtain a different order by comparison with the analogous ruthenium complexes. From his calculations
in the simple ligand field approach a $\underline{k}$ value of 1.98 is obtained.
However, for copper(II) sulfur chelate complexes, it has been shown
that covalency plays a major role and the spin Hamiltonian parameters have been calculated satisfactorily only within a molecular
orbital framework [111,112].

The ESR spectra of various ferricenium complexes have been reported [296,344-351] and interpreted according to the energy-level

order shown in Fig. 17, so that a $^2E_{2g}(a_{1g}^2e_{2g}^3)$ ground state is anticipated. The complex is isoelectronic with the manganese(II) analog, so the treatment is similar for the two ions.

The ESR spectra show high $g_\parallel$ values, in the range 4.3-2.6, and small $g_\perp$ values, which vary correspondingly from 1.3 to 2.0.

3. Iron(III) Spin Equilibria

The spin equilibrium $^6A_1 \rightleftarrows {}^2T_2$ has been observed to occur in certain six-coordinate iron(III) complexes of near octahedral symmetry [352]. This equilibrium has been studied by various techniques, such as magnetic susceptibility, Mössbauer, and of course ESR.

The first iron(III) complexes shown to possess spin equilibrium were the N,N'-dialkyldithiocarbamates [353], and they have been thoroughly studied by ESR. The first problem encountered in the study of these complexes is that of the presence of impurities. As a matter of fact, it has been suggested [354] that the g values seen in the spectra of $Fe(Me_2dtc)_3$ [355] are to some extent due to copper-(II) impurities. Hall and Hendrickson attribute the signals at $g = 3.27$ and 1.83 to transitions within the lowest Kramers doublet of low-spin iron(III), and a signal at $g = 4.3$ to the high-spin iron(III) in the equilibrium [354].

In the case of complexes exhibiting spin equilibrium, it is important to know whether the intermediate magnetic moment is due to a true equilibrium between a high- and a low-spin species or to an intermediate-spin species [356,357]. The ESR spectra show in all reported cases that there are distinct features for molecules in the $^6A_1$ and $^2T_2$ levels. Therefore the spin flipping rate must be slower than 10^7 s^{-1} [354].

Another consideration in spin equilibrium complexes is whether in distorted symmetries there is an S = 3/2 ground state. The complexes $[Fe(S_2C_2O_2)_2X](PPh_4)$ (X = Cl, Br, I) have been reported to be five coordinate with a square pyramidal geometry [358]. The spectra of the bromide and of the iodide derivative at 77 K are typical of S = 3/2 states, and magnetic susceptibility measurements are in accord with this view. The values of the spin Hamiltonian parameters of the bromide, for instance, are $\underline{g}$ = 2.04, D = 4.9 cm^{-1}, E = 0.6 cm^{-1} [358]. A spin equilibrium involving an S = 3/2 state has been reported on the basis of ESR and Mössbauer data [359].

4. *Iron(II)*

Iron(II) is a d^6 ion that can be either low spin, diamagnetic or high spin, paramagnetic in its six-coordinate complexes. ESR spectra of high-spin iron(II) complexes are exceedingly rare. The main reason for this is that the ion has an even number of unpaired electrons and the zero field splitting is in general large. Another factor that makes the observation of the spectra difficult is the fast relaxation time.

For octahedral complexes the ground state is $^5T_{2g}$, which is split by low-symmetry components and spin-orbit coupling. In an axial field the three lowest-lying states are

$$|\psi_1> = a|2, -1> + b|1, 0> + c|0, 1>$$
$$|\psi_0> = d|1, -1> + e|0, 0> + d|-1, 1> \qquad (59)$$
$$|\psi_{-1}> = a|-2, 1> + b|-1, 0> + c|0, -1>$$

in which the kets on the right side are labeled according to M_S and M_L, the eigenvalues of $\underline{S}_z$ and $\underline{L}_z$. In this situation a spin Hamiltonian with a fictitious spin S = 1 can be used and the $|\psi_0>$ state is separated from the other two by an amount D. The $\underline{g}$ values are given by

$$\underline{g}_{\parallel} = 5a^2 + 2b^2 - c^2$$
$$\underline{g}_{\perp} = -(2)^{1/2}(bd + ce) + (6)^{1/2}(be + cd) + 2ad \qquad (60)$$

Since in general D $\gg$ hν, the transitions within $|1>$ and $|-1>$ are not allowed when the static magnetic field is parallel to the $\underline{z}$ axis, but they become allowed when the static magnetic field is in the xy plane. The transition was detected in iron(II)-doped CdCl$_2$, where octahedral MCl$_6$ chromophores (M = Cd, Fe) are present [360]. The line is very broad ($\sim$0.3 T), even at 4.2 K. The crystal field is quite close to octahedral, and the zero field splitting was estimated to be 3.9 ± 0.4 cm^{-1}.

The magnetically perturbed Mössbauer spectroscopy yielded the spin Hamiltonian parameters for Fe[(pyridine-N-oxide)$_6$](ClO$_4$)$_2$: $g_{\parallel}$ = 9.0, $g_{\perp}$ = 0.6 in terms of an S = 1/2 effective spin Hamiltonian [361], since the zero field splitting leaves the $|\psi_1\rangle$, $|\psi_{-1}\rangle$ doublet level well separated from $|\psi_0\rangle$.

F. Cobalt

The naturally occurring isotope of cobalt is ^{59}Co (abundance 100%). It has a nuclear spin quantum number I = 7/2 and a gyromagnetic ratio $\gamma = \mu/I = 1.328$.

1. High-Spin Cobalt(II)

Cobalt(II) is a d^7 ion. In the high-spin state it has three unpaired electrons and very short spin-lattice relaxation time in all the common stereochemistries of interest, so that very low temperatures are needed to detect any signal. This, of course, has limited the number of investigations reported in the literature. Another characteristic of the ESR spectra of high-spin cobalt(II) is that they are extremely sensitive to low-symmetry components, so that lower-than-axial spectra are almost a rule. As a consequence, powder or glassy matrix spectra are rather uninformative.

The ESR theory for octahedral cobalt(II) was first presented in 1951 by Abragam and Pryce [362]. The ground level in this symmetry is $^4T_{1g}$, which is split by low-symmetry components and spin-orbit coupling to give in every case a Kramers doublet as the ground level. The ground level is in general well separated from all the other energy levels in such a way that the ESR spectra can be interpreted using a fictitious spin Hamiltonian with S = 1/2. Since this does not correspond to the true spin and the orbital contributions are not quenched, the expected g values are markedly different from g_e. As a matter of fact, the theory predicts g = 4.33 in octahedral symmetry. In axial symmetry, one of the components of the lowest Kramers doublet is given by

$$|\psi\rangle = c_1|3/2, -1\rangle + c_2|1/2, 0\rangle + c_3|-1/2, 1\rangle \qquad (61)$$

In Eq. (61) the numbers 3/2, 1/2, and -1/2 refer to the eigenvalues of $\underline{S}_z$, and 1, 0, and -1 label the three $\underline{T}_1$ orbital levels. The relations between this labeling and the true states are given in Ref. 9. The $\underline{g}$ values are given by the expressions

$$\underline{g}_{\parallel} = (6 - 2\gamma)c_1^2 + 2c_2^2 + (2\gamma - 2)c_3^2$$

$$\underline{g}_{\perp} = 4c_2^2 + (48)^{1/2}c_1c_3 + 8^{1/2}\gamma c_2 c_3 \qquad (62)$$

where γ is -1 in the strong-field and -3/2 in the weak-field limit. The c_1, c_2, and c_3 coefficients depend on Δ, the low-symmetry splitting of the $^4T_{2g}$ level, and on the spin-orbit coupling constant ξ. The $\underline{g}$ values pattern is $\underline{g}_{\parallel} > \underline{g}_{\perp}$ if ^4E_g lies lowest, whereas the reverse ($\underline{g}_{\parallel} < \underline{g}_{\perp}$) is true if $^4A_{2g}$ has the minimum energy. In Fig. 19 is shown the relation between $\underline{g}_{\parallel}$ and $\underline{g}_{\perp}$. The two curves correspond to the limiting cases of strong and weak field. Several examples of spin Hamiltonian parameters are shown in Table 21.

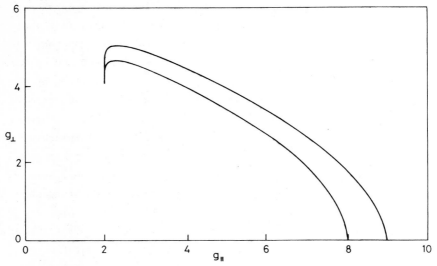

Figure 19 The calculated relation between $\underline{g}_{\parallel}$ and $\underline{g}_{\perp}$ for the high-spin cobalt(II) ion. Upper curve: weak-field case; lower curve: strong-field case. (After Ref. 9.)

In order to use the above expressions correctly, it must be stressed that whereas the observed pattern of g values can be used to guess the nature of the ground level, this does not correspond immediately to axial elongation or compression of the octahedron. As a matter of fact, it is easy to show that as a first approximation the splitting of the ground $^4T_{1g}$ level in tetragonal symmetry is [363]

$$\Delta \equiv E(^4\underline{A}_{2g}) - E(^4\underline{E}_g) \simeq E[d(xy)] - E[d(xz)] \tag{63}$$

Expressing the energies of the d orbitals in the angular overlap formalism gives

$$\Delta = 4e^{eq}_{\underline{\pi}\perp} - 2e^{eq}_{\underline{\pi}\|} - 2e^{ax}_{\underline{\pi}} \tag{64}$$

where $e^{eq}_{\underline{\pi}\perp}$ and $e^{eq}_{\underline{\pi}\|}$ are the parameters relative to π interaction of the equatorial ligands in and out of the tetragonal plane, respectively, and $e^{ax}_{\underline{\pi}}$ is the axial π-bonding parameter. Equation (64) shows that only when the six ligands are identical can it be concluded that the octahedron is compressed or elongated according to the g value pattern. In the case of MA_4B_2 complexes, only the relative π-bonding contributions of the axial and equatorial ligands can be estimated.

An example of how this analysis can be employed is given by the trans-dichlorotetrapyridine and trans-dichlorotetrapyrazole cobalt-(II) complexes [364]. The spectra of the pyridine complex are axial, with $g_\| = 3.33$ and $g_\perp = 4.46$, whereas the spectra of the pyrazole complex yield three g values, $g_1 = 2.14$, $g_2 = 4.55$, and $g_3 = 5.83$. In the two complexes the axial ligands are identical, whereas the equatorial ligands are similar in nature but show different orientation relative to the equatorial plane. In particular, the pyridine planes make an angle of ∿45° to the equatorial plane [365], whereas the pyrazole planes are orthogonal to it [366]. The differences in the g values were satisfactorily analyzed using Eq. (64) and considering that the geometric features of the complexes give different $e^{eq}_{\underline{\pi}\|}$ and $e^{eq}_{\underline{\pi}\perp}$ values.

Table 21   Spin Hamiltonian Parameters for Some High-Spin Cobalt(II) Complexes

Compound	Chromophore[a]	Spectra[b]	g_1[c]	g_2[c]	g_3[c]	A_1[c,d]	A_2[c,d]	A_3[c,d]	Reference
Octahedral complexes									
(Cd,Co)Br$_2$	CoBr$_6$	C	4.692	3.74		159		30	385
(Cd,Co)Cl$_2$	CoCl$_6$	C	4.967	3.06		166.9		42	385
(Mg,Co)Cl$_2$	CoCl$_6$	C	5.032	2.858		161		45	385
K$_4$(Cd,Co)Cl$_6$	CoCl$_6$	C	4.154	4.860		112.2		161.6	385
Cs(Cd,Co)Cl$_3$	CoCl$_6$[e]	C, I	4.459	4.366		144.0		138.3	385
		II	3.782	5.534		91.5		214.6	
(Mg,Co)(ac)$_2$·4H$_2$O	CoO$_6$	C	6.018	4.05	2.518	192	92	31	386
(Zn,Co)SeO$_4$·6H$_2$O	CoO$_6$[f]	C, I	7.45	2.36	1.76	230	40	60	387
		II	5.60	4.27	2.26	110	40	40	
(Cd,Co)(epydo)$_6$(BF$_4$)$_2$	CoO$_6$	P	6.10	4.28	2.38	191	92	$\sim$30	388
(Cd,Co)(epydo)$_6$(ClO$_4$)$_2$	CoO$_6$	P	5.24	5.00	2.58	149	126		388
(Zn,Co)(pyO)$_6$(ClO$_4$)$_2$	CoO$_6$	C	4.77		2.26	384		186	190
[(Zn,Co)(H$_2$O)$_2$(ox)]$_n$	CoO$_6$	P	3.72	2.62	6.28	252			389
[(Zn,Co)(2MeIz)$_2$(ox)]$_n$	CoO$_6$	P	3.62	2.52	6.29	205			389
[(Zn,Co)(DMIz)$_2$(ox)]$_n$	CoO$_6$	P	3.66	3.66	5.51	182			389
[(Zn,Co)(BIz)$_2$(ox)]$_n$	CoO$_6$	P	4.20	2.78	5.68	201			389
(Zn,Co)[HC(pz)$_3$]$_2$(NO$_3$)$_2$	CoN$_6$	C	0.807		8.534	<1		369	369
(Zn,Co)[HB(pz)$_3$]$_2$	CoN$_6$	C	0.97		8.46	<1		362	369
[(Cd,Co)(pz)$_6$](ClO$_4$)$_2$	CoN$_6$	P	4.69		2.34				390
[(Zn,Co)(pz)$_6$](ClO$_4$)$_2$	CoN$_6$	P	4.72		2.38				390
[(Cd,Co)(NMeIz)$_6$](ClO$_4$)$_2$	CoN$_6$	P	4.87		3.06			125	390
[(Zn,Co)(apy)$_6$](ClO$_4$)$_2$	CoN$_6$	P	4.92		3.08	$\sim$43		128	390
[(Cd,Co)(Iz)$_6$](ClO$_4$)$_2$	CoN$_6$	P	3.57		5.50	53		166	390

Compound	Metal and nearest-neighbor atoms		1	2	3		Ref.
$Co(py)_4Cl$	CoN_4Cl_2	C	4.55	4.46	3.33		364
$Co(pz)_4Cl_2$	CoN_4Cl_2	C	5.83		2.14		364
Tetrahedral complexes							
$Cs_3(Zn,Co)Cl_5$	$CoCl_4$	C		4.60	2.40		373
$(NEt_4)_2(Zn,Co)Cl_4$	$CoCl_4$	C	5.44	4.84	3.24		374
$(Zn,Co)[H_2B(pz)_2]_2$	CoN_4	P	0.95	1.13	6.90	~0	391
$(Zn,Co)\{H_2B[3,4,5(CH_3)_3(pz)_2]_2\}_2$	CoN_4	P	0.85	1.02	6.95	~0	391
$(Zn,Co)[(C_6H_5)_2B(pz)_2]_2$	CoN_4	P	2.03	2.28	7.47	160	391
$(Zn,Co)\{H_2B[3,5(CH_3)_2(pz)_2]_2\}_2$	CoN_4	P	2.00	2.30	7.35	130	391
$[(n\text{-}C_4H_9)_4N](Zn,Co)Br_3(quin)$	$CoBr_3N$	C	6.31	2.22	1.60		68
$(Zn,Co)(Ph_3PO)_2Cl_2$	CoO_2Cl_2	C	5.67	3.59	2.16		392
$Co(py)_2Cl_2$	CoN_2Cl_2	P	5.82	2.62	1.77		77
$Co(py)_2Br_2$	CoN_2Br_2	P	5.45	3.77	2.10		393
$Co(\gamma\text{-pic})_2Cl_2$	CoN_2Cl_2	P	6.14	2.12	1.50		393
Five-coordinate complexes							
$Co(Et_4dien)Cl_2$	CoN_3Cl_2	P	7.08	2.53	1.59		394
$Co(terpy)Cl_2$	CoN_3Cl_2	P	5.93	3.53	1.88		394
$[Co(Me_6tren)Cl]Cl$	CoN_4Cl	P		4.25	2.29		394
$(Zn,Co)(MePh_2AsO)_4(NO_3)_2$	CoO_5	C	8.06	1.3	0.9	431	368

a Metal and nearest-neighbor atoms.

b C = single crystal; P = polycrystalline powder.

c (1, 2, 3) refer to (x, y, z) in single-crystal spectra. When two values are reported, the first is the perpendicular component.

d Values in cm^{-1} × 10^{-4}.

e I = D_{3d} site symmetry; II = C_{3v} site symmetry.

f I = C_1 site symmetry; II = C_2 site symmetry.

Another interesting characteristic of the pyrazole compound
spectra is that the symmetry is definitely lower than axial. The
main deviation from D_{4h} symmetry seen in the crystal structure is
that the in-plane bond angles are slightly different from 90°, that
is, 92 and 88° [366]. We believe that this small deviation is able
to account for the observed nonaxial symmetry as shown in Fig. 20.
As a matter of fact, it is possible to calculate that a distortion
of 1° from 90° can cause a splitting of $\underline{g}_\perp$ of the order of 1.4. It
must be mentioned that for this calculation the general ligand field
approach outlined in Sec. VIII was used.

A useful example to show how low-symmetry components can alter
the $\underline{g}$ value pattern as compared to the simple axial case studied by
Abragam and Pryce is given by the five-coordinate square pyramidal
nitratotetrakis(methyldiphenylarsineoxide) cobalt(II) nitrate, whose
structure is sketched in Fig. 21. In the solid the cobalt atom lies
on the axis perpendicular to the O_4 plane, whereas the axial oxygen
is displaced to one side. The powder spectra show three $\underline{g}$ values,

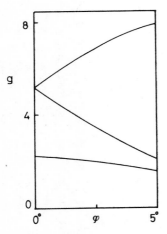

Figure 20   Calculated $\underline{g}$ values for $Co(pz)_4Cl_2$ for various values of
the in-plane bond angles. ϕ indicates the deviation of the in-plane
bond angles from 90°. The other parameters are $B = 800$ cm^{-1}, $\xi = 533$
cm^{-1}, $\underline{e}^{pz}_\sigma = 4333$ cm^{-1}, $\underline{e}^{pz}_\pi = 250$ cm^{-1}, $\underline{e}^{Cl}_\sigma = 1333$ cm^{-1}, $\underline{e}^{Cl}_\pi = 0$. (Reprinted with permission from A. Bencini, C. Benelli, D. Gatteschi,
C. Zanchini, *Inorg. Chem.*, *19*:1301. Copyright 1980 American Chemical
Society.)

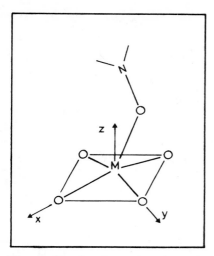

Figure 21 A schematic view of the CoO_5 chromophore in $Co(MePh_2AsO)_4$-$(NO_3)_2$, where M = Co. (After Ref. 367.)

g_1 = 8.6, g_2 = 1.3, and g_3 = 0.9 [368]. Single-crystal studies showed that four magnetically nonequivalent sites are present in the cell. The g_3 axis is close to the perpendicular to the O_4 plane, making an angle of ~12° with it. These spectra were interpreted with the ligand field model of Sec. VIII, by considering that the axial oxygen is not perpendicular to the equatorial plane. The g and A values were reproduced satisfactorily using e_σ and e_π parameters that compare well with the values reported in the literature for similar complexes. Figure 22 shows how the g values are sensitive to a distortion that moves the axial ligand on one side. The angle θ is the angle of the cobalt-oxygen bond to the perpendicular to the basal plane: When θ is 0°, the g values are axial as expected, whereas $g_\perp$ splits into two components as θ increases from zero.

Trigonally distorted octahedral complexes have been studied by Jesson [369]. He found very anisotropic g and A values. In particular, since $A_\perp$ is very small (it was evaluated to be $<1 \times 10^{-4}$ cm^{-1}), a very unusual pattern of spectra was observed when the static magnetic field was close to the molecular plane. This was attributed to second-order corrections in the hyperfine structure perturbation.

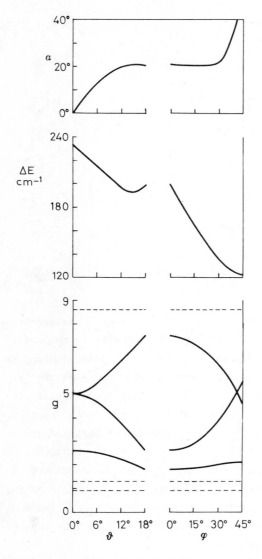

Figure 22 Calculated g values, zero field splitting (ΔE), and angle of $\underline{g}_3$ to the z axis (α) for $Co(MePh_2AsO)_4(NO_3)_2$. θ and φ are the polar angles of the axial oxygen in the reference system of Fig. 21. Left: the effect of varying θ with φ = 0, $\underline{e}_\sigma^{basal}$ = 6000 cm^{-1}, $\underline{e}_\pi^{basal}$ = 1200 cm^{-1}, $\underline{e}_\sigma^{axial}$ = 4000 cm^{-1}, $\underline{e}_\pi^{axial}$ = 960 cm^{-1}, k = 0.8, ξ = 533 cm^{-1}, B = 850 cm^{-1}. Right: the effect of varying φ with θ = 18. Dashed lines correspond to experimental values. (Reprinted with permission from A. Bencini, C. Benelli, D. Gatteschi, and C. Zanchini, *Inorg. Chem.*, *18*:2526. Copyright 1979 American Chemical Society.)

Trigonal bipyramidal complexes have an orbitally nondegenerate ground level, $^4A_2'$ in the limit of D_{3h} symmetry. Since there are several excited levels close to the ground level, the zero field splitting will be larger in general than the energy of the available microwave photons, so that only the transitions within the lowest manifold will be observed. No detailed analysis of this type of complex has yet been reported, but it seems that the powder spectra can be interpreted assuming that they arise from the lowest ±1/2 manifold [370].

The situation is similar for tetrahedral complexes that in T_d symmetry have a $^4A_2$ ground level. In the high-symmetry limit, however, the four levels may be degenerate, so that all the expected transitions can be observed in principle. In general, however, the symmetry will be lower, and transitions will be observed only within the lowest Kramers doublet.

The $[CoCl_4]^{2-}$ chromophore has been studied in several lattices [371-376], which have provided several different geometric environments. In particular, the ±3/2 state was found to be the lowest level in Cs_3CoCl_5 doped into the zinc lattice [373]. The zero field splitting was measured directly using pulsed fields, and the spin Hamiltonian parameters were estimated to be $g_{\parallel}$ = 2.40, $g_{\perp}$ = 2.30, and D = -4.5 cm^{-1}. On the other hand, the ±1/2 state was found to be of lower energy in $(Et_4N)_2CoCl_4$ doped into the zinc analog [374]. Horrocks and Burlone [377] used an angular overlap approach similar to that of Sec. VIII to show that the crossover of the ±1/2 and ±3/2 levels is determined essentially by the bond angles. The former will be of lowest energy for compressed tetrahedral distortions, whereas the latter will be the ground state for elongated octahedra. Confirmation of this comes from the powder spectra of bis(N-t-butylpyrrol-2-carbaldimino)cobalt(II), which at X-band frequency do not show any signal at all [378]. Since the X-ray crystal structure shows that the chromophore is an elongated octahedron with a very narrow N-Co-N angle (84°) [379], the ±3/2 levels must be of lower energy. If the zero field splitting is large as expected, it may be difficult to detect the transition between the ±3/2 states with an ordinary spectrometer. Sample calculations using the parameters that fit the single-crystal polarized optical spectra suggest that $\underline{g}_\parallel$ = 7-8 and $\underline{g}_\perp$ = 0.

Another interesting feature of the $CoCl_4$ spectra is that the $\underline{g}_z$ value in $(Et_4N)_2CoCl_4$ doped into the zinc analog [374] is found to be approximately orthogonal to the room temperature tetragonal axis seen in the crystal structure. Since the cobalt atom has been reported to have C_4 site symmetry [380], it is apparent that some major change occurred in the lattice upon passing from room temperature to 4.2 K. As a matter of fact, phase transitions have been reported for this and other M(II) complexes (M = Mn, Fe, Ni, Cu, Zn) [381]. On the basis of the analysis of the temperature dependence of the ESR spectra of $(Et_4N)_2CuCl_4$ [382], which is not isomorphous but is tetragonal at room temperature, we suggested that the tetragonal symmetry seen at room temperature results from fast hopping of the complex between two equally compressed tetrahedra. At lower temperature this hopping is frozen, and the unique axis is found orthogonal to the room temperature axis.

Another example of the sensitivity of cobalt(II) to low-symmetry components is given by the spectra of $[(C_4H_9)_4N][CoBr_3(quinoline)]$ [68]. The site at the metal atom is C_1 [383], but the electronic spectra were interpreted with some success using C_3 symmetry [384]. However, the ESR spectra show three $\underline{g}$ values and, furthermore, none of the principal directions coincides with the Co-N axis, which would be the trigonal axis. The closest $\underline{g}$ direction is that of $\underline{g}_1$, which makes an angle of $\sim 20°$ with the Co-N direction, showing how polycrystalline powder or glassy solution spectra must be used with much caution to obtain spin Hamiltonian parameters of high-spin cobalt(II) complexes.

2. *Low-Spin Cobalt(II)*

The number of studies of low-spin cobalt(II) complexes is very large, for several reasons. Certainly the fact that it is an S = 1/2 spin system, which is simple to detect and interpret, is one reason; another is biological relevance, since it was discovered that some cobalt(II) complexes can reversibly bind molecular oxygen [395].

Although the system is S = 1/2, the ion possesses seven electrons, making the interpretation of the ESR spectra not as straightforward as in d^1 or d^9 systems. As a consequence, there has been much controversy and confusion in the ESR literature pertaining to square planar cobalt(II) complexes, although in the last few years fundamental steps toward a definitive rationalization of the data have been made.

There are essentially two types of ESR spectra of square planar cobalt(II) complexes, as can be seen from Table 22. The first type (type I), exemplified by cobalt phthalocyanine (CoPc) and cobalt tetraphenylporphine (CoTPP) [396], which have axial spectra, with $g_{\parallel}$ = 1.8-1.9, $A_{\parallel}$ = 160-170 × 10^{-4} cm^{-1}, $g_{\perp}$ = 2.9-3.3, and $\underline{A}_{\perp}$ = 260-320 × 10^{-4} cm^{-1}. An example of the second type (type II) is the CoSALen series of complexes. They possess completely anisotropic $\underline{g}$ values, typically $\underline{g}_1$ = 3.2-3.8, $\underline{A}_1$ = 100-300 × 10^{-4} cm^{-1}, $\underline{g}_2 \simeq \underline{g}_3$ = 1.7-2.0, $\underline{A}_2 \simeq \underline{A}_3$ = 30-50 × 10^{-4} cm^{-1}. In type I spectra the $\underline{g}_{\parallel}$ direction is found as expected perpendicular to the molecular plane, whereas in type II spectra $\underline{g}_1$, which at first guess might be interpreted as $\underline{g}_{\parallel}$, is found in the molecular plane. In all the reported single-crystal studies [397-402], except Co(acacen), the $\underline{g}_{\parallel}$ principal direction bisects the ligand-metal-ligand angle. It must be mentioned, however, that an alternative assignment of the spectra of Co(acacen) [402] has been proposed by Daul et al. [403] in an exhaustive review of the electronic structure of cobalt(II) Schiff base complexes. They argue that the $\underline{g}$ directions should be similar to those in the Co(SALen) complexes, rather than being along the bond axes. When the crystal structure of the diamagnetic host was known, $\underline{g}_1$ was found to bisect the bond angles, which are smaller than 90°.

There had previously been much controversy in the literature concerning the proper ground state for low-spin cobalt(II) complexes possessing type II spectra [404-410]. However, now it seems well established that the ground state responsible for the type II ESR spectra can be described essentially as $d^2(xz)d^2(z^2)d^2(x^2 - y^2)d^1(yz)$,

Table 22 Spin Hamiltonian Parameters for Some Low-Spin Cobalt(II) Complexes

Compound	Chromophore[a]	Spectra[b]	g_1^c	g_2^c	g_3^c	$A_1^{c,d}$	$A_2^{c,d}$	$A_3^{c,d}$	Reference
Square planar complexes									
(Ni,Co)(sacsac)$_2$	CoS$_4$	C	3.280	1.904	1.899	105	35	35	398
(Ni,Co)SALen	CoN$_2$O$_2$	C	3.80	1.66	1.74	291	52	30	399
(Ni,Co)(mesityl)$_2$(PEt$_2$Ph)$_2$	CoP$_2$S$_2$	C	3.72	1.96	1.74	390	50	140	397
(Ni,Co)(mnt)$_2$	CoS$_4$	C	2.798	2.025	1.977	50	28	23	400
(Ni,Co)(amben)	CoN$_4$	C	2.6586	1.9814	2.0068	3.5	29.5	24	401
(Ni,Co)(benacen)	CoN$_2$O$_2$	C	3.372	1.882	1.954	165.4	37	47	401
(Ni,Co)(PhtiazineO)$_2$	CoN$_2$O$_2$	C	2.806	2.062	1.966	16	27	36	450
(Ni,Co)(acacen)	CoN$_2$O$_2$	C	3.26	2.00	1.88	116	34.5	37.5	402
Co(benacacen)	CoN$_4$	G	3.55	1.79	1.89				422
Co(TAAB)(NO$_3$)$_2$	CoN$_4$	G	2.2762		2.0259	12.9		75.3	471
(Ni,Co)(dpe)$_2$(BPh$_4$)	CoP$_4$	P	2.807	2.475	2.001	164	123	129	465
(Ni,Co)(nstsn)	CoN$_2$S$_2$	P	~2.0	~2.0	3.29			164	466
(Ni,Co)(ipe)	CoN$_4$	P	4.256	1.53	1.71	397	55	30	467
Co(cyen)	CoN$_4$	G	2.65	1.975	2.03	<10	28		405
Co(acac)$_2$(-)pm	CoN$_4$	G	3.1619	1.8968	2.0117	75	33	38	428
[Et$_4$N][Zn,Co)(cptc)$_2$	CoS$_4$	P	2.273	2.053	1.988	81	24	20	468

Five-coordinate complexes

Co(CN)$_5$$^{3-}$	CoC$_5$ SQ	P	2.1754		2.0037	-23.9		81.6	469
[CoCl(dpe)$_2$]SnCl$_3$	CoP$_4$Cl SQ	P	2.276	2.357	2.003				454
[Co(NCS)(dpe)$_2$]ClO$_4$	CoP$_4$N SQ	P	2.25		2.01				453
Co(nnp)(NCS)$_2$	CoN$_2$N'$_2$P SQ	P	2.32		2.02				457
[CoCl(dpe)$_2$]SnCl$_3$·C$_6$H$_5$Cl	CoP$_4$Cl TBP	P	2.062		2.258				454
CoBr$_2$(PPhF)$_3$	CoP$_3$Br TBP	C	2.069		2.252	12		102	470
[CoBr(dpe)$_2$]ClO$_4$	CoP$_4$Br TBP	P	2.07		2.26				453
[CoCl(dpe)$_2$]ClO$_4$	CoP$_4$Cl TBP	P	2.06		2.26				453

Six-coordinate complexes

K$_2$BaCo(NO$_2$)$_6$	CoO$_6$	RT P	2.15		2.10				461
		LT	2.15		2.02				
Co(1-Phborabenzene)$_2$		G	2.130	1.915	1.990	142	162	34	464
(Ru,Co)(cp)$_2$		C	1.140	1.219	1.584	25.5	-120.6	-103.4	462
(Ni,Co)(cp)$_2$		C	1.160	1.376	1.939	21.8	-121.6	-103.6	462

[a]Metal and nearest-neighbor atoms. SQ = square pyramidal; TBP = trigonal bipyramidal.

[b]C = single crystal; P = polycrystalline powder; G = frozen solution; RT = room temperature; LT = low temperature. When two values are reported, the first is the perpendicular component.

[c](1, 2, 3) refer to (x, y, z) in the single-crystal spectra.

[d]Values in cm^{-1} × 10^{-4}.

using the coordinate system shown in Fig. 23, whereas for type I spectra the ground state is $d^2(x^2 - y^2)d^1(z^2)d^2(x^2 - y^2)d^2(yz)^2$. It must be stressed that the fact that the ground level originates from the above configurations does not imply that the $d(yz)$ or $d(z^2)$ orbital is higher in energy than, say, $d(x^2 - y^2)$, since electron repulsion terms play a major role in determining the relative energies of the doublet terms originating from different configurations. The electron repulsion contributions to the various doublet levels are shown in Table 23. In square planar complexes the energy of the $d(xy)$ orbital is expected to be much higher than that of the other four, so that it need not be taken into consideration. It is apparent that the configuration with one electron in the $d(x^2 - y^2)$ orbital is 20 B higher in energy than that with the unpaired electron in $d(z^2)$, and 15 B higher than those with the unpaired electron in either $d(xz)$ or $d(yz)$.

Much of the cause of prior controversy in the literature has been due to the use of oversimplified equations relating g and A to electron energy levels. As a matter of fact, the d^3 hole

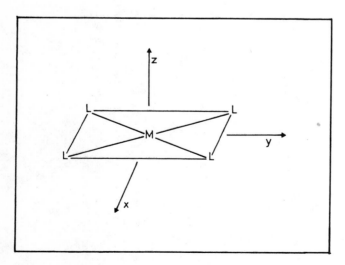

Figure 23   The coordinate system for low-spin square planar cobalt(II) complexes.

Table 23 The Electron Repulsion Contribution to the Doublet Levels of a d^7 Configuration

Configuration $d^i(xz)d^\ell(z^2)d^m(x^2-y^2)d^n(yz)$				Electron repulsion contribution
i	ℓ	m	n	
2	2	2	1	2A - 3B + 4C
2	2	1	2	3A + 12B + 4C
2	1	2	2	3A - 8B + 4C
1	2	2	2	3A - 3B + 4C

configuration has 120 states, and this has set many limitations to the possible calculations up to now. The first approach to the $\underline{g}$ and $\underline{A}$ values was given by Griffith [411], using first-order perturbation theory. He showed that if the unpaired electron is in a $d(z^2)$ orbital as in the case of cobalt phthalocyanine, the spin Hamiltonian parameters are given by

$$\underline{g}_\| = 2$$
$$\underline{g}_\perp = 2 - \frac{6\xi}{\Delta_{xz}}$$
$$\underline{A}_\| = P[-K + \frac{4}{7} + \frac{g_\| - 2}{7}]$$
$$\underline{A}_\perp = P[-\underline{K} - \frac{2}{7} - \frac{15}{14(\underline{g}_\perp - 2)}]$$

(65)

where ξ is the spin-orbit coupling constant, taken as negative, and Δ_{xz} is the energy separation of the excited $d(xz)$ and $d(yz)$ orbitals. He suggested that a $\underline{g}_\|$ value smaller than 2 might be justified by higher-order perturbation terms. In order to fit $\underline{A}_\|$ and $\underline{A}_\perp$, the terms P and $\underline{K}$ were calculated as 0.04 and 0, respectively. The low value of $\underline{K}$ was attributed to the admixture of 4s orbitals into the ground $d(z^2)$ orbital, which makes a positive contribution to the contact term. Later it was recognized that first-order perturbation theory could not be adequate, and second-order corrections must be applied. At the same time, the number of strong field electron configurations included in the calculation was increased, eventually diagonalizing a complete matrix rather than resorting to perturbation [412,413].

McGarvey used a third-order perturbation approach [414], taking into account the effect of low-lying excited quartet states and the mixing of $d(z^2)$ and $d(x^2 - y^2)$ orbitals in lower symmetries. He concluded that it may be difficult to judge the nature of the ground state from $\underline{g}$ and hyperfine structure of cobalt alone, but the magnitude of $\underline{g}_y$ and $\underline{g}_z$ does seem to provide some clue to the correct ground state.

Lin [415,416] has made calculations using different basis sets for cobalt porphyrins, that is, for type I spectra. He found that a nine-function basis set was inadequate to reproduce the $\underline{g}$ values found in the rigorously planar complexes, and that a 16-function basis set, including some doubly excited configurations could nicely reproduce the experimental results. The calculations showed that the ground state is $^2\underline{A}_1$ (D_4 symmetry), which corresponds to placing the unpaired electron into a $d(z^2)$ orbital. He also showed the reasons why previous perturbation treatments, even to second order, were not correct, in that not all the low-lying quartet and doublet states that could mix into the ground state were included. When one considers the formation of adducts the $d(z^2)$ orbital is highly destabilized (Fig. 24), so that the nine-function basis set becomes sufficient.

Hitchman [417], on the other hand, used an angular overlap approach to calculate the $\underline{g}$ values for type II spectra. He used $\underline{e}_\sigma$ and $\underline{e}_\pi$ parameters obtained from the analogous copper complexes [418], and obtained the ground state from the matrix relative to the ligand field and spin-orbit coupling perturbation on 24 states, either doublet or quartet. He considered orthorhombic symmetry C_{2v} within an orthoaxial chromophore. The splitting of the $d(xz)$ and $d(yz)$ orbitals was introduced as an ad hoc artifact, since the angular overlap model predicts for an orthoaxial chromophore that the symmetry is quite close to D_4, because the differences of the O and N ligands on the same axis are averaged by the ligand field perturbation. With

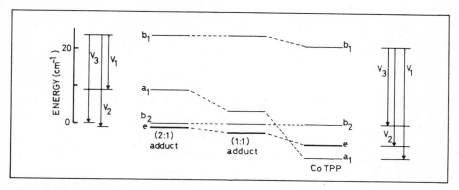

Figure 24 Orbital energy levels for some cobalt porphyrin systems.
Angular overlap expressions for V_1, V_2, and V_3 are given in the
text. (Reprinted with permission from W. C. Lin, *Inorg. Chem.*, *15*:
1114. Copyright 1976 American Chemical Society.)

this model, however, g values are reproduced nicely, corresponding
to a ground state with the unpaired electron in d(yz).

     If one compares the one-electron energy levels in the calcula-
tion by Hitchman with those of Lin, it is apparent that the two
models are very similar. The difference in the orbital energies
calculated by Lin for CoTPP and CoPc and by Hitchman for CoSALen are
shown in Table 24. The energy differences are larger in the case of
CoTPP than for CoSALen. The angular overlap model predicts that for
orthoaxial square planar complexes they are given by

Table 24 The Calculated Differences in Orbital Energies for CoTPP,
CoPc, and CoSALen

Energy differences	CoTPP[a,c]	CoPc[a,c]	CoSALen[b,c]
$V_1 = E\lvert d(xy)\rvert - E\lvert d(z^2)\rvert$	31.8	25	23.3
$V_2 = E\lvert d(xy)\rvert - E\lvert d(xz), d(yz)\rvert$	28.55	27	21.7
$V_3 = E\lvert d(xy)\rvert - E\lvert d(x^2 - y^2)\rvert$	22.5	25	18.9

[a]References 415, 416.
[b]Reference 417.
[c]Values in cm^{-1}.

$$V_1 = 2\underline{e}_\sigma$$

$$V_2 = 3\underline{e}_\sigma - 2\underline{e}_{\pi\parallel} \tag{66}$$

$$V_3 = 3\underline{e}_\sigma - 4\underline{e}_{\pi\perp}$$

where $\underline{e}_{\pi\parallel}$ and $\underline{e}_{\pi\perp}$ mean the interaction parallel and perpendicular to the z axis, respectively. The increased energy differences can be considered to be due to larger interaction (the M-N distance in Ni phthalocyanine is 182 pm [419], whereas the Co-O and Co-N distances in CoSALen are 185 pm [420]). It is worth mentioning that placing the $d(x^2 - y^2)$ orbital higher in energy than $d(xz)$, $d(yz)$, and $d(z^2)$ (i.e., letting V_3 have the smallest value) does not imply that the metal-ligand interaction is ionic, but only that the π interaction in the molecular plane is very strong. A similar approach was used by Nishida and Kida [421], who obtained good agreement with the experimental values of the spin Hamiltonian parameters for type II spectra using a basis set of eight functions.

The low-symmetry components, which split the $d(xz)$ and $d(yz)$ orbitals, are also relevant in the determination of the appearance of type I or type II spectra. For the complexes that have these spectra, the angles in the plane are a little different from 90°. For such an arrangement one expects that the $d(xz)$ and $d(yz)$ orbitals split, the $d(yz)$ lying higher in energy by $4 \cos \alpha \underline{e}_{\pi\parallel}$, where α is the small angle between two ligands.

The order of the orbitals becomes definitely simpler when one considers five- and six-coordinate adducts. In that case the energy differences are given as shown in Table 25 according to the angular overlap formalism. The energy of the $d(z^2)$ orbital is increased by the σ interaction with the axial ligands, and the unpaired electron remains in this orbital. Also, the larger spread of the levels makes the first-order treatment of Griffith valid, and g values of the adducts in Table 22 show the expected pattern $\underline{g}_\parallel \simeq 2$ and $\underline{g}_\perp \simeq 2.2$.

Following the above discussion, it is apparent that any formula that gives the $\underline{g}$ and $\underline{A}$ values as a function of ligand field parameters must be looked at with suspicion, and that the best way to

Table 25 Angular Overlap Energy Differences of Five- and Six-Coordinate Adducts of Planar Low-Spin Cobalt(II) Complexes

Energy differences[a]	Five-coordinate adducts[b]	Six-coordinate adducts[b]
V_1	$2e_\sigma^{eq} - e_\sigma^{ax}$	$2e_\sigma^{eq} - 2e_\sigma^{ax}$
V_2	$3e_\sigma^{eq} - 2e_{\pi\parallel}^{eq} - e_\pi^{ax}$	$3e_\sigma^{eq} - 2e_{\pi\parallel}^{eq} - 2e_\pi^{ax}$
V_3	$3e_\sigma^{eq} - 4e_{\pi\perp}^{eq} - e_\pi^{ax}$	$3e_\sigma^{eq} - 4e_{\pi\perp}^{eq} - 2e_\pi^{ax}$

[a] V_1, V_2, and V_3 are defined in Table 24.

[b] eq = equatorial ligands, ax = axial ligand(s). The symbols $\pi\parallel$ and $\pi\perp$ refer to the component of the π interaction parallel and perpendicular to the normal to the basal plane.

compare experimental values against theoretical models is to resort to complete calculations.

For most of the above compounds there have been numerous studies in different solvents, and extended correlations between the nature of the solvent and the g values have been made [422-438]. In general, however, no unambiguous trend has emerged that allows one to distinguish the presence of five- or six-coordinate complexes.

Among the adducts the most interesting are those with dioxygen [438]. The ESR spectra in general show g values quite close to 2.00, and a small cobalt hyperfine coupling constant [430,439-445]. Elegant studies with ^{17}O-enriched oxygen have revealed that the two oxygen atoms are equivalent, with a large coupling constant a_O = 21.6 G. This is comparable in magnitude to that observed for symmetrical superoxo groups [446,447]. The equivalence can be explained either by a π-bonded symmetric structure or by rapid hopping of the two oxygen atoms between two bent positions. There has been some dispute in the literature regarding the nature of the bond in these complexes. The ESR spectra show beyond doubt that the unpaired electron is essentially localized on the oxygen atoms. However, this does not imply an ionic $Co^{III}-O_2^-$ bond, but can also be rationalized in a molecular orbital scheme, without the transfer of one electron from cobalt to oxygen [448]. In this model the cobalt hyperfine splitting is attributed to a spin polarization mechanism [448,449].

At the end of this section on square planar complexes, it is worth mentioning the detailed single-crystal study of bix[1-methyl-3-(2-chloro-6-methyl)phenyltiazine-1-oxide]cobalt(II) doped into the diamagnetic nickel(II) analog [450]. In this complex, of unknown crystal structure, very large quadrupole effects were detected, which yielded very intense "forbidden" lines. By complete diagonalization of the Hamiltonian matrix, one of the largest quadrupole coupling constants ever found for cobalt was reported.

Beside the large number of adducts of square planar complexes, five-coordinate complexes of other geometries have been reported.

Square pyramidal complexes are expected to have a ground term originating from $d^2(xz)d^1(z^2)d^2(x^2 - y^2)d^2(yz)$ [451,452] well separated from excited states, so that the normal $\underline{g}$ value pattern is $\underline{g}_\perp > \underline{g}_\parallel \simeq 2.00$, and all the complexes reported so far conform to this prediction. For trigonal bipyramidal complexes, in the limit of D_{3h} symmetry, the unpaired electron would be expected to yield a $^2E'$ ground term, which is unstable under Jahn-Teller distortions [451]. As a matter of fact, all such complexes studied so far show marked deviations from regular D_{3h} symmetry, and the unpaired electron is either in a $d(x^2 - y^2)$ or $d(xy)$ orbital yielding $\underline{g}_\parallel > \underline{g}_\perp > 2.00$, a pattern not dissimilar from that found for copper(II) complexes in which the unpaired electron is in a $d(x^2 - y^2)$ or $d(xy)$ orbital.

One of the characteristics of five-coordinate low-spin cobalt(II) is the ease by which the complexes distort and give rise to intermediate geometries. For instance, the complexes $[CoCl(dpe)_2]^+$ can crystallize as either a genuine square pyramid or a distorted bipyramid, depending on the lattice forces. Thus, when the anion is $SnCl_3^-$ the $\underline{g}$ value pattern suggest a $d(z^2)$ ground state, whereas when C_6H_5Cl enters the lattice, the $\underline{g}$ value pattern is that of a trigonal bipyramid [453]. The same pattern is observed for the perchlorate and other phosphine ligands [454,455].

The complex $Co(nnp)(NCS)_2$ [456] exhibits a spin equilibrium so that at room temperature there is about an equimolar mixture of the high-spin and low-spin forms, whereas at 77 K the complex is practically all in the low-spin form. The ESR spectra at 300 and 77 K show the presence of the low-spin form [457], without any appreciable change, and a g pattern in accord with the presence of a square pyramidal complex. Variable-temperature X-ray data [457,458] have shown that this complex is indeed present at low temperature, whereas at room temperature the structure is intermediate between a square pyramid and a trigonal bipyramid. The ESR data definitely show that there is an equilibrium between two forms, one low, the other high spin, whose interconversion rate at room temperature is slower than 10^9-10^{10} s^{-1}.

In a regular octahedral geometry the ground state of low-spin cobalt(II) should be ^2E_g, which is unstable toward Jahn-Teller distortions. If a compression of the octahedron takes place as a result of vibronic coupling, the unpaired electron will be in the $d(x^2 - y^2)$ orbital, whereas it will be in the $d(z^2)$ orbital in the case of elongation of the octahedron. $K_2BaCo(NO_2)_6$ was reported to be exactly octahedral at room temperature [459], whereas it becomes rhombic at lower temperatures [460], with a substantial elongation of the octahedron. The ESR spectra [461] confirm these results with isotripic spectra at room temperature, and $g_{\parallel}$ = 2.02, $g_{\perp}$ = 2.15 at 143 K.

Very accurate single-crystal work has been done on cobaltocene in various lattices and glassy matrices [296,462,463]. The ligand field one-electron orbital scheme for cobaltocene is shown in Fig. 17, and the unpaired electron is expected to be in the e_{1g} orbitals. No signal could be detected for the pure complex, whereas reasonably well-featured spectra were observed in the ruthenium and nickel host single crystals. The line width was found to be extremely orientation dependent, yielding a well-resolved cobalt hyperfine structure parallel to z, whereas the x-axis spectrum is just one broad line with no hyperfine structure. The data were interpreted with the model described in Sec. VIII.

Also, the similar complex bis(1-phenylborabenzene)cobalt(II) has been studied, but only in a glassy matrix [464]. The ground state has been described as a mixture of $d(x^2 - y^2)$ and $d(z^2)$, but

the arguments must be considered with some suspicion, since the
principal directions are not experimentally available.

G. Nickel

The only stable isotope of nickel having nonzero nuclear spin in ^{61}Ni
(abundance 1.25%, I = 3/2, $\gamma = \mu_N/I = 0.497$). Because of the small
abundance of ^{61}Ni, no hyperfine structure is seen in the ESR spectra
of paramagnetic nickel ions.

1. Nickel(III)

Nickel(III) is a d^7 ion. Complexes that have been studied by
ESR spectroscopy are in distorted octahedral environments with the
metal ion in the low-spin state, so that an S = 1/2 spin Hamiltonian
is used to interpret the spectra. In this situation the spin-lattice
relaxation time is sufficiently long to yield observable ESR spectra
at room temperature.

In Table 26, spin Hamiltonian parameters for well-characterized
complexes are reported.

The spectra of octahedral complexes usually show $g_\perp > g_\parallel \simeq 2.0023$,
suggesting that the unpaired electron is essentially in a $d(z^2)$ orbital.

Bernstein and Gray [472] reported the single-crystal and powder
spectra of [Ni(diars)$_2$Cl$_2$]Cl doped into [Co(diars)$_2$Cl$_2$]ClO$_4$. Although the two complexes are not isomorphous and the site symmetry of
nickel is C$_i$, spectra show axial symmetry and have been interpreted
assuming C$_{2h}$ symmetry. The $\underline{g}$ values and the superhyperfine interaction with the chlorine atoms have been interpreted using a molecular
orbital approach outlined by McGarvey [15] and have been shown to be
consistent with a ^2A_g $d(z^2)$ ground state and the d-level ordering
$d(x^2 - y^2) < d(xz) \simeq d(yz) < d(z^2) < d(xy)$. The nonaxial symmetry of
the chlorine superhyperfine splitting tensor has been interpreted as
resulting from mixing of the 3p(x) orbital of chlorine into the
ground-state molecular orbital. The powder spectra of the pure
nickel complex showed a completely anisotropic g tensor, and this has
been interpreted by a mixing of d(xy) into the $\underset{\sim}{g}$round state due to

the assumed larger Ni-Cl distance in the pure compound as compared
to the doped one.

The same energy-level order has been found to be consistent with
the spin Hamiltonian parameters of single crystals of $[Ni(DP)_2X_2]^+$
(X = Cl, Br) doped into the corresponding Co(III) lattices [473].

Reinen et al. [474] studied the ESR spectra of octahedral NiF_6^{3-}
doped into various cubic lattices. They found that the low-spin
configuration $t_{2g}^6 e_g$ is stabilized versus the high-spin alternative
$t_{2g}^5 e_g^2$ by a Jahn-Teller splitting of the ^2E_g state of about 7000 cm^{-1}.
From the experimental g values, they calculated that the first ex-
cited quartet state $^4A_{2g}$ $(t_{2g}^5 e_g^2)$ is about 1000 cm^{-1} higher in energy
than the $^2A_{2g}$ ground state and observed that the quartet states can
give significant third-order contributions to g.

The ESR spectra of nickel(III) in trigonal cubic lattices have
been interpreted assuming that a coupling of the metal ion with lattice phonons is operative, and a detailed study of the Jahn-Teller
effect in these lattices has been published [475,476].

Various authors [338,477-480] have characterized the oxidation
state of several nickel complexes both in solid state and in solution
by their ESR spectra. An average g value of about 2.1 has been considered indicative of low-spin d^7 nickel(II) systems rather than
nickel(II) complexes of stabilized cation radicals as acting ligands.

2. Nickel(II)

Nickel(II) has a d^8 electronic configuration, and its ESR spectra can be interpreted using an S = 1 spin Hamiltonian. Because of
the even number of electrons, it does not possess a Kramers doublet
as the lowest state in a magnetic field, so zero field splitting can
be large enough to lift the threefold degeneracy of the S = 1 spin
state and give no observable ESR signal. Usually ESR spectra can be
recorded for octahedral complexes.

In Table 27, spin Hamiltonian parameters for well-characterized
nickel(II) complexes are reported. Spectra have usually been interpreted with nearly isotropic g tensors having principal values in
the range 2-2.3. Octahedral $[Ni(H_2O)_6]^{2+}$ ions have observable ESR

Table 26 Spin Hamiltonian Parameters for Some Low-Spin Nickel(III) Complexes

Compound	Chromophore[a]	Spectra[b]	g_1[c]	g_2[c]	g_3[c]	$\underline{A}_1(N)$[c,d]	$\underline{A}_2(N)$[c,d]	$\underline{A}_3(N)$[c,d]	Reference
[Ni(diars)$_2$Cl$_2$]Cl	NiAs$_4$Cl$_2$	P	2.0539	2.0913	2.1421				472
[(Co,Ni)(diars)$_2$Cl$_2$]Cl	NiAs$_4$Cl$_2$	P	2.142		2.008				472
[(Co,Ni)(diars)$_2$Cl$_2$]ClO$_4$	NiAs$_4$Cl$_2$	C	2.142		2.008	-8.5 (^{75}As)	-6.3 (^{75}As)	-30 (^{75}As)	472
[(Co,Ni)(DP)$_2$Cl$_2$]ClO$_4$	NiP$_4$Cl$_2$	C	2.1123	2.1157	2.0089	-32 (Cl) -5.1 (^{31}P)	-49 (Cl) -13 (^{31}P)	-27 (Cl) -43.6 (^{31}P)	473
[(Co,Ni)(DP)$_2$Br$_2$]PF$_6$	NiP$_4$Br$_2$	C	2.0961	2.1413	1.9936	-15.3 (Cl) 20 (^{31}P) -27 (Br)	-18.3 (Cl) 20 (^{31}P) -61.7 (Br)	-17.6 (Cl) -19.4 (^{31}P) -169 (Br)	473
Cs$_2$KNiF$_6$	NiF$_6$	C (77 K)	2.54		2.12				474
Rb$_2$KNiF$_6$	NiF$_6$	C (77 K)	2.39		2.12				474
Cs$_2$NaNiF$_6$	NiF$_6$	C (77 K)		2.28					474
[Ni(Me$_2$[14]aneN$_4$)Br$_2$]ClO$_4$	NiN$_4$Br$_2$	P	2.171		2.022				477

Compound	Donor atoms[a]	[b]	g[c]			A[d]		Ref.
[Ni(Me₂[14]aneN₄)Cl₂]ClO₄	NiN₄Cl₂	P	2.181		2.025			477
[Ni(Me₂[14]aneN₄)NO₃]ClO₄	NiN₄O₂	P	2.182	2.221	2.103			477
[Ni(Me₂[14]aneN₄)]³⁺·2CH₃CN	NiN₄N₂	G	2.205		2.020	17 (¹⁴N)	20.1 (¹⁴N)	477
[Ni(Me₆[14]4,11-dieneN₄)]³⁺·2CH₃CN	NiN₄N₂	G	2.199		2.024		21.3 (¹⁴N)	478
[Ni(Me₆[14]1,4,11,tetraeneN₄)]³⁺·2CH₃CN	NiN₄N₂	G	2.186	2.015	2.018	17.3 (¹⁴N)	22.8 (¹⁴N)	478
(Co,Ni)(cp)₂PF₆		P	1.969		1.812			297
[NiBr₂(cyclam)]⁺	NiN₄Br₂	G	2.17		2.014			338
[NiCl₂(cyclam)]⁺	NiN₄Cl₂	G	2.18		2.022			338
Ni(H₋₂G₃)	NiN₆	G	2.242	2.295	2.015			479
Ni(H₋₃G₃a)	NiN₆	G	2.310	2.281	2.006			479
Ni(BPO)₃	NiN₃N₃	P	2.08	2.10	2.14			480

[a] Metal and nearest-neighbor atoms.

[b] C = single crystal; P = polycrystalline powder; G = frozen solution.

[c] (1, 2, 3) refer to (x, y, z) in single-crystal spectra. When two values are reported, the first is the perpendicular component.

[d] Values in cm⁻¹ × 10⁻⁴. N is the nucleus with which the unpaired electron is coupled.

Table 27 Spin Hamiltonian Parameters for some Nickel(II) Complexes

Compound	Chromophore[a]	Spectra[b]	g_1[c]	g_2[c]	g_3[c]	D[d]	E/D	Reference
$NiSiF_6 \cdot 6H_2O$	NiO_6	C		2.25		-0.51	0.	482
$(Zn,Ni)SiF_6 \cdot 6H_2O$	NiO_6	C		2.26		-0.64	0.	483
$Ni(BrO_3)_2 \cdot 6H_2O$	NiO_6	C		2.29		-1.97	0.	485
$NiSO_4 \cdot 6H_2O$	NiO_6	C		2.25		4.85	0.01	494
$(Fe,Ni)SiF_6 \cdot 6H_2O$	NiO_6	C		2.255		-3.05	0.056	495
$Cs(Mg,Ni)Cl_3$	$NiCl_6$	C	2.241		2.257	2.000	0.	239
$Cs(Mg,Ni)Br_3$	$NiBr_6$	C	2.23		2.23	1.70	0.	239
$Cs(Mg,Ni)I_3$	NiI_6	C	2.16		2.16	1.03	0.	239
$Cs(Cd,Ni)Br_3$	$NiBr_6$	C	2.22		2.22	1.28	0.	497
$Cs(Cd,Ni)Cl_3$	$NiCl_6$	C	2.29		2.28	0.905	0.	237
$Ni(Iz)_6(ClO_4)_2$	NiN_6	P		2.20		0.46	0.005	74
$Ni(N-MeIz)_6(ClO_4)_2$	NiN_6	P		2.180		0.820	0.004	74
$Ni(pz)_6(ClO_4)_2$	NiN_6	P		2.19		0.07	0.001	74
$Ni(CH_3CN)_6SbCl_6$	NiN_6	P		2.190		0.260	0.10	74
$Ni(CH_3CN)_6InBr_4$	NiN_6	P		2.198		0.270	0.17	74

Ni(CH$_3$CN)$_6$(ClO$_4$)$_2$	NiN$_6$	P	2.195	0.380	0.047	74
Ni(5-Mepz)$_6$(ClO$_4$)$_2$	NiN$_6$	P	2.178	0.400	0.00	74
Ni(4-Mepz)$_6$(ClO$_4$)$_2$	NiN$_6$	P	2.190	0.510	0.333	74
Ni(4-Clpz)$_6$(ClO$_4$)$_2$	NiN$_6$	P	2.180	0.26	0.00	74
(Zn,Ni)(NH$_3$)$_6$(NO$_3$)$_2$	NiN$_6$	P	2.17	0.606		488
(Cd,Ni)(NH$_3$)$_6$Cl$_2$	NiN$_6$	P	2.18	0.265		491
(Cd,Ni)(NH$_3$)$_6$Br$_2$	NiN$_6$	P	2.17	0.30		491
(Zn,Ni)(NH$_3$)$_6$Cl$_2$	NiN$_6$	P	2.16	0.30		491
(Zn,Ni)(NH$_3$)$_6$Br$_2$	NiN$_6$	P	2.18	0.30		491
Ni(dbsc)$_2$(py)$_2$	NiSe$_4$N$_2$	G	2.087	0.00443		496

[a]Metal and nearest-neighbor atoms.

[b]C = single crystal; P = polycrystalline powder; G = frozen solution.

[c](1, 2, 3) refer to (x, y, z) in single-crystal spectra. When two values are reported, the first is the perpendicular component.

[d]Values in cm^{-1}.

spectra at room temperature, and the sign of D places the $|\pm1>$ doublet lower and the $|0>$ singlet higher in energy at zero field [481]. The D values for a series of nickel(II) hexaaquoions have been found to decrease with decreasing temperature [482-486], and this dependence has been attributed to electron-phonon interactions [484,487]. Data reported in Table 27 refer to room-temperature spectra.

Single-crystal studies of $Ni(BrO_3)_2 \cdot 6H_2O$ have been interpreted assuming an antiferromagnetic interaction between the nickel(II) ions with Weiss temperature of -0.4 K [485]. A ferromagnetic interaction has been reported [482] for pure $NiSiF_6 \cdot 6H_2O$ leading to a Curie temperature of about 0.1 K.

The $[Ni(NH_3)_6]^{2+}$ ion has been studied for a long time [15] because of the interesting temperature variation of its ESR spectra [488-491]. The spectra are isotropic at room temperature with $\underline{g} \simeq$ 2.2 both in pure and in diluted samples. Below a critical temperature, which varies from one complex to another, the line suddenly broadens in the pure compound while the spectra of the diluted compounds become anisotropic. A typical spectrum is shown in Fig. 25. This behavior has been interpreted using a theory proposed by Bates

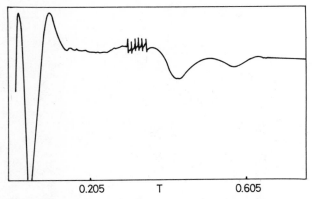

0.205   T   0.605

Figure 25   X-band ESR spectrum of a polycrystalline powder sample of $(Cd,Ni)(NH_3)_6Cl_2$ at 163 K.   The spectrum shows manganese(II) impurities at $\underline{g} \simeq 2$.   (After Ref. 491.)

and Stevens [492,493], in which the low-symmetry effect derives from the freezing out of the rotation of the ammonia molecules.

Methods of extracting the spin Hamiltonian parameters from powder spectra of S = 1 ions have been discussed by Reedijk and Nieuwenhuijse [74] and applied to a series of NiN_6 chromophores. Some results are reported in Table 27. Formulas are reported for the resonance line positions in the case where an asymmetric g tensor is present.

3. Nickel(I)

Nickel(I) is a d^9 ion, and its ESR spectra can be interpreted using an S = 1/2 spin Hamiltonian. The spin-lattice relaxation time is generally sufficiently long to give observable spectra at room temperature. In the case of pseudotetrahedral $Ni(p_3)X$ complexes (X = Cl, Br, I), the spectra were observed only at temperatures lower than 30 K [52]. This was justified by ligand field calculations, which put the levels derived from the tetrahedral ^2E ground state quite close in energy, thus shortening the spin-lattice relaxation time. The pure compounds are exchange narrowed, and the g values reported in Table 28 refer to the a, b, and c crystallographic axes of the orthorhombic cell. In the copper(I)-doped compounds, super-hyperfine structure due to the coupling of the unpaired electron with two equivalent phosphorus atoms has been observed with almost an isotropic A tensor with principal value in the range 0.0072- 0.0080 cm^{-1}. In some orientations, superhyperfine splitting due to the coupling with ^{79}Br and ^{81}Br of about 0.0019 cm^{-1} has been observed.

The coupling to two phosphorus atoms is surprising, since the three Ni-P distances seen in the structure are identical within error. Also, the principal directions of g, mentioned in Sec. III, are unusual, so that the presence of a Jahn-Teller distortion of the ^2E ground level, expected in C_3 symmetry, has been suggested.

A series of trigonal bipyramidal five-coordinate complexes of formula $Ni(np_3)X$ (X = Cl, Br, I) with C_3 site symmetry have been studied as powders [498]. The g value pattern $g_\perp > g_\parallel \simeq 2.0023$ has

Table 28 Spin Hamiltonian Parameters for Some Nickel(I) Complexes

Compound	Chromophore[a]	Spectra[b]	g_1^c	g_2^c	g_3^c	Reference
Ni(np$_3$)Cl	NiNP$_3$Cl	P		2.210	2.001	499
Ni(np$_3$)Br	NiNP$_3$Br	P		2.184	2.004	499
Ni(np$_3$)I	NiNP$_3$I	P		2.151	2.004	499
(Cu,Ni)(p$_3$)Cl	NiP$_3$Cl	C	2.344	2.124	2.040	52
(Cu,Ni)(p$_3$)Br	NiP$_3$Br	C	2.391	2.102	2.006	52
Ni(p$_3$)Cl	NiP$_3$Cl	C	2.14	2.12	2.24	52
Ni(p$_3$)Br	NiP$_3$Br	C	2.12	2.16	2.24	52
Ni(p$_3$)I	NiP$_3$I	C	1.99	2.10	2.38	52
(Ph$_3$P)$_2$(Cu,Ni)Cl	NiP$_2$Cl	C	2.111	2.167	2.446	501
(Ph$_3$P)$_2$(Cu,Ni)Br	NiP$_2$Br	C	2.112	2.209	2.435	501

[a]Metal and nearest-neighbor atoms.

[b]C = single crystal; P = polycrystalline powder.

[c](1, 2, 3) refer to (x, y, z) in single-crystal spectra. When two values are reported, the first is the perpendicular component.

also been observed in trigonal bipyramidal copper(II) complexes, and is indicative of a $d(z^2)$ ground state. The small deviations of $\underline{g}$ from the free-electron value has been rationalized through simple molecular orbital calculations [499].

Unusual three-coordinate complexes $(Ph_3P)_2NiX$ (X = Cl, Br) doped into the copper(I) analogs (site symmetry C_s [500]) have been characterized by Belford et al. by single-crystal spectra [501]. The coupling of the unpaired electron with the two nonequivalent ^{31}P atoms has been described through tensors that take the principal values $\underline{A}_1 = 0.0067$, $\underline{A}_2 = \underline{A}_3 = 0.0055$ cm^{-1} and $\underline{A}_1 = 0.0041$, $\underline{A}_2 = 0.0039$, $\underline{A}_3 = 0.0035$ cm^{-1} for the chloride complex; $\underline{A}_1 = 0.0072$, $\underline{A}_2 = \underline{A}_3 = 0.0053$ cm^{-1}, and $\underline{A}_1 = \underline{A}_2 = \underline{A}_3 = 0.0035$ cm^{-1} for the bromide. In the bromide complex, the coupling with ^{79}Br and ^{81}Br nuclei ($\underline{A}_1 = -0.0007$, $\underline{A}_2 = 0.0031$, $\underline{A}_3 = -0.0014$ cm^{-1}) has been noted, the quadrupole coupling constant has been determined.

H. Copper

The two stable isotopes of copper, ^{63}Cu (abundance 69.09%) and ^{65}Cu (abundance 30.91%), have nuclear spin quantum number I = 3/2 and gyromagnetic ratios $\gamma = \mu_N/I = 1.486$ and 1.590, respectively.

1. Copper(II)

The number of reports on the ESR spectra of copper(II) complexes is much too large to be considered here. Therefore we shall limit our attention to three selected classes of complexes that appear to be of particular interest. Some excellent reviews have been published, which the interested reader may consult in order to have the necessary information on the appearance of the ESR spectra of copper-(II) complexes in practically every conceivable symmetry, with all kinds of donor atoms [161,502,503].

We shall treat here the ESR spectra of octahedral copper(II) complexes experiencing Jahn-Teller distortions, as well as five-coordinate and tetrahedral complexes. The first topic has been studied frequently in the last years, especially in magnetically concentrated materials, whereas the last two have become increasingly

popular due to the impact of ESR spectroscopy on bioinorganic
chemistry.

   Copper(II) complexes are unlikely to have a regular octahedral
geometry, since the resulting ^2E_g ground state is unstable to Jahn-
Teller distortions. For many years the presence of a static Jahn-
Teller effect has been inferred from known crystal structures of
hexacoordinate copper(II) complexes wherein bond length and angle
distortions have been observed. There also are now some well-
established cases in which the ESR spectra (see Table 29) and sometimes also the X-ray crystal structures, have shown dynamic
Jahn-Teller distortions. One of the first instances in which ESR
showed unequivocally the operation of a Jahn-Teller effect was for
copper(II) doped into $ZnSiF_6 \cdot 6H_2O$. There the metal ion possesses
trigonal symmetry, which yields a ^2E ground state as in octahedral
symmetry [504]. A similar case is that of copper(II) doped into
$La_2Mg_3(NO_3)_{12} \cdot 24H_2O$ [505,506]. At high temperatures (above 20 K),
the spectra are quasiisotropic, with $\underline{g}_\parallel$ = 2.219, $\underline{g}_\perp$ = 2.218, $\underline{A}_\parallel$ =
29.0×10^{-4} cm^{-1}, $\underline{A}_\perp$ = 27.5×10^{-4} cm^{-1}, whereas at lower temperatures the spectra become anisotropic, with $\underline{g}_\parallel$ = 2.465, $\underline{g}_\perp$ = 2.099,
$\underline{A}_\parallel$ = -122×10^{-4} cm^{-1}, $\underline{A}_\perp$ = 16×10^{-4} cm^{-1}. This can be easily understood in terms of an elongated tetragonal distortion of the
octahedron. These data were interpreted assuming that strong vibronic coupling determines three equivalent minima in the potential
energy surface corresponding to elongation of a regular octahedron
along three orthogonal axes. The three wells are separated by a
barrier whose height is small compared to thermal energy at room
temperature, so that each chromophore will be hopping rapidly among
the three minima, yielding average spectra. At low temperature,
however, the distortion will be localized, so that the $\underline{g}$ and $\underline{A}$ values will be typical of a tetragonally elongated copper(II) complex.
The room-temperature $\underline{g}$ and $\underline{A}$ values correspond quite well to the
average of the $\underline{g}$ and $\underline{A}$ values at low temperature, in accord with the
suggested mechanism.

   Many other examples have been reported concerning Jahn-Teller
effects in hexaaquo cations. In the doped complexes the transition

from the isotropic spectra of high temperature to the anisotropic
spectra of low temperatures occurs in a range of temperatures in
which a motional broadening of the ESR signal can be observed. From
this it is possible to follow the dynamics of the process, in particular obtaining information on the height of the barrier, that separ-
ates the three equivalent potential wells, and on the mean residence
time of the cation in the potential wells [507,508].

Jahn-Teller distortions were also observed in some pure compounds, such as $K_2PbCu(NO_2)_6$ [509-512], $Cu(en)_3SO_4$ [513,514], and
$Cu(OMPA)_3(ClO_4)_2$ [515], in which the X-ray structures showed high-
symmetry sites for the metal ions, in contrast to the Jahn-Teller
theorem [516,517]. In these cases cooperative effects operating
among the distorting ions determine a phase transition, above which
almost isotropic spectra are observed, while anisotropic spectra
appear below it. The low-temperature spectra of $Cu(OMPA)_3(ClO_4)_2$
[515] yield g and A values typical of elongated octahedra, and as
such they are readily interpreted, since, apparently, exchange interactions between the copper ions are not strong enough to average the
signals of magnetically nonequivalent centers. In $K_2PbCu(NO_2)_6$ [509,
510] and $Cu(en)_3SO_4$ [514,518,519], the low-temperature phase shows
exchange-narrowing effects, so that the observed g values do not
correspond to molecular g values, but are the crystal average. The
failure to realize this point produced some literature reports that
claimed the stabilization of compressed octahedral chromophores, but
it was later realized that this was not the case, and the usual
elongated octahedral chromophores are actually formed. The matter
has been treated at length elsewhere [520]. The Cu(pyridine-N-
oxide)$_6X_2$ complexes are interesting in this respect (X = ClO_4, BF_4)
[521,522]. Here a dynamic Jahn-Teller interaction is observed for
the pure compounds, but the g values at low temperatures are highly
different, being $g_\parallel$ = 2.08, $g_\perp$ = 2.22 for the perchlorate and $g_\parallel$ =
2.37, $g_\perp$ = 2.07 for the tetrafluoroborate derivative. In the latter
it was suggested that the elongated octahedra are aligned parallel to
each other in the low-temperature lattice [521], whereas the unique

Table 29 Spin Hamiltonian Parameters for Some Octahedral Copper(II) Complexes

Compound	Chromophore[a]	Spectra[b]	g_1^c	g_2^c	g_3^c	$A_1^{c,d}$	$A_2^{c,d}$	$A_3^{c,d}$	Reference
Cu^{2+} in $La_2Mg_3(NO_3)_{12} \cdot 24D_2O$	CuO_6	HT	2.218		2.219	27.5		29	505
		LT	2.465		2.099	16		-112	506
		IIT		2.13					548
$(Zn,Cu)(phen)_3(NO_3)_2 \cdot 2H_2O$	CuN_6	LT	2.064		2.273	<7		160	514
		HT	2.129		2.153	50		68	
$(Zn,Cu)(en)_3(NO_3)_2$	CuN_6	LT	2.082		2.248			-168	514
		HT	2.110		2.126				
$Cu(en)_3SO_4$	CuN_6	LT	2.134	2.159	2.053				547

Compound	Chromophore							
Cu(OMPA)$_3$(ClO$_4$)$_2$	· CuN$_6$	HT			2.22		90	515
		LT	2.06	2.12		2.52		
K$_2$PbCu(NO$_2$)$_6$	CuN$_6$	HT			2.12			509
		LT	2.143	2.152		2.060		
Cu(pyO)$_6$(ClO$_4$)$_2$	CuO$_6$	HT			2.18			521
		LT	2.22	2.23		2.08		
Cu(pyO)$_6$(BF$_4$)$_2$	CuO$_6$	HT			2.18			521
		LT	2.07	2.07		2.37		

[a] Metal and nearest-neighbor atoms.

[b] C = single crystal; P = polycrystalline powder. HT = high temperature; LT = low temperature.

[c] (1, 2, 3) refer to (x, y, z) in single-crystal spectra. When two values are reported, the first is the perpendicular component; when one value is reported, it is the isotropic value.

[d] Values in cm^{-1} × 10^{-4}.

axis of one chromophore is rotated 90° from that of the neighboring complex in the former.

It must be mentioned that dynamic effects due to vibronic coupling can be observed in cases where the metal ion is in a low-symmetry site not amenable to Jahn-Teller distortions [523-526]. One example is that of copper Tutton salts [527,528], in which the copper(II) ion shows three different bond distances to oxygen atoms [529]. The $\underline{g}$ values were found to be temperature dependent, and so was the ^{63}Cu hyperfine splitting (Fig. 26) [528]. The data were interpreted in terms of a dynamic Jahn-Teller effect, according to which three nonequivalent potential energy minima are available for the copper(II) ion, according to the scheme in Fig. 27.

Five-coordinate copper(II) complexes with trigonal bipyramidal symmetry have the unpaired electron in the $d(z^2)$ orbital, so the $\underline{g}$ and $\underline{A}$ values should conform to the relations [530]

$$\underline{g}_\| = \underline{g}_e \qquad\qquad A_\| = P'[-\underline{K} + \frac{4}{7} - \frac{1}{7}(\underline{g}_\perp - 2)]$$

$$\underline{g}_\perp = \underline{g}_e - \frac{6\xi}{\Delta_{xz}} \qquad A_\perp = P'[-\underline{K} - \frac{2}{7} + \frac{15}{14}(\underline{g}_\perp - 2)] \tag{67}$$

where Δ_{xz} is the energy separation of the $d(z^2)$ and $d(xz)$ orbitals. The value of $\underline{g}_\|$ is expected to be slightly lower than $\underline{g}_e$ because of the presence of quadratic terms containing the spin-orbit coupling constant. The well-characterized cases of complexes having at least C_3 site symmetry are [Cu(Me$_6$tren)X]X (X = Cl, Br, I) [531,499], Cu(NH$_3$)$_2$Ag(SCN)$_3$ [532,499], and [Co(NH$_3$)$_6$]CuCl$_5$ [533]. The last complex would be interesting, but unfortunately all attempts to prepare dilute single crystals have failed. The spectra of the pure compound, which possesses a cubic lattice [534], yield only one isotropic $\underline{g}$ value, which results from the exchange average of magnetically nonequivalent chromophores. The observed $\underline{g}_{av}$ = 2.157 has been analyzed [533] in terms of $\underline{g}_\|$ and $\underline{g}_\perp$ assuming that $\underline{g}$ = 2.00. Under this assumption, $\underline{g}_\|$ = 2.23, which might be reasonable, but demands experimental verification.

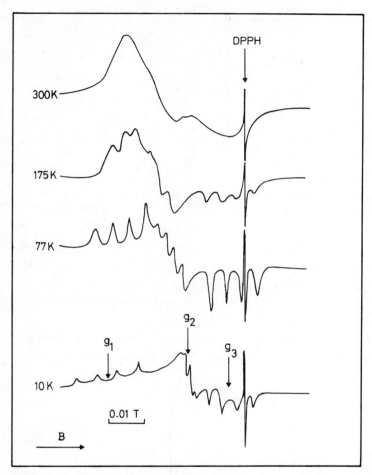

Figure 26 X-band ESR spectra of a polycrystalline powder sample of zinc(II) Tutton salt doped with copper(II) at various temperatures. $\underline{g}_1$ = 2.42, $\underline{g}_2$ = 2.15, $\underline{g}_3$ = 2.03. (After Ref. 528.)

In $Cu(NH_3)_2Ag(SCN)_3$ the copper chromophore has the structure [535] shown in Fig. 28, with each thiocyanate ion bridging through its sulfur atom to the silver atom. The site symmetry is D_3, and the $\underline{g}$ values conform to the prediction of Eq. (67) [532]. However, at room temperature, $\underline{g}_\parallel$ = 2.006 is slightly larger than $\underline{g}_e$, and it is temperature dependent [499]. In order to justify a positive

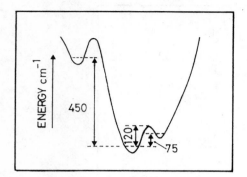

Figure 27 Schematic section of the potential energy surface for copper(II) Tutton salt showing the three Jahn-Teller minima. (After Ref. 528.)

deviation from g_e, it is necessary to invoke some vibronic coupling that allows the mixing of different excited levels into the ground state, which mixing would be forbidden in D_3 symmetry. On this ground it is thus possible to justify the temperature dependence of the g values [499].

A large deviation of g below the spin-only value is observed for [Cu(Me$_6$tren)X]X, where X = I [499]. The reason for this is the

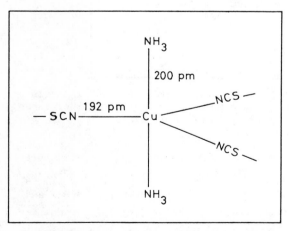

Figure 28 A schematic view of the molecular structure of Cu(NH$_3$)$_2$Ag(SCN)$_3$. (After Ref. 535.)

contribution of the spin-orbit coupling constant of the halogen
ligand, no longer negligible [Eq. (37)].

For square pyramidal complexes, the unpaired electron is in the
$d(x^2 - y^2)$ orbital, so that the g and A value patterns are similar
to those observed for square planar and tetragonally elongated octahedra:

$$g_{\parallel} = g_e - \frac{8\xi}{\Delta_{xy}} \qquad A_{\parallel} = P'[-K - \frac{4}{7} + (g_{\parallel} - 2) + \frac{3}{7}(g_{\perp} - 2)]$$

$$\tag{68}$$

$$g_{\perp} = g_e - \frac{2\xi}{\Delta_{xz}} \qquad A_{\perp} = P'[-K + \frac{2}{7} + \frac{11}{14}(g_{\perp} - 2)]$$

where Δ_{xy} and Δ_{xz} are the energy separations of the d(xy) and d(xz)
orbitals relative to the ground state. As compared to the square
planar case, the deviation of g values from g_e tend to be larger,
and the A values to be smaller. Typically $g_{\parallel}$ is in the range 2.25-
2.40 and $A_{\parallel}$ 120-150 × 10^{-4} cm^{-1} (see Table 30).

Many complexes, however, have symmetries that are intermediate
between these two extremes (trigonal bipyramidal or square pyramidal).
The Florence group [536] gave a simple model for the interpretation
of the spin Hamiltonian parameters of five-coordinate complexes with
geometries intermediate between a square pyramid and a trigonal
bipyramid. A square pyramid can be converted into a trigonal bipyramid through a pathway that preserves C_{2v} symmetry as shown in Fig.
29. In Fig. 30 it is shown how the energies of the electronic levels, as well as the g and A values, are affected by the geometric
distortion. It is apparent that in a large range of distortions
completely anisotropic g and A values are expected. Indeed, a good
correlation was found between this qualitative pattern and the g and
A values of the distorted (Zn,Cu)(SALMe) and (Zn,Cu)(SALMeDPT) complexes. In the latter complex the five donor atoms are all bound to
the metal atom [537], whereas the former is a dinuclear complex with
the structure shown in Fig. 31 [538]. At a low doping level only
one copper atom will enter the dinuclear unit.

Also, in other cases very anisotropic g values were reported.
In one case very highly anisotropic g values, associated with very

Table 30 Spin Hamiltonian Parameters for Some Five-Coordinate Copper(II) Complexes

Compound	Chromophore[a]	Spectra[b]	g_1[c]	g_2[c]	g_3[c]	A_1[c,d]	A_2[c,d]	A_3[c,d]	Reference
Square pyramidal complexes									
$Cu(apy)_5(ClO_4)_2$	CuO_5	C		2.079	2.400	35		126	549
$(Zn,Cu)(ac)_2 \cdot H_2O$	CuO_4O	C	2.082	2.052	2.344	<23	<18	147	550
$(Zn,Cu)(pyO)_2Cl_2$	CuO_3Cl_2	C	2.080	2.056	2.323	<10	<10	129	551
Trigonal bipyramidal complexes									
$[(Zn,Cu)(Me_6tren)Br]Br$	CuN_4Br	C	2.182		1.956	100		80	499
$[(Zn,Cu)(Me_6tren)I]I$	CuN_4I	C	2.226		1.895	103		97	499
$Cu(NH_3)_2Ag(SCN)_3$	CuN_3N_2	C (300 K)	2.201		2.006				532
		C (4.2 K)	2.199		1.998				499
Intermediate geometries									
$[Cu(bipy)_2I]I$	CuN_4I_2	C	2.169	2.169	2.033				552
$[Cu(bipy)_2tu]ClO_4$	CuN_4S	C	2.167	2.141	2.020				553

Compound									
[Cu(bipy)₂NH₃](BF₄)₂	C	CuN₄N	2.228	2.145	2.015				554
[Cu(phen)₂H₂O](NO₃)₂	C	CuN₄O	2.227	2.125	2.022				555
[dchpCuCl]CuCl₂	C	CuN₄Cl	2.232	2.202	2.007				554
Cu(py)₃(NO₃)₂	C	CuN₃O₂	2.266	2.166	2.025				554
Cu(dpq)Cl₂	C	CuN₃Cl₂	2.239	2.166	2.062				539
Cu(dmp)Cl₂H₂O	C	CuN₂Cl₂O	2.275	2.132	2.030				556
(Zn,Cu)SALMe	C	CuN₃O₂	2.25	2.13	2.02	131	65		536
(Zn,Cu)SALMeDPT	C	CuN₃O₂	2.23	2.07	2.04	150	38	47	536
[Cu(phen)₂CN]NO₃·H₂O	P	CuN₄C	2.242	2.149	2.054			66	557
[Cu₂(Et₅dien)₂ox](BPh₄)₂	P	CuN₃O₂	2.242	2.119	2.020				558

[a] Metal and nearest-neighbor atoms.

[b] C = single crystal; P = polycrystalline powder spectra.

[c] (1, 2, 3) refer to (x, y, z) in single-crystal spectra. When two values are reported, the first is the perpendicular component.

[d] Values in $cm^{-1} \times 10^{-4}$.

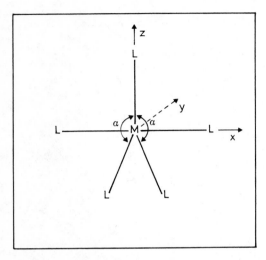

Figure 29   The interconversion pathway of a square pyramid into a
trigonal bipyramid. As α moves from 90 to 120°, the chromophore
passes from a square pyramid to a trigonal bipyramid, preserving an
overall C_{2v} symmetry. (Reprinted with permission from A. Bencini,
I. Bertini, D. Gatteschi, A. Scozzafava, *Inorg. Chem.*, *17*:3194.
Copyright 1978 American Chemical Society.)

small $\underline{A}$ constants, have been reported [539] due to a five-coordinate
complex. They do not seem to fit the above model, which suggests
that when anisotropic $\underline{g}$ values occur, anisotropic $\underline{A}$ values are to be
expected.

Pseudotetrahedral copper(II) complexes are not very numerous,
but an interest in their ESR spectra has resulted because of the
presence of this coordination geometry in certain metalloenzymes
[540]. Except for two cases of trigonally distorted copper(II) doped
into ionic lattices [541,542], all the reported cases are of tetragonally distorted tetrahedra. Some representative examples are shown in
Table 31. For these complexes the unpaired electron should be located in an essentially $d(xy)$ metal orbital. Accordingly, the $\underline{g}$ and
$\underline{A}$ values should conform to Eq. (68). However, the energies of the
excited states are in general lower for tetrahedral geometry than for
planar or octahedral, so that the $\underline{g}$ values tend to be larger. The
most puzzling feature of the ESR spectra of tetrahedral copper(II),

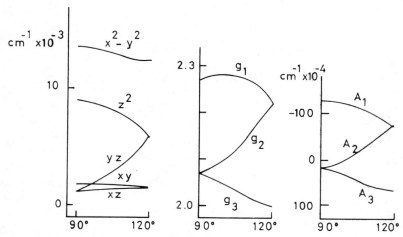

Figure 30 The effect of passing from a square pyramid to a trigonal bipyramid on the electronic energies and the spin Hamiltonian parameters for a copper(II) ion. From left to right: the effect of varying α, defined in Fig. 29, on the electronic energies, the g values, and the A values. (Reprinted with permission from A. Bencini, I. Bertini, D. Gatteschi, A. Scozzafava, *Inorg. Chem.*, *17*:3194. Copyright 1978 American Chemical Society.)

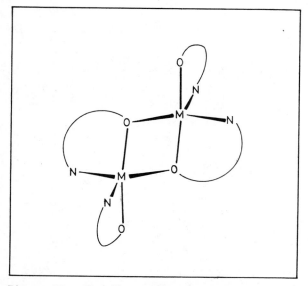

Figure 31 The dinuclear unit of M(SALMe) (M = Cu, Zn). (Reprinted with permission from P. L. Orioli, M. DiVaira, and L. Sacconi, *Inorg. Chem.*, *5*:400. Copyright 1966 American Chemical Society.)

Table 31 Spin Hamiltonian Parameters for Some Tetrahedral Copper(II) Complexes

Compound	Chromophore[a]	Spectra[b]	g_1[c]	g_2[c]	g_3[c]	A_1[c,d]	A_2[c,d]	A_3[c,d]	Reference
(Be,Cu)O	CuO_4	C	2.379		1.709	108		50	541
(Zn,Cu)O	CuO_4	C	1.531		0.74	231		195	542
(Zn,Cu)(dapy)$_2$(ClO$_4$)$_2$	CuO_4	P	2.101		2.467	11.7		128.3	561
Cs$_2$(Zn,Cu)Cl$_4$	$CuCl_4$	C	2.100	2.083	2.445	46	51	23	544
[(CH$_3$)$_4$N]$_2$(Zn,Cu)Cl$_4$	$CuCl_4$	C	2.101	2.078	2.462	51	51	17	545
Cu(Ph$_3$PO)$_2$Cl$_2$	CuO_2Cl_2	C	2.071	2.064	2.408				559
Cu(dpeO$_2$)Cl$_2$	CuO_2Cl_2	C	2.073	2.067	2.413				560
(Zn,Cu)(phen)$_2$Cl$_2$	CuN_2Cl_2	C	2.062	2.058	2.297	9		123	546
(Zn,Cu)[(-)sparteine]Cl$_2$	CuN_2Cl_2	P	2.075	2.050	2.299	9	14	97	562
Cu(N-t-butylpyrrole-2-carboxaldimine)$_2$	CuN_4	C	2.090	2.065	2.266				563
(Zn,Cu)(py)$_2$(NCS)$_2$	CuN_2N_2	P	2.162		2.406	<16		68	564
Cu[Me$_2$Ga(dmpz)$_2$]$_2$	CuN_4	P	2.041		2.316	50		95	565
(Zn,Cu)(SALip)$_2$	CuN_2O_2	C	2.095	2.014	2.280	~30	~15	117	566
(Zn,Cu)(SALtBu)$_2$	CuN_2O_2	C	2.08	2.03	2.29	~20	~12	117	567
Cu(1-ad-SAL)$_2$	CuN_2O_2	G	2.089		2.262	22		145	568
Cu(DMAEP)Cl$_2$	CuN_2Cl_2	G	2.118	2.057	2.305				569

[a]Metal and nearest-neighbor atoms.

[b]C = single crystal; P = polycrystalline powder; G = frozen solution

[c](1, 2, 3) refer to (x, y, z) in single crystal spectra. When two values are reported, the first is the perpendicular component.

[d]Values in cm^{-1} × 10^{-4}.

however, is given by the metal hyperfine coupling constant. In general, $\underline{A}$ is smaller than in planar and octahedral complexes, whereas $\underline{A}$ tends to be slightly larger. The limiting case for this is given by the $CuCl_4^{2-}$ ion, present in $Cs_2(Zn,Cu)Cl_4$, for which $\underline{A}_{\parallel} \simeq 25 \times 10^{-4}$ cm^{-1}, $\underline{A}_{\perp} \simeq 50 \times 10^{-4}$ cm^{-1} have been reported [543-545]. It is apparent from Eq. (68) that it is not possible to reproduce the observed $\underline{A}$ values. Using the experimental $\underline{g}$ values, McGarvey calculated a Fermi contact term that compared well with that observed for other copper(II) complexes, but the dipolar term was too small [163]. A molecular orbital treatment also gave too small a value for the molecular orbital coefficient of the ground state. Sharnoff [544] considered the admixture of metal 4p states into the ground state and was able to reproduce consistently both the $\underline{g}$ and $\underline{A}$ values. However, he obtained rather large admixtures of the 4p orbitals into the ground level, according to which the unpaired electron spends $\sim15\%$ of its time in the 4p orbitals. It must also be mentioned that in the complex $(Zn,Cu)(phen)Cl_2$, a detailed analysis of the optical and ESR data induced Kokoszka et al. to conclude that the ground-state metal orbital contains less than 3% 4p character [546]. On the other hand, as shown by Table 31, the hyperfine pattern of $CuCl_4^{2-}$ is rather unusual, and only certain copper(II) enzymes show similar values. All the other pseudotetrahedral copper(II) complexes have $\underline{A}$ values that are similar to those of $Cu(phen)Cl_2$; that is, they resemble the usual tetragonal values, with $\underline{A}_{\parallel}$ being smaller essentially due to the large $\underline{g}_{\parallel}$ values. More experimental data are required for a full understanding of the spin Hamiltonian parameters in tetrahedral copper(II) complexes [587].

Some experimental correlations have been found among d-d band energies, $\underline{A}_{\parallel}$, $\underline{g}_{\parallel}$, $\underline{A}_0$ and $\underline{g}_0$ values [570,571], which can be of help in the interpretation of the spectral properties of copper(II) proteins.

NOTE ADDED IN PROOF

Recently we reported the single crystal EPR spectra of $Cu(Ph_3PO)_2Cl_2$ doped into the zinc analogue [587]. From the analysis of the spectra

we concluded that the small value of the copper hyperfine coupling
constant can be justified by the covalency of the copper-chlorine
interaction and the large value of the spin orbit coupling constant
for chlorine. According to this model, the role of d-p mixing is
minor, the reduction in the dipolar contribution being due to the
above-mentioned effects. It was suggested that the model can be
applied also to sulphur-containing chromophores.

ACKNOWLEDGMENTS

It is a pleasure to thank Professor L. Sacconi for his constant
encouragement and interest during the preparation of this chapter.
Thanks are also due to Professor I. Bertini for many helpful discussions, and to Professor M. Wicholas for revising the manuscript.
The help of Dr. C. Zanchini in the literature scanning is gratefully
acknowledged. Mr. F. Cecconi is thanked for his help in preparing
the manuscript.

ABBREVIATIONS

ac: acetato

acac: acetylacetonato

acacen: N,N'-ethylenebisacetylacetonato

acen: bis-acetylacetonatoethylenediimine

acpn: bis-acetylacetonato-1,2-propylenediimine

1-ad-SAL: N-1-adamantylsalycilaldiminato

amben: N,N'-ethylenebis(o-aminobezylideneiminato)

apy: 2,3-dimethyl-1-phenyl-Δ^3-pyrazolin-t-one

B: 2,13-dimethyl-3,6,9,12,18-pentaazabicyclo-12,3,1-octadeca-1(18),
      2,12,14,16-pentane

benacacen: N,N'-ethylenebis(benzoylacetoneiminato)

benzac: benzoylacetonato

biet: biuret

bipy: 2,2'-bipyridine

BIz: benzimidazole

BPO: syn-2-benzoylpyridineoximato

bzen: benzoylacetoneethylenediimine

bzpn: benzoylacetone-1,2-propylenediimine

$(C_6H_5)_2B(pz)_2$:   diphenylbis(1-pyrazolyl)borato

4-Clpz: 4-chloropyrazole

cp: cyclopentadiene

cptc: cyclopentadienedithiocarboxilate

cyclam: 1,4,8,11-tetraazacyclo[14]ane

cyen: N,N'-ethylenebis(o-aminobenzylideneiminato)

dapy: diantipyrylmethane

dbsc: di-n-butyl-diselenocarbamate

dchp: dodeca(dimethylamino)cyclohexaphosphazene-NNNN

diars: o-phenylenebis(dimethylarsine)

dipyam: di-2-pyridyl-amine

dithiox: dithiooxalato

DMAEP: 2-(2-dimethylaminoethyl)pyridine

4,4'-dmb: 4,4'-(dimethyl)bipyridine

5,5'-dmb: 5,5'-(dimethyl)bipyridine

DMFc: 1,1'-dimethylferricenium

DMIz: 1,2-dimethylimidazole

dmp: 2,9-dimethyl-1,10-phenanthroline

dmpz: 3,5-dimethylpyrazolyl

DP: o-phenylenebis(dimethylphosphine)

dpe: 1,2-bis(diphenylphosphino)ethane

$dpeO_2$:   1,2-diphenyl-phosphinyl-ethane

dqp: 2,6-di(2'-quinolyl)pyridine

dtb: dithiobenzoato

dtc: diethyldithiocarbamato

dtp: diethyldithiophosphinato

dtt: dithio-p-toluate

EDTA: ethylendiamminetetraacetato

en: 1,3-diamminopropane

epydo: 1-methyl-2-pyridone

Et: ethyl

Et_5dien: 1,1,4,7,7-pentaethyldiethylenetriamine

G_3:   glycylglycylglycine

G_{3a}:   glycylglycylglycinamide

$HB(pz)_3$:   hydrotris(1-pyrazolyl)borato

$H_2B(pz)_2$: dihydrobis(1-pyrazolyl)borato

$H_2B\ 3,5-(CH_3)_2(pz)_2$: dihydrobis(3,5-dimethylpyrazolyl)borato

$H_2B\ 3,4,5-(CH_3)_3(pz)_2$: dihydrobis(3,4,5-trimethylpyrazolyl)borato

$HC(pz)_3$: hydrotris(1-pyrazolyl)methane

HL: 1,4-diazabicyclo[2.2.2]octane cation

HMPA: hexamethylphosphoramide

ipe: 6,13-diphenyl-1,8-dihydrodibenzo[b,i][1,4,8,11]tetraazacyclo-
      tetradecene

Iz: imidazole

Me: methyl

Mecp: methylcyclopentadienyl

Me_2dtc: dimethyldithiocarbamato

3-Meisoquin: 3-methylisoquinoline

2-MeIz: 2-methylimidazole

4-MepyO: 4-methylpyridine-N-oxide

4-Mepz: 4-methylpyrazole

5-Mepz: 5-methylpyrazole

Mesalim: N-methylsalicylaldiminato

Me_6tren: tris(2-dimethylaminoethyl)amine

mnt: maleonitriledithiolato

$N_4hexaene$: 5,7,12,14-tetramethyldibenzo[b,i][1,4,8,11]tetraazacyclo-
      tetradecahexaene

NIPA: nonamethylimidodiphosphoramide

N-MeIz: N-methylimidazole

nnp: N-(2-(diphenylphosphino)ethyl)-N',N'-diethylethylenediamine

np_3: tris(2-diphenylphosphinoethyl)ethane

nstn: N,N'-bis(1-phenyl-2-thiobenzoylvinyl)trimethylenediamine

OMPA: octamethylpyrophosphoramide

ox: oxalato

P_3: 1,1,1-tris(diphenylphosphinomethyl)ethane

pc: phtalocyanine

pfp: perfluoropinacolate

Ph: phenyl

phen: 1,10-phenanthroline

phtiazineO: 1-methyl-3-(2-chloro-6-methyl)phenyltriazine-1-oxide

γ-pic: γ-picoline

pn: 1,2-diaminopropane

py: pyridine

pyO: pyridine-N-oxide

pz: pyrazole

quin: quinoline

quinO: 2-methyl-8-quinolinate

sacsac: dithioacetylacetonate

SALen: N,N'-ethylenebissalicylaldiminato

SALip: N-isopropylsalicylaldiminato

SALMe: N-methylsalicylaldiminato

SALMedpt: (3-salicylaldiminatopropyl)methylamine

SALtb: N-t-butylsalicylaldiminato

TAAB: tetrabenzo[b,f,j,n][1,5,9,13]tetraazacyclohexadecine

TCA⁻: trichloroacetato

TCAA: trichloroacetic acid

tfen: bis-1,1,1-trifluoroacetylacetonatoethylenediimine

tmu: tetramethylurea

TPP: tetraphenylporphine

ttd: trithioperoxy-p-toluate

tu: thiourea

APPENDIX A: CALCULATION OF SINGLE-CRYSTAL AND POWDER SPECTRA

The transition probability in an ESR experiment is given by the variation in the magnetic dipole moment induced by the oscillating magnetic field B_1 in the static magnetic field B.

For a general orientation of B and B_1 in the principal axis system of g, the probability P_{ij} of a transition between the states $|i>$ and $|j>$ when only the Zeeman term is considered is given by

$$P_{ij}(B, B_1) \propto M_{ij}(B, B_1)f_{ij}(\omega) \tag{69}$$

in which $M_{ij}(B, B_1)$ is

$$M_{ij}(B, B_1) \propto |<i|\mu_B B_1 \cdot g \cdot S|j>|^2 \tag{70}$$

and $f_{ij}(\omega)$ is a function indicating the extent to which the resonance condition (71),

$$\omega \equiv \left| E_i(\underset{\sim}{B}) - E_j(\underset{\sim}{B}) \right| - h\nu = 0 \tag{71}$$

must be fulfilled in order to have absorption of microwave radiation. When $f_{ij}(\omega)$ is normalized to unity, it is called a line-shape function. By means of this function the various mechanisms responsible for the broadening of the absorption lines are taken into account (see Sec. VI).

If (ℓ_1, m_1, n_1) are the direction cosines of $\underset{\sim}{B}_1$, Eq. (70) can be written

$$M_{ij}(\underset{\sim}{B}, \underset{\sim}{B}_1) \propto \mu_B^2 B_1^2 \left| <i| \ell_1 \underline{g}_x \underline{S}_x + m_1 \underline{g}_y \underline{S}_y + n_1 \underline{g}_z \underline{S}_z |j> \right|^2 \tag{72}$$

This general relation may be used for the computation of the transition probability between the $|i>$ and $|j>$ states in every case. However, it often happens that a simpler formula can be used [63,572]. This is obtained by quantizing the spin operators in Eq. (72) along the direction of the static magnetic field:

$$M_{ij}(\underset{\sim}{B}, \underset{\sim}{B}_1) \propto \mu_B^2 B_1^2 \underline{g}_1^2 \sum_k \left| <i| \underline{S}_k |j> \right|^2 \tag{73}$$

where the sum is over all the spin components for which $<i| \underline{S}_k |j>$ is different from zero and $\underline{g}_1$ is [573,574]

$$\underline{g}_1^2 = \{ \underline{g}_x^2 \underline{g}_y^2 \cos^2 \psi \sin^2 \theta + \underline{g}_z^2 [\sin^2 \psi \sin^2 (\eta - \theta)(\underline{g}_y^2 \sin^2 \phi +$$
$$\underline{g}_x^2 \cos^2 \phi) + \cos^2 \psi \cos^2 \theta (\underline{g}_y^2 \cos^2 \phi + \underline{g}_x^2 \sin^2 \phi) +$$
$$2 \sin \psi \cos \psi \sin \phi \cos \phi \sin(\eta - \theta) \cos \theta (\underline{g}_y^2 - \underline{g}_x^2)] \}/\underline{g}^2 \tag{74}$$

θ, ϕ, ψ, η are the angular coordinates defined in Fig. 32. When $\underset{\sim}{B}_1$ is perpendicular to $\underset{\sim}{B}$, $(\eta - \theta) = \pi/2$ and $\underline{g}_1$ has maximum value. The transition probabilities for specific systems involving zero field splitting, hyperfine, and nuclear quadrupole coupling have been considered [57,63,64,572].

Because of the experimental difficulties of recording ESR spectra at fixed fields, generally the microwave frequency is kept constant and the field is varied. Equation (69) can be changed to account for this as follows [101,575]:

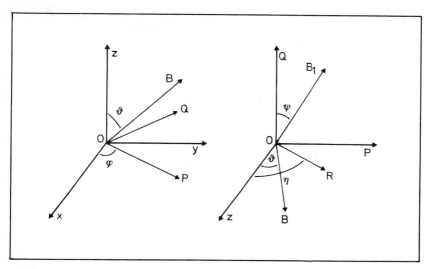

Figure 32 Polar coordinates for $\underset{\sim}{B}$ (left) and B_1 (right). OP and OQ are in the xy plane and OQ is perpendicular to the zBP plane. η is the angle that the projection of B_1 on the zBP plane, OR, makes with the z axis. (After Ref. 573].

$$P_{ij}(\underset{\sim}{B}, \underset{\sim}{B}_1) \propto M_{ij}(\underset{\sim}{B}, \underset{\sim}{B}_1) \left| \frac{\partial B_{ij}}{\partial h\nu} \right| f_{ij}(\underline{B} - \underline{B}_{ij}) \tag{75}$$

in which $\underline{B}_{ij}$ is the value of the magnetic field at which the transition between the $|i>$ and $|j>$ levels occurs at frequency ν. The importance of including $|\partial B_{ij}/\partial h\nu|$ in the simulation of the spectra has been discussed [77].

By summing Eq. (75) over all possible transitions, we obtain the absorption spectrum of a single crystal:

$$I(\underset{\sim}{B}) \propto \sum_a P_a(\underset{\sim}{B}, \underset{\sim}{B}_1) = \sum_a M_a(\underset{\sim}{B}, \underset{\sim}{B}_1) \left| \frac{\partial B_a}{\partial h\nu} \right| f_a(\underline{B} - \underline{B}_a) \tag{76}$$

where a is a concise notation for indicating individual transitions.

The intensity of absorption from a polycrystalline powder sample at the field B [W(B)] is simulated as a superimposition of spectra of single crystals with different orientations relative to the applied magnetic field [576,577]. Assuming a uniform distribution of the orientations of the crystallites over the unit sphere, W(B) is given by Eq. (77):

$$W(\underline{B}) \propto \int_0^\pi d\psi \int_0^\pi d\phi \int_0^\pi \sum_a M_a(\underline{B}, \underline{B}_1) \left| \frac{\partial B_{-a}}{\partial h\nu} \right| f_a(\underline{B} - \underline{B}_a) \sin \theta \ d\theta \quad (77)$$

or, equivalently,

$$W(\underline{B}) \propto \sum_a \int_0^\pi d\phi \int_0^\pi <M_a> \left| \frac{\partial B_{-a}}{\partial h\nu} \right| f_a(\underline{B} - \underline{B}_a) \sin \theta \ d\theta \quad (78)$$

and for the derivative spectrum,

$$\frac{dW(\underline{B})}{d\underline{B}} \propto \sum_a \int_0^\pi d\phi \int_0^\pi <M_a> \left| \frac{\partial B_{-a}}{\partial h\nu} \right| \frac{d}{d\underline{B}} f_a(\underline{B} - \underline{B}_a) \sin \theta \ d\theta \quad (79)$$

In Eqs. (78) and (79), we have defined

$$<M_a> = \int_0^\pi M_a(\underline{B}, \underline{B}_1) \ d\psi \quad (80)$$

When B and B_1 are kept orthogonal, as in the usual ESR experiment, the three Euler angles θ, ϕ, ψ are sufficient to determine the position of $\underset{\sim}{B}$ and $\underset{\sim}{B}_1$. In this case $\ell_1 = \cos \theta \cos \phi \cos \psi -$ $\sin \theta \sin \psi$, $m_1 = \cos \theta \sin \phi \cos \psi + \cos \phi \sin \psi$, and $n_1 = - \sin \theta \cos \psi$, and averaging Eq. (72) over ψ, one obtains

$$<M_a> = \left(\frac{\pi}{2}\right) \mu_B^2 B_{-1}^2 [\ | <i| \underline{g}_x \cos \theta \cos \phi \ \underline{S}_x + \underline{g}_y \cos \theta \sin \theta \ \underline{S}_y -$$

$$\underline{g}_z \sin \theta \ \underline{S}_z |j>|^2 + |<i| \underline{g}_x \sin \phi \ \underline{S}_x - \underline{g}_y \cos \phi \ \underline{S}_y |j>|^2] \quad (81)$$

A simpler expression for Eq. (80) is obtained averaging $\underline{g}_1^2$ in Eq. (73) over ψ [573]:

$$<\underline{g}_1^2> = \{ \underline{g}_x^2 \underline{g}_y^2 \sin^2 \theta + \underline{g}_y^2 \ \underline{g}_z^2 [\sin^2 \phi \ \sin^2(\eta - \theta) +$$

$$\cos^2 \theta \ \cos^2 \phi] + \underline{g}_z^2 \underline{g}_x^2 [\cos^2 \phi \ \sin^2(\eta - \theta) +$$

$$\cos^2 \theta \ \sin \phi] \} / (2\underline{g}^2) \quad (82)$$

In order to apply the general formula (79), saving computer time, the angular dependence of $<M_a>$, $|\partial B_{-a}/\partial h\nu|$, and f_a can be neglected because in general they do not show strong angular variations [100]. In this framework it is possible to define the function

$$S_a(\underline{B}) = \int_0^\pi d\phi \int_0^\pi f_a(\underline{B} - \underline{B}_a) \sin\theta \, d\theta \tag{83}$$

and Eq. (79) becomes

$$\frac{dW(\underline{B})}{d\underline{B}} \propto \sum_a <M_a> \left|\frac{\partial \underline{B}_a}{\partial h\nu}\right| \frac{d}{d\underline{B}} S_a(\underline{B}) \tag{84}$$

The mathematical form of the function $f_a(\underline{B} - \underline{B}_a)$ has been briefly discussed in Sec. VI; for powder spectra, however, $f_a(\underline{B} - \underline{B}_a)$ can be considered similar to a δ function in that it differs significantly from zero only in the region $\underline{B}_a - \Gamma_a < \underline{B} < \underline{B}_a + \Gamma_a$, where Γ_a is the peak-to-peak amplitude of the transition a. Thus $S_a(B)$ can be considered as the area of a narrow contour at the surface of the unit sphere whose points (θ, ϕ) satisfy the condition $\underline{B}_a(\theta, \phi) \simeq \underline{B}$.

Equation (84) shows that it is sufficient to determine the principal features of the $S_a(\underline{B})$ function in order to have the approximate features of the spectrum. In particular, peaks in the derivative spectrum correspond to the turning points of the angular dependence of the resonance fields $\underline{B}_a$. At these points $S_a(\underline{B})$ changes from a finite value to zero. In Fig. 33 the relation between the turning points of $\underline{B}_a$ and the ESR spectrum is shown for an S = 3/2 spin system.

Van Veen [101] has briefly discussed the validity of Eq. (84) and has shown that for those turning points of $\underline{B}_a(\theta, \phi)$ for which $\theta \neq 0$, the following expression for $S_a(\underline{B})$ can be used:

$$S_a(\underline{B}) \simeq \sin\theta_0 \left| \left(\frac{\partial^2 \underline{B}_a}{\partial\theta^2}\right)\left(\frac{\partial^2 \underline{B}_a}{\partial\phi^2}\right) - \left(\frac{\partial^2 \underline{B}_a}{\partial\theta\partial\phi}\right)^2 \right|^{-1/2} \tag{85}$$

in which $\theta_0 \neq 0$.

The calculation of the derivatives of the transition fields occurring in Eqs. (79) and (85) can be done by standard methods. The relevant equations needed in the computation are briefly shown in App. B. In App. B a brief survey of the methods for calculating the transition fields is also given.

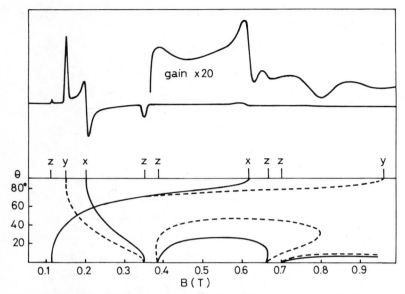

Figure 33   The relation between the ESR spectrum and the turning
points in the angular dependence of the resonance fields.   Upper:
the frozen solution spectra of $[Cr(en)_2Cl_2]Cl$ (S = 3/2).   Lower:
resonance fields as a function of magnetic field orientation in the
xz (full lines) and yz (dashed lines) planes, computed with g_x =
g_y = g_z = 1.99, D = 0.489 cm^{-1}, E/D = 0.10.   (Reprinted with permission from J. C. Hempel, L. O. Morgan, and W. Burton Lewis, *Inorg.*
Chem., 9:2064.   Copyright 1970 American Chemical Society.)

APPENDIX B: CALCULATION OF TRANSITION FIELDS

1.   The Indirect Method

The most widely used method for computing the transition fields can
be called the indirect method, in contrast to the method of Belford
and Belford [578,579], which is called the direct method or the
eigenfield method.

      The field at which the transition between the levels |i> and
|j> occurs at a fixed orientation of the static magnetic field is
computed, when the perturbation expressions of Sec. IV are not
applicable, by calculating the expression

$$\Delta_{ij}(\underline{B}) = \left| E_i(\underline{B}) - E_j(\underline{B}) \right| - h\nu \tag{86}$$

for a series B_m of field strength values. In Eq. (86) the energies of the states at the field $\underline{B}$ [$E_i(\underline{B})$ and $E_j(\underline{B})$] are obtained by diagonalization of the appropriate spin Hamiltonian (see Sec. V). When the product

$$\Delta_{ij}(\underline{B}_n) \times \Delta_{ij}(\underline{B}_{n+1}) \tag{87}$$

becomes negative, it is clear that the required resonance field $\underline{B}_a$ is contained in the interval $(\underline{B}_n, \underline{B}_{n+1})$. An interpolation procedure can then be used to approach $\underline{B}_a$.

This method, although simple in principle, is very time consuming when applied to the simulation of powder spectra requiring hundreds of diagonalizations. Moreover, when more than one resonance is contained between $\underline{B}_n$ and $\underline{B}_{n+1}$, some solutions may be overlooked.

Eigenvectors obtained from the diagonalization of the spin Hamiltonian matrix for $\underline{B} = \underline{B}_a$ can be used in Eq. (81) to compute $<M_a>$.

2. The Eigenfield Method

The basic concept of the eigenfield method is to find an operator whose eigenvalues are the transition frequencies [580]. For a system characterized by N levels in the magnetic field, N^2 energy differences can be defined, even if N^2 - N of them, at most, can correspond to actual transitions. The required operator is defined on an N^2-dimensional vector space. It has been shown that this operator acts on a vector space formed by operators, and it has been called a superoperator [580]. The operators that span the N^2-dimensional vector space on which the required superoperator is defined are called level-shift operators and are defined by Eq. (88):

$$\underset{\sim}{S}_{ij}|k> = \delta_{jk}|i> \quad (i, k = 1, \ldots, N) \tag{88}$$

The operator that acts on them, giving as eigenvalues the transition energies, is called the derivation superoperator $\underset{\sim}{h}^D$, and it is defined through [580]

$$h^D_\sim(S_{\sim ij}) = [\underline{H}, \; S_{\sim ij}] = \big|E_i(\underline{B}) - E_j(\underline{B})\big| S_{\sim ij} \tag{89}$$

in which $[\underline{H}, \; S_{\sim ij}]$ is the commutator of the operators $\underline{H}$ and $S_{\sim ij}$ and the other symbols have the usual meaning.

Expressing $\underline{H}$ and $S_{\sim ij}$ in matrix form, using the N-dimensional spin space as representation space, Eq. (89) can be written as

$$\underline{HL} - \underline{LH} = \omega L \tag{90}$$

where $\underline{L} = \underline{pq}^+$ is the $N \times N$ matrix representing $S_{\sim ij}$, $\underline{p}$ and $\underline{q}$ are the column matrices representing $|i\rangle$ and $|j\rangle$; $\omega = h\nu$ is the microwave quantum, and $\underline{H}$ is the spin Hamiltonian matrix. It can be written as

$$\underline{H} = \underset{\sim}{F} + \underset{\sim}{BG} \tag{91}$$

where $\underset{\sim}{F}$ and $\underset{\sim}{G}$ are $N \times N$ Hermitian matrices representing the field-independent (zero field, hyperfine, etc.) and the linear Zeeman term, respectively. Substituting Eq. (91) into Eq. (90), the $N^2 \times N^2$ generalized eigenvalue equation (92) is obtained after some manipulation:

$$\underset{\sim\sim}{AZ} = \underset{\sim\sim}{BKZ} \tag{92}$$

in which $\underset{\sim}{Z} = \underset{\sim}{p} \otimes \underset{\sim}{q}$ is the N^2 column vector formed by the $N \times N$ matrix elements of $\underset{\sim}{L}$ and A, and K are the $N^2 \times N^2$ matrices defined by Eqs. (93) and (94):

$$A = \underset{\sim}{1} \otimes \underset{\sim}{1} - \underset{\sim}{F} \otimes \underset{\sim}{1} + \underset{\sim}{1} \otimes \underset{\sim}{F} \tag{93}$$

$$K = \underset{\sim}{G} \otimes \underset{\sim}{1} - \underset{\sim}{1} \otimes \underset{\sim}{G} \tag{94}$$

$\underline{B}$ in Eq. (92) is the transition field (eigenfield) at the frequency $\omega/h\nu$. $\underset{\sim}{A}$ and $\underset{\sim}{K}$ can be computed when the system is defined. The resolution of Eq. (92) yields $\underline{B}$ and $\underset{\sim}{Z}$.

The generalized eigenvalue equation (92) can be solved by standard methods [581,582] by reducing both A and K (which are Hermitian) simultaneously to lower triangular form. Some suggested strategies of computation can be found in Ref. 578.

In this way the eigenfield can be computed as

$$\underset{\sim}{B}_i = \frac{\alpha_{ii}}{\kappa_{ii}} \tag{95}$$

where α_{ii} and κ_{ii} are the diagonal matrix elements of the triangular $\underset{\sim}{A}$ and $\underset{\sim}{K}$ matrices.

Also, the transition probabilities can be computed through the relation

$$<\mu>_{ij} \propto \underset{\sim}{\mu}\underset{\sim}{Z} \tag{96}$$

where $<\mu>_{ij}$ is the transition moment, $\underset{\sim}{\mu}$ is a row matrix containing all the rows of the transition operator matrix, one after the other, and $\underset{\sim}{Z}$ is defined in Eq. (92). $\underset{\sim}{M}_{ij}$ in Eq. (72) is proportional to the square of $<\mu>_{ij}$.

Applications of the eigenfield method to the calculation of the angular variation of the transition fields for various spin systems have been published by McGregor et al. [583].

Although the method requires less computer time than the direct one, for calculations involving fine and hyperfine terms the dimensions of the problem are defined by $N = (2S + 1)(2I + 1)$, and the storage of $\underset{\sim}{A}$, $\underset{\sim}{B}$, and $\underset{\sim}{Z}$ matrices can be a serious problem.

3. Derivatives of the Transition Fields

In order to obtain expressions for the derivatives of the transition fields with respect to the spin Hamiltonian parameters (as well as with respect to $h\nu$ or to the angles defining the orientation of $\underset{\sim}{B}$) that are required in the simulation of the spectra, one needs the expressions for the derivatives of the energy levels. These are straightforward when perturbation expressions of the energy levels are used.

Analytical expressions can be obtained more generally by perturbative expansion of the eigenvalue equation $\underline{H}|j> = \underline{E}_j|j>$ obtained by expanding $\underline{E}_j$ into a Taylor series. The relevant equations for the first and second derivatives are

$$\frac{\partial E_j}{\partial x} = <j|\frac{\partial H}{\partial x}|j>$$

$$\frac{\partial^2 E_j}{\partial x\,\partial y} = <j|\frac{\partial^2 H}{\partial x\,\partial y}|j> + 2\sum_{i\neq j}\frac{<i|\frac{\partial H}{\partial x}|j><j|\frac{\partial H}{\partial y}|i>}{E_i - E_j}$$

$$(97)$$

where i and j indicate energy levels, and x and y are variables of the spin Hamiltonian. The derivatives of $\underline{H}$ with respect to x and y are generally simple to compute.

The derivatives of the transitions field relative to the $|i> \rightarrow |j>$ transition are related to the derivatives of the difference $\Delta = |\underline{E}_j - \underline{E}_i|$ by the equations [101]

$$\frac{\partial \underline{B}_a}{\partial x} = -\frac{\partial\Delta/\partial x}{\partial\Delta/\partial\underline{B}} \tag{98}$$

and with the conditions $\partial \underline{B}_a/\partial x = 0$ and $\partial \underline{B}_a/\partial y = 0$, which apply to Eq. (85):

$$\frac{\partial^2 \underline{B}_a}{\partial x\partial y} = -\frac{\partial^2\Delta/\partial x\,\partial y}{\partial\Delta/\partial\underline{B}} \tag{99}$$

Other methods are reported that do not give explicit expressions [584-586].

REFERENCES

1. A. Abragam and B. Bleaney, *Electron Paramagnetic Resonance of Transition Metal Ions*, Clarendon Press, Oxford, 1970.

2. J. E. Wertz and J. R. Bolton, *Electron Spin Resonance. Elementary Theory and Practical Applications*, McGraw-Hill, New York, 1972.

3. C. P. Poole, Jr., and H. A. Farach, *The Theory of Magnetic Resonance*, Wiley-Interscience, New York, 1972.

4. P. B. Ayscough, *Electron Paramagnetic Resonance in Chemistry*, Barnes & Noble, New York, 1967.

5. A. Carrington and A. D. McLachlan, *Introduction to Magnetic Resonance*, Harper & Row, New York, 1967.

6. S. A. Al'tshuler and B. M. Kozyrev, *Electron Paramagnetic Resonance in Compounds of Transition Elements*, 2nd ed., Nauka, Moscow, 1972 (Halsted Press, Wiley, New York, 1974).

7. L. A. Sorin and M. V. Vlasova, *Electron Spin Resonance of Paramagnetic Crystals*, Plenum Press, New York, 1973.

8. G. Pake and T. Estle, *The Physical Principles of E.P.R.*, Addison-Wesley, Reading, Mass., 1973.

9. J. S. Griffith, *The Theory of Transition-Metal Ions*, 2nd ed., Cambridge University Press, Cambridge, 1971.

10. J. E. Harriman, *Theoretical Foundations of Electron Spin Resonance*, Pergamon Press, London, 1978.

11. R. S. Drago, *Physical Methods in Chemistry*, Saunders, Philadelphia, 1977.

12. N. M. Atherton, *Electron Spin Resonance: Theory and Applications*, Ellis Harwood, Wiley, Chichester, 1973.

13. R. S. Alger, *Electron Paramagnetic Resonance, Techniques and Applications*, New York, 1968.

14. A. D. McLachlan, *Electron Spin Resonance*, Harper & Row, New York, 1969.

15. B. R. McGarvey, in *Transition Metal Chemistry* (R. L. Carlin, ed.), Vol. 3, Dekker, New York, 1966.

16. B. A. Goodman and J. B. Raynor, in *Advances in Inorganic Chemistry and Radiochemistry* (H. J. Emeléus and A. G. Sharpe, eds.), Vol. 13, Academic Press, 1970.

17. H. A. Kuska and M. T. Rogers, in *Spectroscopy in Inorganic Chemistry* (C. N. R. Rao and J. R. Ferraro, eds.), Vol. II, Academic Press, New York, 1971.

18. S. Geschwind, ed., *Electron Paramagnetic Resonance*, Plenum Press, New York, 1972.

19. Teh Fu Yen, ed., *Electron Spin Resonance of Metal Complexes*, Adam Hilger, Condon, 1969.

20. H. A. Buckmaster and D. B. Delay, *Magnetic Res. Rev.*, 3:127 (1974).

21. J. R. Wasson and P. J. Cowen, *Anal. Chem.*, 50:92R (1978).

22. G. F. Kokoszka and G. Gordon, in *Transition Metal Chemistry* (R. L. Carlin, ed.), Vol. 5, Dekker, New York, 1969.

23. J. P. Fackler, Jr., in *Progress in Inorganic Chemistry* (S. J. Lippard, ed.), Vol. 21, Wiley-Interscience, New York, 1977.

24. J. Catterick and P. Thornton, in *Advances in Inorganic Chemistry and Radiochemistry* (H. J. Emeléus and A. G. Sharpe, eds.), Vol. 20, Academic Press, New York, 1977.

25. G. F. Kokoszka and R. W. Duerst, *Coord. Chem. Rev.*, 5:68 (1970).

26. D. J. Hodgson, in *Progress in Inorganic Chemistry* (S. J. Lippard, ed.), Vol. 19, Wiley-Interscience, New York, 1975.

27. G. L. Eichorn, ed., *Inorganic Biochemistry*, Elsevier, Amsterdam, 1973.

28. W. E. Blumberg and J. Peisach, in *Bioinorganic Chemistry* (R. Dessy, J. Dillard, and L. Taylor, eds.), American Chemical Society Special Publication No. 100, Washington, D. C., 1971.

29. B. G. Malmstrom, L. Andreassen, and B. Reinhammar, in *The Enzymes* (P. Boyer, ed.), Academic Press, New York, 1975.

30. R. C. Bray, in *The Enzymes* (P. Boyer, ed.), Academic Press, New York, 1975.

31. G. Palmer, in *The Enzymes* (P. Boyer, ed.), Academic Press, New York (1975).

32. P. F. Knowles, *Essays Biochem.*, *8*:79 (1972).

33. P. J. Quilly and G. A. Webb, *Coord. Chem. Rev.*, *12*:407 (1974).

34. M. H. L. Pryce, *Proc. Phys. Soc.*, *A63*:25 (1950).

35. A. Abragam and M. H. L. Pryce, *Proc. Roy. Soc.*, *A205*:135 (1951).

36. T. Ray, *Proc. Roy. Soc.*, *A277*:76 (1964).

37. F. S. Ham, G. W. Ludwig, G. D. Watkins, and H. H. Woodbury, *Phys. Rev. Lett.*, *5*:468 (1960).

38. I. Love, *Mol. Phys.*, *4*:1217 (1975).

39. G. F. Foster and H. Statz, *Phys. Rev.*, *113*:445 (1969).

40. W. J. C. Grant and M. W. P. Strandberg, *J. Phys. Chem. Solids*, *25*:635 (1964)

41. H. A. Buckmaster, R. Chatterjee, and Y. H. Shing, *Phys. Stat. Solidi A*, *13*:9 (1972).

42. J. C. Geusic and L. Carlton-Brown, *Phys. Rev.*, *112*:64 (1958).

43. D. S. Shonland, *Proc. Phys. Soc.*, *73*:788 (1959).

44. D. E. Billing and B. J. Hathaway, *J. Chem. Phys.*, *50*:2258 (1969).

45. W. G. Waller and M. T. Rogers, *J. Magnetic Res.*, *9*:92 (1973).

46. W. G. Waller and M. T. Rogers, *J. Magnetic Res.*, *18*:39 (1975).

47. C. P. Keijzers, G. F. Paulussen, and E. de Boer, *Mol. Phys.*, *29*:973 (1975)

48. J. A. Weil, T. Buch, and J. E. Clapp, in *Advances in Magnetic Resonance* (J. S. Waugh, ed.), Vol. 6, Academic Press, New York, 1973.

49. A. Lund and T. Vänngård, *J. Chem. Phys.*, *42*:2979 (1965).

50. L. Sacconi and S. Midollini, *J. Chem. Soc. Dalton*, 1213 (1972).

51. P. Dapporto, G. Fallani, and L. Sacconi, *Inorg. Chem.*, *13*:2847 (1974).

52. A. Bencini, C. Benelli, D. Gatteschi, and L. Sacconi, *Inorg. Chim. Acta*, *37*:195 (1979).

53. B. Bleaney, *Phil. Mag.*, *42*:441 (1951).

54. G. H. Azarbayejani, *Phys. Lett. A*, *25*:767 (1967).

55. R. E. D. McClung, *Can. J. Phys.*, *46*:2271 (1968).

56. R. M. Golding, *Applied Wave Mechanics*, van Nostrand, London, 1969.

57. U. Sakaguchi, Y. Arata, and S. Fujiwara, *J. Magnetic Res.*, *9*: 118 (1973).

58. W. C. Lin, *Mol. Phys.*, *25*:247 (1973).

59. R. Kirmse, S. Wartewig, W. Windsh, and E. Hoyer, *J. Chem. Phys.*, *56*:5273 (1972).

60. J. R. Pilbrow and M. E. Winfield, *Mol. Phys.*, *25*:1073 (1973).

61. R. M. Golding and W. C. Tennant, *Mol. Phys.*, *25*:1163 (1973).

62. A. Rockenbauer and P. Simon, *J. Magnetic Res.*, *11*:217 (1973).

63. A. Rockenbauer and P. Simon, *Mol. Phys.*, *28*:1113 (1974).

64. R. M. Golding and W. C. Tennant, *Mol. Phys.*, *28*:167 (1974).

65. J. A. Weil, *J. Magnetic Res.*, *18*:113 (1975).

66. R. Skinner and J. A. Weil, *J. Magnetic Res.*, *21*:271 (1976).

67. R. Skinner and J. A. Weil, *J. Magnetic Res.*, *29*:223 (1978).

68. A. Bencini and D. Gatteschi, *Inorg. Chem.*, *16*:2141 (1977).

69. E. Wasserman, L. C. Snyder, and W. A. Yager, *J. Chem. Phys.*, *41*:1763 (1964).

70. J. Baranowski, T. Cukierda, B. Jezowska-Trzebiatowska, and H. Kozlowski, *Chem. Phys. Lett.*, *39*:606 (1976).

71. J. R. Wasson, C. Shyr, and C. Trapp, *Inorg. Chem.*, 7:469 (1968).

72. J. H. van der Waals and M. S. de Groot, *Mol. Phys.*, 2:333 (1959).

73. J. H. van der Waals and M. S. de Groot, *Mol. Phys.*, 3:130 (1960).

74. J. Reedijk and B. Nieuwenhuijse, *Rec. Trav. Chim.*, *91*:533 (1972).

75. P. Kottis and R. Lefebvre, *J. Chem. Phys.*, *39*:393 (1963).

76. E. Pedersen and H. Toftlund, *Inorg. Chem.*, *13*:1603 (1974).

77. J. R. Pilbrow, *J. Magnetic Res.*, *31*:479 (1978).

78. J. T. Vallin and G. D. Watkins, *Phys. Rev.*, *9*:2051 (1974).

79. U. Kaufmann, *Phys. Rev.*, *14*:1848 (1976).

80. C. Rudowicz, *Acta Phys. Pol.*, *A51*:515 (1977).

81. D. M. Duggan and D. N. Hendrickson, *Inorg. Chem.*, *14*:955 (1975).

82. J. Baranowski, T. Cukierda, B. Jezowska-Trzebiatowska, and H. Kozlowski, *J. Magnetic Res.*, *33*:585 (1979).

83. T. Castner, G. S. Newell, W. C. Holton, and C. P. Slichter, *J. Chem. Phys.*, *32*:668 (1969).

84. J. S. Griffith, *Mol. Phys.*, *8*:213 (1964).

85. J. S. Griffith, *Mol. Phys.*, *8*:217 (1964).

86. W. E. Blumberg, in *Magnetic Resonance in Biological Systems* (A. Ehrenberg, B. E. Malmstrom, and T. Vanngard, eds.), Pergamon Press, London, 1967.

87. R. M. Golding, T. Singhasuwich, and W. C. Tennant, *Mol. Phys.*, *34*:1343 (1977).

88. R. D. Dowsing and J. F. Gibson, *J. Chem. Phys.*, *50*:294 (1969).

89. R. Aasa, *J. Chem. Phys.*, *52*:3919 (1970).

90. F. W. Warner and A. J. McPhate, *J. Magnetic Res.*, *24*:125 (1976).

91. R. D. Dowsing, *J. Comput. Phys.*, *6*:326 (1970).

92. J. D. Swalen and H. M. Gladney, *IBM J. Res. Dev.*, *8*:515 (1964).

93. H. M. Gladney, *IBM Res. Rep.*, RJ 318, October 16 (1964).

94. B. Fox, F. Holuj, and W. E. Baylis, *J. Magnetic Res.*, *10*:347 (1973).

95. A. K. Jain and G. C. Upreti, *Mol. Phys.*, *34*:273 (1977).

96. J. Uhrin, *Czech. J. Phys. B*, *23*:551 (1973).

97. H. A. Buckmaster, R. Chatterjee, J. C. Dering, D. J. I. Fry, Y. H. Shing, and B. Venkatesan, *J. Magnetic Res.*, *4*:113 (1971).

98. S. K. Misra and G. R. Sharp, *J. Magnetic Res.*, *23*:191 (1976).

99. M. R. Smith, H. A. Buckmaster, and D. J. I. Fry, *J. Magnetic Res.*, *23*:103 (1976).

100. P. C. Taylor, J. F. Baugher, and H. M. Kriz, *Chem. Rev.*, *75*: 203 (1975).

101. G. van Veen, *J. Magnetic Res.*, *30*:91 (1978).

102. J. H. Mackey, M. Kopp, E. C. Tynan, T. F. Yen, in *Electron Spin Resonance of Metal Complexes* (Teh Fu Yen, ed.), Adam Hilger, London, 1969.

103. K. Shimokoshi, *Bull. Chem. Soc. Jap.*, *51*:635 (1978).

104. W. Meisel, *Phys. Status Solidi B*, *43*:k129 (1971).

105. G. Elste, *Z. Astrophys.*, *33*:39 (1953).

106. R. E. Coffmann, *J. Phys. Chem.*, *79*:1129 (1975).

107. A. H. Maki and B. R. McGarvey, *J. Chem. Phys.*, *29*:31 (1958).

108. A. H. Maki, N. Edelstein, A. Davison, and R. H. Holm, *Inorg. Chem.*, *3*:4580 (1964).

109. W. Windson and M. Welter, *Z. Naturforsch.*, *22a*:1 (1967).

110. D. Kivelson and R. Neiman, *J. Chem. Phys.*, *35*:149 (1961).

111. C. P. Keijzers and E. de Boer, *Mol. Phys.*, *29*:1007 (1975).

112. C. P. Keijzers and E. de Boer, *J. Chem. Phys.*, *57*:1277 (1972).

113. A. J. Stone, Thesis, Cambridge University, Cambridge, 1963.

114. D. Attanasio, C. P. Keijzers, J. P. van de Berg, and E. de Boer, *Mol. Phys.*, *31*:501 (1976).

115. J. I. Zink and R. S. Drago, *J. Amer. Chem. Soc.*, *94*:4550 (1972).

116. D. W. Smith, *J. Chem. Soc.*, *A*, 3108 (1970).

117. L. Shields, *J. Chem. Soc.*, *A*, 1048 (1971).

118. J. Owen and J. H. M. Thornley, *Rep. Prog. Phys.*, *29*:675 (1966).

119. R. F. Fenske, K. G. Caulton, D. D. Redke, and C. C. Sweeney, *Inorg. Chem.*, *5*:951 (1966).

120. R. D. Brown and P. G. Burton, *Theor. Chim. Acta*, *18*:309 (1970).

121. D. W. Clack, N. S. Hush, and J. R. Yandle, *J. Chem. Phys.*, *57*: 3503 (1972).

122. J. C. Slater, *J. Chem. Phys.*, *35*:228 (1965).

123. K. H. Johnson, *J. Chem. Phys.*, *45*:3085 (1966).

124. K. H. Johnson, *Int. J. Quant. Chem.*, *15*:361 (1967).

125. F. A. Cotton and B. J. Kalbacher, *Inorg. Chem.*, *16*:2389 (1977).

126. J. B. Johnson and W. G. Klemperer, *J. Amer. Chem. Soc.*, *99*: 7132 (1977).

127. J. P. Jasinski and S. L. Holt, *Inorg. Chem.*, *14*:1267 (1975).

128. S. Larsson and J. W. D. Connolly, *J. Chem. Phys.*, *60*:1514 (1974).

129. S. Larsson, *Theor. Chim. Acta*, *39*:173 (1975).

130. R. S. Mulliken, *J. Chem. Phys.*, *23*:1833, 1841, 2338, 2341 (1955).

131. H. Bethe, *Ann. Phys.*, *3*:133 (1929).

132. J. H. van Vleck, *Phys. Rev.*, *41*:208 (1932).

133. K. W. H. Stevens, *Proc. Roy. Soc.*, *A219*:542 (1953).

134. J. Owen, *Disc. Faraday Soc.*, *19*:127 (1955).

135. J. Owen, *Proc. Roy. Soc.*, *A227*:183 (1955).

136. C. K. Jørgensen, R. Pappalardo, and H. H. Schmidtke, *J. Chem. Phys.*, *39*:1422 (1963).

137. C. E. Schaffer, in *Structure and Bonding* (C. K. Jørgensen, J. B. Neilands, R. S. Nyholm, D. Reinen, and R. J. P. Williams, eds.), Vol. 5, Springer-Verlag, Berlin, 1968.

138. C. E. Schaffer, in *Wave Mechanics* (W. C. Price, S. S. Chissick, and T. Ravensdale, eds.), Butterworths, London, 1973.

139. M. Gerloch and R. C. Slade, *Ligand Field Parameters*, Cambridge University Press, London, 1973.

140. D. W. Smith, in *Structure and Bonding* (J. D. Dunitz, J. B. Goodenough, P. Hemmerich, J. A. Ibers, C. K. Jørgensen, J. B. Neilands, D. Reinen, and R. J. P. Williams, eds.), Vol. 35, Springer-Verlag, Berlin, 1978.

141. I. Bertini, D. Gatteschi, and A. Scozzafava, *Isr. J. Chem.*, *15*:188 (1977).

142. M. Gerloch and R. F. McMeeking, *J. Chem. Soc. Dalton*, 2443 (1975).

143. W. de W. Horrocks and D. A. Burlone, *J. Amer. Chem. Soc.*, *98*: 6513 (1976).

144. A. Bencini and D. Gatteschi, *J. Magnetic Res.*, *34*:653 (1979).

145. C. K. Jørgensen, *Modern Aspects of Ligand Field Theory*, North Holland, Amsterdam, 1971.

146. A. D. van Ingen Schenau, G. C. Vershoor, and C. Romers, *Acta Crystallogr.*, *B30*:1686 (1974).

147. T. S. Bergendahl and J. S. Wood, *Inorg. Chem.*, *14*:338 (1975).

148. C. J. O'Connor and R. L. Carlin, *Inorg. Chem.*, *14*:291 (1975).

149. R. L. Carlin, C. J. O'Connor, and R. S. Bathia, *J. Amer. Chem. Soc.*, *98*:3523 (1976).

150. R. L. Carlin, C. J. O'Connor, and R. S. Bathia, *J. Amer. Chem. Soc.*, *98*:685 (1976).

151. D. J. Mackey and R. F. McMeeking, *J. Chem. Soc. Dalton*, 2186 (1977).

152. D. J. Mackey, S. V. Evans, and R. F. McMeeking, *J. Chem. Soc. Dalton*, 160 (1978).

153. J. S. Wood, personal communication.

154. B. N. Figgis and J. Lewis, in *Progress in Inorganic Chemistry* (S. J. Lippard, ed.), Vol. 6, Wiley-Interscience, New York, 1964.

155. M. Tinkham, *Proc. Roy. Soc.*, *A236*:535 (1956).

156. M. Tinkham, *Proc. Roy. Soc.*, *A236*:549 (1956).

157. W. Low, in *Paramagnetic Resonance in Solids* (F. Seitz and D. Turnbull, eds.), Academic Press, New York, 1960.

158. M. Gerloch and J. R. Miller, in *Progress in Inorganic Chemistry* (S. J. Lippard, ed.), Vol. 10, Wiley-Interscience, New York, 1968.

159. C. K. Jørgensen, in *Progress in Inorganic Chemistry* (S. J. Lippard, ed.), Vol. 4, Wiley-Interscience, New York, 1962.

160. C. K. Jørgensen, *Absorption Spectra and Chemical Bonding in Complexes*, Pergamon Press, Oxford, 1962.

161. B. J. Hathaway and D. E. Billing, *Coord. Chem. Rev.*, 5:143 (1970).

162. D. E. Billing and B. J. Hathaway, *J. Chem. Soc.*, A, 1516 (1968).

163. B. R. McGarvey, *J. Phys. Chem.*, 71:51 (1967).

164. J. A. McMillan and T. Halpern, *J. Chem. Phys.*, 55:33 (1971).

165. H. M. Gladney and J. D. Swalen, *J. Chem. Phys.*, 42:1999 (1965).

166. R. W. Schwartz and R. L. Carlin, *J. Amer. Chem. Soc.*, 92:6763 (1970).

167. T. H. Cottrell and C. R. Quade, *Bull. Amer. Phys. Soc.*, 9:502 (1964).

168. B. Bleaney, *Proc. Phys. Soc.*, A63:407 (1950).

169. B. Bleaney, G. S. Bogle, A. H. Cooke, R. J. Duffus, M. C. M. O'Brien, and K. W. H. Stevens, *Proc. Phys. Soc.*, A68:57 (1955).

170. J. A. McKinnon and G. F. Dionne, *Can. J. Phys.*, 44:2329 (1966).

171. G. F. Dionne, *Can. J. Phys.*, 42:2419 (1964).

172. J. A. McKinnon and M. Shannon, *Can. J. Phys.*, 53:841 (1975).

173. G. F. Dionne and J. A. McKinnon, *Phys. Rev.*, 172:325 (1968).

174. P. I. Premovic and P. R. West, *Can. J. Chem.*, 53:1630 (1975).

175. B. R. McGarvey, *J. Chem. Phys.*, 38:388 (1963).

176. N. Rumin, C. Vincent, and D. Walsh, *Phys. Rev.*, 7B:1811 (1973).

177. F. S. Ham, in *Electron Paramagnetic Resonance* (S. Geschwind, ed.), Plenum Press, New York, 1972.

178. R. M. McFarlane, *Phys. Rev.*, 184:B603 (1969).

179. P. I. Premovic and P. R. West, *Can. J. Chem.*, 53:1593 (1975).

180. J. G. Kenworthy, J. Myatt, and M. C. R. Symons, *J. Chem. Soc.*, A, 3428 (1971).

181. J. G. Kenworthy, J. Myatt, and M. C. R. Symons, *J. Chem. Soc.*, A, 1020 (1971)

182. G. Henrici-Olive and S. Olive, *Angew. Chem. Int. Ed. Engl.*, *6*: 790 (1967).

183. J. J. Solzmann, *Helv. Chim. Acta*, *51*:526 (1968).

184. D. R. Lorenz and J. R. Wasson, *J. Inorg. Nucl. Chem.*, *37*:2265 (1975).

185. C. J. Ballhausen and H. B. Gray, *Inorg. Chem.*, *1*:11 (1962).

186. J. Selbin, *Chem. Rev.*, *65*:153 (1965).

187. J. Selbin, *Coord. Chem. Rev.*, *1*:293 (1966).

188. T. R. Ortolano, J. Selbin, and S. P. McGlynn, *J. Chem. Phys.*, *41*:262 (1964).

189. C. J. Ballhausen, B. B. F. Djminskij, and K. J. Watson, *J. Amer. Chem. Soc.*, *90*:3305 (1968).

190. J. R. Wasson and H. J. Stoklosa, *J. Inorg. Nucl. Chem.*, *36*:227 (1974).

191. A. K. Viswanath, *J. Chem. Phys.*, *67*:3744 (1977).

192. B. N. Misra and R. Kripal, *Chem. Phys. Lett.*, *46*:536 (1977).

193. B. N. Misra and R. Kripal, *Chem. Phys.*, *31*:95 (1978).

194. M. Roberts, W. S. Koski, and W. S. Caughey, *J. Chem. Phys.*, *34*: 591 (1961).

195. D. Kivelson and S. K. Lee, *J. Chem. Phys.*, *41*:1896 (1964).

196. K. V. S. Rao and M. D. Sastry, *J. Chem. Phys.*, *49*:4984 (1968).

197. P. F. Bramman, T. Lund, J. B. Raynor, and C. J. Willis, *J. Chem Soc. Dalton*, 45 (1975).

198. H. Kon and N. E. Sharpless, *J. Phys. Chem.*, *70*:105 (1966).

199. J. J. Altenau, M. T. Pope, R. A. Prados, and H. So, *Inorg. Chem.*, *14*:417 (1975).

200. M. Otake, Y. Komiyama, and T. Otaki, *J. Phys. Chem.*, *77*:2896 (1973).

201. D. P. Smith, H. So, J. Benden, and M. T. Pope, *Inorg. Chem.*, *12*:685 (1973).

202. H. So, C. M. Flynn, Jr., and M. T. Pope, *J. Inorg. Nucl. Chem.*, *36*:329 (1974).

203. M. A. Hitchman, B. W. Moores, and R. L. Belford, *Inorg. Chem.*, *8*:1817 (1969).

204. M. A. Hitchman and R. L. Belford, *Inorg. Chem.*, *8*:958 (1969).

205. M. A. Hitchman, C. D. Olson, and R. L. Belford, *J. Chem. Phys.*, *50*:1195 (1969).

206. H. J. Stoklosa, J. R. Wasson, and B. J. McCormick, *Inorg. Chem.*, *13*:592 (1974).

207. C. P. Stewart and A. L. Porte, *J. Chem. Soc. Dalton*, 1661 (1972).

208. H. A. Kuska and P. H. Yang, *Inorg. Chem.*, *13*:1090 (1974).

209. J. R. Wasson, *Inorg. Chem.*, *10*:1531 (1971).

210. B. Jezowska-Trzebiatowska and J. Jezierska, *Chem. Phys. Lett.*, *34*:237 (1975).

211. H. J. Stoklosa, G. L. Seebach, and J. R. Wasson, *J. Phys. Chem.*, *78*:962 (1974).

212. H. J. Stoklosa and J. R. Wasson, *Inorg. Nucl. Chem. Lett.*, *10*: 377 (1974).

213. J. McCormick, J. L. Featherhouse, H. J. Stoklosa, and J. R. Wasson, *Inorg. Chem.*, *12*:692 (1973).

214. D. R. Lorenz, D. K. Johnson, H. J. Stoklosa, and J. R. Wasson, *J. Inorg. Nucl. Chem.*, *36*:1184 (1974).

215. H. J. Stoklosa and J. R. Wasson, *Inorg. Nucl. Chem. Lett.*, *10*: 401 (1974).

216. J. L. Petersen and L. F. Dahl, *J. Amer. Chem. Soc.*, *97*:6416 (1975).

217. J. L. Petersen and L. F. Dahl, *J. Amer. Chem. Soc.*, *97*:6422 (1975).

218. D. P. Bakalik and R. J. Hayes, *Inorg. Chem.*, *11*:1734 (1972).

219. C. P. Stewart and A. L. Porte, *J. Chem. Soc. Dalton*, 722 (1973).

220. D. C. Bradley, R. H. Moss, and K. D. Sales, *Chem. Commun.*, 1255 (1969).

221. G. F. Kokoszka, H. C. Allen, Jr., G. Gordon, *Inorg. Chem.*, *5*: 91 (1966).

222. E. C. Alyea, D. C. Bradley, *J. Chem. Soc.*, *A*, 2330 (1969).

223. C. E. Holloway, F. E. Mabbs, and W. R. Smail, *J. Chem. Soc.*, *A*, 2980 (1968).

224. J. L. Petersen, D. L. Lichtenberger, R. F. Fenske, and L. F. Dahl, *J. Amer. Chem. Soc.*, *97*:6433 (1975).

225. W. L. Kwik and E. I. Stiefel, *Inorg. Chem.*, *12*:2337 (1973).

226. V. G. Krishnan, *Mol. Phys.*, *34*:297 (1977).

227. N. M. Atherton and J. C. Winscom, *Inorg. Chem.*, *12*:383 (1973).

228. A. Manoogian and J. A. McKinnon, *Can. J. Phys.*, *45*:2769 (1967).

229. R. G. Cavell, E. D. Day, W. Byers, and P. M. Watkins, *Inorg. Chem.*, *11*:1591 (1972).

230. L. J. Boucher, E. C. Tynan, and T. F. Yen, in *Electron Spin Resonance of Metal Complexes* (T. F. Yen, ed.), Adam Hilger, London, 1969.

231. G. D. Simpson, R. L. Belford, and R. Biagioni, *Inorg. Chem.*, *17*:2424 (1978).

232. J. M. Flowers, J. C. Hempel, W. E. Hatfield, and H. H. Dearman, *J. Chem. Phys.*, *58*:1479 (1973).

233. J. N. McElearney, R. W. Schwartz, A. E. Siegle, and R. L. Carlin, *J. Amer. Chem. Soc.*, *93*:4333 (1971).

234. J. Pontman and R. Adde, *Phys. Rev.*, *14*:3778 (1976).

235. J. G. Clarke and J. A. McKinnon, *Can. J. Phys.*, *49*:1539 (1971).

236. G. L. McPherson and M. R. Freedman, *Inorg. Chem.*, *15*:2299 (1976).

237. Y. R. Chang, G. L. McPherson, and J. L. Atwood, *Inorg. Chem.*, *14*:3079 (1975).

238. J. Y. Buzaré, A. Leble, and J. C. Fayet, *Phys. Stat. Solidi B*, *67*:455 (1975).

239. G. L. McPherson, R. C. Koch, and G. D. Stucky, *J. Chem. Phys.*, *60*:1424 (1974).

240. B. R. McGarvey, *J. Chem. Phys.*, *41*:3743 (1964).

241. W. T. M. Andriessen, *Inorg. Chem.*, *14*:792 (1975).

242. B. B. Garrett, K. de Armond, and H. S. Gutowsky, *J. Chem. Phys.*, *44*:3393 (1966).

243. L. E. Mohrmann, Jr., and B. B. Garrett, *Inorg. Chem.*, *13*:357 (1974).

244. R. M. McFarlane, *J. Chem. Phys.*, *39*:3118 (1963).

245. R. M. McFarlane, *J. Chem. Phys.*, *42*:442 (1965).

246. R. M. McFarlane, *J. Chem. Phys.*, *47*:2066 (1967).

247. E. E. Pedersen and S. Kallesoe, *Inorg. Chem.*, *14*:85 (1975).

248. M. T. Holbrook and B. B. Garrett, *Inorg. Chem.*, *15*:150 (1976).

249. W. T. M. Andriessen and J. Meuldijk, *Inorg. Chem.*, *15*:2044 (1976).

250. W. T. M. Andriessen, *J. Magnetic Res.*, *23*:339 (1975).

251. A. Manoogian and A. Leclerc, *J. Chem. Phys.*, *63*:4450 (1975).

252. A. Leclerc and A. Manoogian, *J. Chem. Phys.*, *63*:4456 (1975).

253. F. J. Owens, *Phys. Stat. Solidi B*, *79*:623 (1977).

254. B. B. Garrett, M. T. Holbrook, and J. A. Stanko, *Inorg. Chem.*, *16*:1159 (1977).

255. E. W. Stout, Jr., and B. B. Garrett, *Inorg. Chem.*, *12*:2565 (1973).

256. D. C. Doetschmann, *J. Chem. Phys.*, *60*:2647 (1974).

257. R. A. Bernheim and E. F. Reichenbecher, *J. Chem. Phys.*, *51*: 996 (1969).

258. J. H. M. Mooy, H. J. de Jong, M. Glasbeek, and J. D. W. van Voorst, *Chem. Phys. Lett.*, *18*:51 (1973).

259. J. H. M. Mooy, *J. Phys. Chem. Solids*, *38*:103 (1977).

260. J. C. Hempel, D. Klassen, W. E. Hatfield, and H. H. Dearman, *J. Chem. Phys.*, *58*:1487 (1973).

261. D. C. Bradley, R. G. Copperthwaite, S. A. Cotton, K. D. Sales, and J. F. Gibson, *J. Chem. Soc. Dalton*, 191 (1973).

262. J. Schmidt, *Phys. Lett.*, *28A*:419 (1968).

263. J. C. Hempel, L. O. Morgan, and W. Burton Lewis, *Inorg. Chem.*, *9*:2064 (1970).

264. W. L. Klotz and M. K. de Armond, *Inorg. Chem.*, *14*:3125 (1975).

265. S. J. Baker and B. B. Garrett, *Inorg. Chem.*, *13*:2683 (1974).

266. Y. Kawasaki and L. S. Forster, *J. Chem. Phys.*, *50*:1010 (1969).

267. W. E. Estes, D. Y. Jeter, J. C. Hempel, and W. E. Hatfield, *Inorg. Chem.*, *10*:2074 (1971).

268. W. T. M. Andriessen and M. P. Groenewege, *Inorg. Chem.*, *15*:621 (1976).

269. J. S. van Wieringen, *Disc. Faraday Soc.*, *19*:118 (1955).

270. G. M. Woltermann and G. R. Wasson, *J. Magnetic Res.*, *9*:486 (1973).

271. G. M. Woltermann and J. R. Wasson, *Chem. Phys. Lett.*, *16*:92 (1972).

272. R. D. Dowsing, J. F. Gibson, D. M. L. Goodgame, M. Goodgame, and P. J. Hayward, *Nature*, *219*:1037 (1968).

273. R. M. Golding and T. Singhasuwich, *Mol. Phys.*, *29*:1847 (1975).

274. R. B. Birdy and M. Goodgame, *J. Chem. Soc. Dalton*, 461 (1977).

275. G. M. Woltermann and J. R. Wasson, *Inorg. Chem.*, *12*:2366 (1973).

276. G. M. Woltermann and J. R. Wasson, *J. Phys. Chem.*, *78*:45 (1974).

277. M. W. G. de Bolster, B. Nieuwenhuijse, and J. Reedijk, *z. Naturforsch.*, *28b*:104 (1973).

278. R. D. Dowsing, J. F. Gibson, M. Goodgame, and P. J. Hayward, *J. Chem. Soc.*, *A*, 187 (1969).

279. W. J. Nicholson and G. Burns, *Phys. Rev.*, *129*:2490 (1963).

280. H. Watanabe, *Progr. Theor. Phys.*, *18*:405 (1957).

281. H. J. Stoller, G. Rist, and H. H. Gunthard, *J. Chem. Phys.*, *57*:4651 (1972).

282. C. D. Burbridge and D. M. L. Goodgame, *J. Chem. Soc.*, A, 237 (1968).

283. R. D. Dowsing, J. F. Gibson, M. Goodgame, and P. J. Hayward, *J. Chem. Soc.*, A, 1133 (1970).

284. R. B. Birdy and M. Goodgame, *Inorg. Chem.*, *18*:472 (1979).

285. R. D. Dowsing, J. F. Gibson, D. M. L. Goodgame, M. Goodgame, and P. J. Hayward, *J. Chem. Soc.*, A, 1242 (1969).

286. D. M. L. Goodgame, M. Goodgame, P. J. Hayward, *J. Chem. Soc.*, A, 1352 (1970).

287. R. B. Birdy, G. Brun, D. M. L. Goodgame, and M. Goodgame, *J. Chem. Soc. Dalton*, 149 (1979).

288. D. Vivien and J. F. Gibson, *J. Chem. Soc. Faraday II*, *71*:1640 (1975).

289. J. Chandrasekhar, N. Chandrakumar, and J. Subramanian, *J. Magnetic Res.*, *18*:129 (1975).

290. R. S. Saraswat and G. C. Upreti, *Chem. Phys.*, *23*:97 (1977).

291. G. M. Woltermann and J. R. Wasson, *J. Phys. Chem.*, *77*:945 (1973).

292. J. Reedijk and J. A. Smit, *Rec. Trav. Chim.*, *91*:681 (1972).

293. J. Reedijk, H. Schrijver, and J. A. Welleman, *Rec. Trav. Chim.*, *94*:40 (1975).

294. M. Goodgame and P. J. Hayward, *J. Chem. Soc.*, A, 3406 (1971).

295. R. D. Dowsing, B. Nieuwenhuijse, and J. Reedijk, *Inorg. Chim. Acta*, *5*:301 (1971).

296. J. H. Ammeter, *J. Magnetic Res.*, *30*:299 (1978).

297. J. H. Ammeter, N. Oswald, and R. Bucher, *Helv. Chim Acta*, *58*: 671 (1975).

298. S. Evans, M. L. H. Green, B. Jewitt, G. M. King, and A. F. Orchard, *J. Chem. Soc. Faraday II*, *70*:356 (1974).

299. M. E. Switzer, R. Wang, M. F. Rettig, and A. H. Maki, *J. Amer. Chem. Soc.*, *96*:7669 (1974).

300. J. H. Ammeter, R. Bucher, N. Oswald, *J. Amer. Chem. Soc.*, *96*: 7833 (1974).

301. K. D. Warren, in *Structure and Bonding* (J. D. Dunitz, J. B. Goodenough, P. Hemmerich, J. A. Ibers, C. K. Jørgensen, J. B. Neilands, D. Reinen, and R. J. P. Williams, eds.), Vol. 27, Springer Verlag, Berlin, 1978.

302. W. Low, *Phys. Rev.*, *105*:792 (1957).

303. R. R. Sharma, T. P. Das, and R. Orbach, *Phys. Rev.*, *149*:257 (1966).

304. R. R. Sharma, T. P. Das, and R. Orbach, *Phys. Rev.*, *155*:338 (1967).

305. G. R. Gabriel, D. F. Johnston, and M. J. D. Powell, *Proc. Roy. Soc.*, *264*:503 (1961).

306. I. Fidone and K. W. H. Stevens, *Proc. Phys. Soc.*, *73*:116 (1959).

307. R. Lacroix and G. Eruch, *Helv. Phys. Acta*, *35*:592 (1962).

308. S. A. Cotton and J. F. Gibson, *J. Chem. Soc.*, *A*, 2105 (1970).

309. G. M. Cole, Jr., and B. B. Garrett, *Inorg. Chem.*, *13*:2680 (1974).

310. S. K. Misra and G. R. Sharp, *J. Chem. Phys.*, *66*:4172 (1977).

311. S. A. Cotton and J. F. Gibson, *J. Chem. Soc.*, *A*, 859 (1971).

312. S. A. Cotton and J. F. Gibson, *J. Chem. Soc.*, *A*, 1696 (1971).

313. D. C. Doetschman, B. J. McCool, *Chem. Phys.*, *8*:1 (1975).

314. W. V. Sweeney, D. Coucouvanis, and R. E. Coffman, *J. Chem. Phys.*, *59*:369 (1973).

315. M. Cox, B. W. Fitzsimmons, A. W. Smith, L. F. Larkworthy, and K. A. Rogers, *J. Chem. Soc.*, *A*, 2158 (1971).

316. C. J. Radnell, J. R. Pilbrow, S. Subramanian, and M. T. Rogers, *J. Chem. Phys.*, *62*:4948 (1975).

317. J. Peisach and W. E. Blumberg, *Proc. Nat. Acad. Sci. U.S.*, *67*: 172 (1970).

318. R. Tsai, C. A. Yu, I. G. Gunsalus, J. Peisach, W. E. Blumberg, W. H. Orme-Johnson, and B. H. Beinert, *Proc. Nat. Acad. Sci. U.S.*, *66*:1157 (1970).

319. M. Sato and H. Kon, *Inorg. Chem.*, *14*:2016 (1975).

320. M. Sato and H. Kon, *Chem. Phys.*, *12*:199 (1976).

321. S. C. Tang, S. Koch, G. C. Papaefthymiou, S. Foner, R. B. Frankel, J. A. Ibers, and R. H. Holm, *J. Amer. Chem. Soc.*, *98*: 2414 (1976).

322. H. Uenoyama and K. Sakai, *Spectrochim. Acta*, *31A*:1517 (1975).

323. Y. H. Ya, *J. Chem. Phys.*, *57*:3020 (1972).

324. J. M. Gaite and J. Michoulier, *J. Chem. Phys.*, *59*:488 (1973).

325. R. C. Kemp, *J. Phys. C*, *5*:3566 (1972).

326. J. Michoulier and J. M. Gaite, *J. Chem. Phys.*, *56*:5205 (1972).

327. S. A. Cotton and J. F. Gibson, *J. Chem. Soc.*, *A*, 1690 (1971).

328. S. A. Cotton and J. F. Gibson, *J. Chem. Soc.*, *A*, 1693 (1971).

329. S. A. Cotton, *Chem. Phys. Lett.*, *41*:606 (1976).

330. J. S. Griffith, *Mol. Phys.*, *21*:135 (1971).

331. B. Bleaney and M. C. M. O'Brien, *Proc. Phys. Soc. B*, 69:1216 (1956).

332. C. Ezzeh and B. R. McGarvey, *J. Magnetic Res.*, *15*:183 (1974).

333. P. B. Merrithew, C. C. Lo, and A. J. Modestino, *Inorg. Chem.*, *14*:242 (1975).

334. J. Baker, L. M. Engelhardt, B. N. Figgis, and A. H. White, *J. Chem. Soc. Dalton*, 530 (1975).

335. R. L. Collins, *J. Chem. Phys.*, *42*:1072 (1965).

336. W. M. Reiff, *Chem. Phys. Lett.*, *8*:297 (1971).

337. E. Munck and B. H. Huynh, in *ESR and NMR of Paramagnetic Species in Biological and Related Systems* (I. Bertini and R. S. Drago, eds.), D. Reidel, Dordrecht, 1979.

338. A. Desideri, J. B. Raynor, and C. K. Poon, *J. Chem. Soc. Dalton*, 2051 (1977).

339. S. A. Cotton and J. F. Gibson, *J. Chem. Soc.*, *A*, 803 (1971).

340. R. Rickards, C. E. Johnson, and H. A. O. Hill, *J. Chem. Soc.*, *A*, 1755 (1971).

341. K. Knauer, P. Hemmerich, and J. D. W. van Voorst, *Angew. Chem. Int. Ed. Engl.*, *6*:262 (1967).

342. R. Beckett, J. A. Heath, B. F. Hoskins, B. P, Kelly, R. L. Martin, A. G. Roos, and P. L. Weickardt, *Inorg. Nucl. Chem. Lett.*, *6*:257 (1970).

343. R. E. DeSimone, *J. Amer. Chem. Soc.*, *95*:6238 (1973).

344. M. D. Rowe and A. J. McCaffery, *J. Chem. Phys.*, *59*:3786 (1973).

345. R. Prins and F. J. Reinders, *J. Amer. Chem. Soc.*, *91*:4929 (1969).

346. R. Prins and A. G. T. G. Kortbeek, *J. Organometal. Chem.*, *33*:C33 (1971).

347. R. Prins and A. R. Korswagen, *J. Organometal. Chem.*, *25*:C74 (1970).

348. W. M. Reif and R. E. DeSimone, *Inorg. Chem.*, *12*:1793 (1973).

349. R. E. DeSimone and R. S. Drago, *J. Amer. Chem. Soc.*, *92*:2343 (1970).

350. M. Duggan and D. N. Hendrickson, *Inorg. Chem.*, *14*:955 (1975).

351. S. E. Anderson and R. Rai, *Chem. Phys.*, *2*:216 (1973).

352. R. L. Martin and A. H. White, in *Transition Metal Chemistry* (R. L. Carlin, ed.), Vol. 4, Dekker, New York, 1968.

353. L. Cambi and A. Cagnasso, *Atti Accad. Naz. Lincei*, *13*:809 (1951).

354. G. R. Hall and D. N. Hendrickson, *Inorg. Chem.*, *15*:607 (1976).

355. R. Rickards, C. E. Johnson, and H. A. O. Hill, *J. Chem. Phys.*, *53*:3118 (1970).

356. G. Harris, *Theor. Chim. Acta*, *10*:119 (1968).

357. G. Harris, *Theor. Chim. Acta*, *10*:155 (1968).

358. D. Niarchos, A. Kostikas, A. Simopoulos, D. Coucouvanis, D. Piltingsrud, and R. E. Coffman, *J. Chem. Phys.*, *69*:4411 (1978).

359. D. P. Rininger, N. V. Duffy, R. C. Weir, E. Gelerinter, J. Stanford, and D. L. Uhrich, *Chem. Phys. Lett.*, *52*:102 (1977).

360. C. Trapp, D. Smith, and R. Johnson, *J. Chem. Phys.*, *55*:5195 (1971).

361. J. R. Sams and T. B. Tsin, *Chem. Phys.*, *15*:209 (1976).

362. A. Abragam and M. H. L. Pryce, *Proc. Roy. Soc.*, *A206*:173 (1951).

363. M. Gerloch, R. F. McMeeking, and A. White, *J. Chem. Soc. Dalton*, 655 (1976).

364. A. Bencini, C. Benelli, D. Gatteschi, and C. Zanchini, *Inorg. Chem.*, *19*:1301 (1980).

365. M. A. Porai-Koshits, *Trudy Inst. Krist. Akad. Nauk S.S.S.R.*, 19 (1955).

366. C. W. Reimann, A. D. Mighell, and F. Mauer, *Acta Crystallogr.* *23*:135 (1967).

367. P. Pauling, G. B. Robertson, and G. A. Rodley, *Inorg. Chim. Acta*, *5*:341 (1971).

368. A. Bencini, C. Benelli, D. Gatteschi, and C. Zanchini, *Inorg. Chem.*, *18*:2526 (1979).

369. J. P. Jesson, *J. Chem. Phys.*, *45*:1049 (1968).

370. A. Desideri, L. Morpurgo, J. B. Raynor, and G. Rotilio, *Biophys. Chem.*, *8*:267 (1978).

371. H. G. Beljers, P. F. Bongers, R. P. van Stapele, and H. Zijlstra, *Phys. Lett.*, *12*:81 (1964).

372. R. P. van Stapele, H. G. Beljers, P. F. Bongers, and H. Zijlstra, *J. Chem. Phys.*, *44*:3719 (1966).

373. J. C. M. Henning, *Z. Angew. Phys.*, *24*:281 (1969).

374. G. E. Shankle, J. N. McElearney, R. W. Schwartz, A. R. Kemp, and R. L. Carlin, *J. Chem. Phys.*, *56*:3750 (1972).

375. R. P. van Stapele, J. C. M. Henning, G. E. G. Hardeman, and P. F. Bongers, *Phys. Rev.*, *150*:310 (1967).

376. K. D. Bowers and J. Owen, *Rep. Progr. Phys.*, *18*:348 (1955).

377. W. DeW. Horrocks and D. A. Burlone, *J. Amer. Chem. Soc.*, *98*:6512 (1976).

378. Unpublished results of our Laboratory.

379. C. H. Wei, *Inorg. Chem.*, *11*:1100 (1972).

380. G. D. Stucky, J. B. Folkers, T. J. Distenmacher, *Acta Crystal-logr.*, *23*:1064 (1967).

381. T. P. Melia and R. Merrifield, *J. Chem. Soc.*, *A*, 1258 (1971).

382. A. Bencini, C. Benelli, and D. Gatteschi, *Inorg. Chem.*, *19*:1632 (1980).

383. W. DeW. Horrocks, D. H. Templeton, and A. Zalkin, *Inorg. Chem.*, *7*:2303 (1968).

384. I. Bertini, D. Gatteschi, and F. Mani, *Inorg. Chim. Acta*, *7*: 717 (1973).

385. A. Edgar, *J. Phys. C*, *9*:4303 (1976).

386. J. A. Cowen and G. T. Johnston, *J. Chem. Phys.*, *44*:2217 (1965).

387. S. Subramanian, Z. Rahman, and J. Whittery, *J. Chem. Phys.*, *49*: 473 (1968).

388. F. B. Hulsberger, J. A. Welleman, and J. Reedijk, *Delft Progr. Rep.*, *Al*:137 (1976).

389. C. G. van Kralingen, J. A. C. van Ooijen, and J. Reedijk, *Transition Metal Chem.*, *3*:90 (1978).

390. J. Reedijk and P. J. van der Put, *Proc. Int. Conf. Coord. Chem.*, *16*:227b (1974).

391. L. J. Guggenberger, C. T. Prewitt, P. Meakin, S. Trofimenko, and J. P. Jesson, *Inorg. Chem.*, *12*:508 (1973).

392. A. Bencini, C. Benelli, D. Gatteschi, and C. Zanchini, *Inorg. Chem.*, *18*:2137 (1979).

393. Y. V. Yablokov, V. K. Voronkova, V. F. Shishkov, A. V. Ablov, and Z. H. Weisbeim, *Sov. Phys. Solid State*, *13*:831 (1971).

394. F. S. Kennedy, H. A. O. Hill, T. A. Kaden, and B. L. Vallee, *Biochem. Biophys. Res. Commun.*, *6*:1533 (1972).

395. R. H. Bailes and M. Calvin, *J. Amer. Chem. Soc.*, *69*:1886 (1947).

396. F. A. Walker, *J. Magnetic Res.*, *15*:201 (1974).

397. R. B. Bentley, F. Mabbs, W. R. Smail, M. Gerloch, and J. Lewis, *J. Chem. Soc.*, *A*, 3003 (1970).

398. A. K. Gregson, R. L. Martin, and S. Mitra, *Chem. Phys. Lett.*, *5*:310 (1970).

399. A. von Zelewsky and H. Fierz, *Helv. Chim. Acta*, *56*:977 (1973).

400. A. H. Maki, N. Edelstein, A. Davison, and R. H. Holm, *J. Amer. Chem. Soc.*, *86*:4580 (1964).

401. V. Malatesta and B. R. McGarvey, *Can. J. Chem.*, *53*:5791 (1975).

402. F. Cariati, F. Morazzoni, C. Busetto, G. DelPiero, and A. Zazzetta, *J. Chem. Soc. Dalton*, 342 (1976).

403. C. Daul, C. W. Sclapfer, and A. von Zelewsky, in *Structure and Bonding* (J. D. Dunitz, J. B. Goodenough, P. Hemmerich, J. A. Ibers, C. K. Jørgensen, J. B. Neilands, D. Reinen, and R. J. P. Williams, eds.), Vol. 36, Springer Verlag, Berlin, 1979.

404. E. I. Ochiai, *J. Chem. Soc., Chem. Commun.*, 489 (1972).

405. L. M. Engelhardt, J. D. Duncan, and M. Green, *Inorg. Nucl. Chem. Lett.*, *8*:725 (1972).

406. C. J. Hipp, W. A. Baker, Jr., *J. Amer. Chem. Soc.*, *92*:792 (1970).

407. C. Busetto, F. Cariati, P. C. Fantucci, D. Galizzioli, F. Morazzoni, *Inorg. Nucl. Chem. Lett.*, *9*:313 (1973).

408. C. Busetto, F. Cariati, A. Fusi, M. Gullotti, F. Morazzoni, A. Pasini, R. Ugo, and V. Valenti, *J. Chem. Soc. Dalton*, 754 (1973).

409. K. Migita, M. Iwaizuru, and T. Isobe, *J. Amer. Chem. Soc.*, *97*: 4228 (1975).

410. P. Fantucci, V. Valenti, F. Cariati, *Inorg. Nucl. Chem. Lett.*, *11*:585 (1975).

411. J. S. Griffith, *Disc. Faraday Soc.*, *26*:81 (1958).

412. F. Cariati, A. Sgamellotti, and V. Valenti, *Atti Accad. Naz. Lincei*, *45*:344 (1968).

413. L. M. Engelhardt and M. Green, *J. Chem. Soc. Dalton*, 724 (1972).

414. B. R. McGarvey, *Can. J. Chem.*, *53*:2498 (1975).

415. W. C. Lin, *Inorg. Chem.*, *15*:1114 (1976).

416. W. C. Lin, *Mol. Phys.*, *31*:657 (1976).

417. M. Hitchman, *Inorg. Chem.*, *16*:1985 (1977).

418. R. L. Belford and W. A. Yeranos, *Mol. Phys.*, *6*:121 (1963).

419. S. E. Harrison and J. M. Assour, in *Paramagnetic Resonance* (W. Low, ed.), Academic Press, New York, 1963.

420. W. P. Schaeffer and R. E. Marsh, *Acta Crystallogr.*, *B25*:1675 (1969).

421. Y. Nishida and S. Koda, *Bull. Chem. Soc. Jap.*, *51*:143 (1978).

422. A. Pezeshk, F. T. Greenaway, and G. Voncow, *Inorg. Chem.*, *17*: 422 (1978).

423. N. S. Hush and J. S. Woolsey, *J. Chem. Soc. Dalton*, 24 (1974).

424. W. C. Lin and P. W. Lau, *J. Amer. Chem. Soc.*, *98*:1447 (1976).

425. J. M. Assour and W. R. Kahn, *J. Amer. Chem. Soc.*, *87*:207 (1965).

426. C. Busetto, F. Cariati, P. Fantucci, D. Galizzioli, and F. Morazzoni, *J. Chem. Soc. Dalton*, 1712 (1973).

427. A. Pezeshk, F. T. Greenaway, J. Dabrowiak, and G. Vincow, *Inorg. Chem.*, *17*:1717 (1978).

428. F. L. Urbach, R. D. Bereman, J. A. Topich, M. Hariharu, and B. J. Kalbacher, *J. Amer. Chem. Soc.*, *96*:5063 (1974).

429. B. B. Wayland and M. E. Abd-Elmageed, *J. Amer. Chem. Soc.*, *96*: 4809 (1974).

430. F. A. Walker, *J. Amer. Chem. Soc.*, *92*:4235 (1970).

431. J. M. Assour, *J. Amer. Chem. Soc.*, *87*:4701 (1965).

432. B. M. Hoffman, F. Basolo, and D. L. Diemente, *J. Amer. Chem. Soc.*, *95*:6497 (1973).

433. J. F. Endicott, J. Lilie, J. M. Kuszaj, B. S. Ramaswamy, W. G. Schmonsees, M. G. Simic, M. D. Glick, and D. P. Rillema, *J. Amer. Chem. Soc.*, *99*:429 (1977).

434. B. B. Wayland, M. E. Abd-Elmageed, and L. F. Mehne, *Inorg. Chem.*, *14*:1456 (1975).

435. M. Chikira, T. Kawakita, and T. Isobe, *Bull. Chem. Soc. Jap.*, *47*:1283 (1974).

436. H. A. O. Hill, P. J. Sadler, and R. J. P. Williams, *J. Chem. Soc. Dalton*, 1663 (1973).

437. L. D. Rollmann and S. I. Chan, *Inorg. Chem.*, *10*:1978 (1971).

438. R. D. Jones, D. A. Summerville, and F. Basolo, *Chem. Rev.*, *79*: 139 (1979).

439. F. A. Walker, *J. Amer. Chem. Soc.*, *95*:1150 (1973).

440. A. L. Crumbliss and F. Basolo, *J. Amer. Chem. Soc.*, *92*:55 (1970).

441. B. M. Hoffmann, D. L. Diemente, and F. Basolo, *J. Amer. Chem. Soc.*, *92*:61 (1970).

442. J. H. Bayston, N. K. King, F. D. Looney, and M. E. Winfield, *J. Amer. Chem. Soc.*, *91*:2775 (1969).

443. B. B. Wayland, J. V. Minkiewicz, and M. E. Abd-Elmageed, *J. Amer. Chem. Soc.*, *96*:2795 (1974).

444. B. Jezowska-Trzebiatowska, A. Vogt, H. Kozlowski, and A. Jezierski, *Bull. Acad. Pol. Sci.*, *XX*:187 (1972).

445. B. Jezowska-Trzebiatowska, A. Vogt, *Pure Appl. Chem.*, *38*:367 (1974).

446. E. Melamud, B. L. Silver, and Z. Dori, *J. Amer. Chem. Soc.*, *96*: 4689 (1974).

447. D. Getz, E. Melamud, B. L. Silver, and Z. Dori, *J. Amer. Chem. Soc.*, *97*:3846 (1975).

448. B. S. Tovrog, D. J. Kitko, and R. S. Drago, *J. Amer. Chem. Soc.*, *98*:5144 (1976).

449. J. B. Raynor, *Inorg. Nucl. Chem. Lett.*, *10*:867 (1974).

450. V. P. Chacko and P. T. Manoharan, *J. Magnetic Res.*, *22*:7 (1976).

451. R. Morassi, I. Bertini, and L. Sacconi, *Coord. Chem. Rev.*, *11*: 343 (1973).

452. M. Ciampolini, in *Structure and Bonding* (C. K. Jørgensen, J. B. Neilands, R. S. Nyholm, D. Reinen, and R. J. P. Williams, eds.), Vol. 6, Springer Verlag, Berlin, 1969.

453. Y. Nishida and S. Kida, *Bull. Chem. Soc. Jap.*, *48*:1045 (1975).

454. Y. Nishida and H. Shimohori, *Bull. Chem. Soc. Jap.*, *46*:2406 (1973).

455. R. S. Drago, personal communication.

456. R. Morassi, F. Mani, and L. Sacconi, *Inorg. Chem.*, *12*:1246 (1973).

457. D. Gatteschi, C. A. Ghilardi, A. Orlandini, and L. Sacconi, *Inorg. Chem.*, *17*:3023 (1978).

458. A. Bianchi Orlandini, C. Calabresi, C. A. Ghilardi, P. L. Orioli, and L. Sacconi, *J. Chem. Soc. Dalton*, 1383 (1973).

459. J. A. Bertrand and D. A. Carpenter, *Inorg. Chem.*, *5*:514 (1966).

460. J. A. Bertrand, D. A. Carpenter, and A. R. Kalyanaraman, *Inorg. Chim. Acta*, *5*:113 (1971).

461. G. Backes and D. Reinen, *Z. Anorg. Allg. Chem.*, *418*:217 (1975).

462. J. H. Ammeter and J. D. Swalen, *J. Chem. Phys.*, *57*:678 (1972).

463. J. H. Ammeter and J. M. Brom, Jr., *Chem. Phys. Lett.*, *27*:380 (1974).

464. G. E. Herberich, T. Lund, and J. B. Raynor, *J. Chem. Soc. Dalton*, 985 (1975).

465. D. Attanasio, *Chem. Phys. Lett.*, *49*:547 (1977).

466. Y. Nishida, A. Sumita, and S. Kida, *Bull. Chem. Soc. Jap.*, *50*: 759 (1977).

467. Y. Nishida, A. Sumita, and S. Kida, *Bull. Chem. Soc. Jap.*, *50*: 2485 (1977).

468. B. J. Kalbacher and R. D. Bereman, *Inorg. Chem.*, *12*:2997 (1973).

469. R. J. Booth and W. C. Lin, *J. Chem. Phys.*, *61*:1226 (1974).

470. O. Stelzer, W. S. Sheldrick, and J. Subramanian, *J. Chem. Soc. Dalton*, 966 (1977).

471. A. Reuveni, V. Malatesta, and B. R. McGarvey, *Can. J. Chem.*, *55*:70 (1977).

472. P. K. Bernstein and H. B. Gray, *Inorg. Chem.*, *11*:3035 (1972).

473. C. N. Sethulakshmi, S. Subramanian, M. A. Bennett, and P. T. Manoharan, *Inorg. Chem.*, *18*:2520 (1979).

474. D. Reinen, C. Friebel, and V. Propach, *Z. Anorg. Allg. Chem.*, *408*:187 (1974).

475. M. Abou-Ghantous, C. A. Bates, I. A. Clark, J. R. Fletcher, P. J. Jaussaud, and W. S. Moore, *J. Phys. C*, *7*:2707 (1974).

476. M. Abou-Ghantous, C. A. Bates, J. R. Fletcher, and P. J. Jaussaud, *J. Phys. C*, *8*:3641 (1975).

477. E. S. Gore and D. H. Busch, *Inorg. Chem.*, *12*:1 (1973).

478. F. V. Lovecchio, E. S. Gore, and D. H. Busch, *J. Amer. Chem. Soc.*, *96*:3109 (1974).

479. A. G. Lappin, C. K. Murray, and D. W. Margerum, *Inorg. Chem.*, *17*:1630 (1978).

480. R. S. Drago and E. I. Bancom, *Inorg. Chem.*, *11*:2064 (1972).

481. J. H. E. Griffiths and J. Owen, *Proc. Roy. Soc.*, *213*:458 (1952).

482. R. S. Rubins, J. D. Clark, and S. K. Jani, *J. Chem. Phys.*, *67*:893 (1977).

483. R. S. Rubins and S. K. Jani, *J. Chem. Phys.*, *66*:3297 (1977).

484. K. N. Shrivastava, *Chem. Phys. Lett.*, *20*:106 (1973).

485. J. F. Suassuna, C. Rettori, H. Vargas, G. E. Barberis, C. E. Hennies, and N. F. Oliveira, Jr., *J. Phys. Chem. Solids*, *38*:1075 (1977).

486. J. Czakon, *Acta Phys. Pol.*, *A53*:419 (1978).

487. W. M. Walsh, Jr., *Phys. Rev.*, *114*:1473 (1959).

488. J. A. Ochi, W. Sano, S. Isotani, and C. E. Hennies, *J. Chem. Phys.*, *62*:2115 (1975).

489. B. Sczaniecki, L. Larys, U. Gruszczynska, *Acta Phys. Pol.*, *A46*:759 (1974).

490. J. Stankowski, J. M. Janik, A. Dezor, and B. Sczaniecki, *Phys. Status Sol.*, *A16*:K167 (1973).

491. C. Trapp and C. I. Shyr, *J. Chem. Phys.*, *54*:196 (1971).

492. A. R. Bates and K. W. H. Stevens, *J. Phys. C*, *2*:1573 (1969).

493. A. R. Bates, *J. Phys. C*, *3*:1825 (1970).

494. W. M. Pontuschka, A. Piccini, C. J. A. Quadros, and S. Isotani, *Phys. Lett.*, *44A*:57 (1973).

495. R. S. Rubins, *J. Chem. Phys.*, *60*:4189 (1974).

496. R. Kirmse, S. Wartewig, and R. Bottcher, *Chem. Phys. Lett.*, *22*:427 (1973).

497. G. L. McPherson and K. O. Devaney, *Inorg. Chem.*, *16*:1565 (1977).

498. P. Dapporto and L. Sacconi, *Inorg. Chim. Acta*, *39*:61 (1980).

499. R. Barbucci, A. Bencini, and D. Gatteschi, *Inorg. Chem.*, *16*: 2117 (1977).

500. P. H. Davis, R. L. Belford, and I. C. Paul, *Inorg. Chem.*, *12*: 213 (1973).

501. M. J. Nilges, E. K. Barefield, R. L. Belford, and P. H. Davis, *J. Amer. Chem. Soc.*, *99*:755 (1977).

502. W. E. Hatfield and R. Whyman, in *Transition Metal Chemistry* (R. L. Carlin, ed.), Vol. 5, Dekker, New York, 1969.

503. B. J. Hathaway and A. A. G. Tomlinson, *Coord. Chem. Rev.*, *5*:1 (1970).

504. A. Abragam and M. H. L. Pryce, *Proc. Phys. Soc.*, *A63*:409 (1950).

505. B. Bleaney, K. D. Bowers, and R. S. Treman, *Proc. Roy. Soc.*, *A228*:157 (1955).

506. D. P. Brean, D. C. Krupka, and F. I. B. Williams, *Phys. Rev.*, *179*:241 (1969).

507. M. C. M. O'Brien, *Proc. Roy. Soc.*, *A281*:323 (1964).

508. A. Hudson, *Mol. Phys.*, *10*:575 (1966).

509. C. Friebel, *Z. Anorg. Allg. Chem.*, *417*:197 (1975).

510. C. Friebel, *Z. Naturforsch. B*, *28*:295 (1974).

511. B. V. Harrowfield and J. P. Pilbrow, *J. Phys. C*, *6*:755 (1973).

512. I. Bertini, D. Gatteschi, P. Paoletti, and A. Scozzafava, *Inorg. Chim. Acta*, *13*:L5 (1975).

513. R. Rajan and T. R. Reddy, *J. Chem. Phys.*, *39*:1140 (1963).

514. I. Bertini, D. Gatteschi, and A. Scozzafava, *Inorg. Chem.*, *16*: 1973 (1977).

515. R. C. Koch, M. D. Joesten, and J. H. Venable, Jr., *J. Chem. Phys.*, *59*:6312 (1973).

516. D. L. Cullen and E. L. Lingafelter, *Inorg. Chem.*, *9*:1858 (1970).

517. S. Takagi, M. D. Joesten, and P. G. Lenhert, *J. Amer. Chem. Soc.*, *96*:6606 (1974).

518. I. Bertini, P. Dapporto, D. Gatteschi, and A. Scozzafava, *Solid State Commun.*, *26*:749 (1978).

519. I. Bertini, P. Dapporto, D. Gatteschi, and A. Scozzafava, *J. Chem. Soc. Dalton*, 1409 (1979).

520. I. Bertini, D. Gatteschi, and A. Scozzafava, *Coord. Chem. Rev.*, *29*:64 (1979).

521. D. Reinen and S. Krause, *Solid State Commun.*, *29*:691 (1979).

522. E. de Boer, C. P. Keijzers, and J. S. Wood, *Proc. XIX ICCC*, Prague, 1978, p. 17a.

523. J. H. Ammeter, H. B. Burgi, E. Gamp, V. Meyer-Sandrin, and W. P. Jensen, *Inorg. Chem.*, *18*:733 (1979).

524. R. Allmann, W. Henke, and D. Reinen, *Inorg. Chem.*, *17*:378 (1978).

525. J. Pradilla-Sorzano and J. P. Fackler, *Inorg. Chem.*, *13*:38 (1974).

526. B. J. Hathaway, P. G. Hodgson and P. C. Power, *Inorg. Chem.*, *13*:2009 (1974).

527. D. Getz and B. L. Silver, *J. Chem. Phys.*, *61*:630 (1974).

528. B. L. Silver and D. Getz, *J. Chem. Phys.*, *61*:638 (1974).

529. D. J. Robinson and C. H. L. Kennard, *Crystal Struct. Commun.*, *1*:185 (1972).

530. G. A. Senyukova, I. D. Mikheikhin, and K. I. Zamaraev, *Russ. J. Struct. Chem.*, *11*:18 (1970).

531. R. Barbucci and M. J. M. Campbell, *Inorg. Chim. Acta*, *15*:L15 (1975).

532. B. J. Hathaway, D. E. Billing, R. J. Dudley, R. J. Fereday, and A. A. G. Tomlinson, *J. Chem. Soc.*, *A*, 186 (1970).

533. W. E. Hatfield and T. S. Piper, *Inorg. Chem.*, *3*:841 (1964).

534. K. N. Raymond, D. W. Meek, and J. A. Ibers, *Inorg. Chem.*, *7*:1111 (1968).

535. H. Jin-Ling, L. Jien-Ming, L. Jia-Xi, *Hua Hsueh Hsueh Pao*, *32*:162 (1966).

536. A. Bencini, I. Bertini, D. Gatteschi, and A. Scozzafava, *Inorg. Chem.*, *17*:3194 (1978).

537. M. DiVaira, P. L. Orioli, and L. Sacconi, *Inorg. Chem.*, *10*:553 (1971).

538. P. L. Orioli, M. DiVaira, and L. Sacconi, *Inorg. Chem.*, *5*:400 (1966).

539. J. R. Wasson, D. M. Klassen, H. W. Richardson, and W. E. Hatfield, *Inorg. Chem.*, *16*:1906 (1977).

540. R. C. Rosenberg, C. A. Root, P. K. Bernstein, and H. B. Gray, *J. Amer. Chem. Soc.*, *97*:2092 (1975).

541. R. E. Dietz, H. Kamimura, M. D. Sturge, and A. Yariv, *Phys. Rev.*, *132*:1559 (1963).

542. M. de Wit and A. R. Reinberg, *Phys. Rev.*, *163*:261 (1967).

543. M. Sharnoff, *J. Chem. Phys.*, *41*:2203 (1964).

544. M. Sharnoff, *J. Chem. Phys.*, *42*:3383 (1965).

545. M. Sharnoff and C. W. Reimann, *J. Chem. Phys.*, *43*:2993 (1965).

546. G. F. Kokoszka, C. W. Reimann, and H. C. Allen, Jr., *J. Phys. Chem.*, *71*:121 (1967).

547. I. Bertini, D. Gatteschi, and A. Scozzafava, *Inorg. Chim. Acta*, *11*:L17 (1974).

548. G. F. Kokoszka, C. W. Reimann, H. C. Allen, Jr., and G. Gordon, *Inorg. Chem.*, *6*:1657 (1967).

549. R. Srinivasan and C. K. Subramanian, *Indian J. Pure Appl. Phys.*, *8*:817 (1970).

550. G. F. Kokoszka, H. C. Allen, Jr., and G. Gordon, *J. Chem. Phys.*, *42*:3693 (1965).

551. G. F. Kokoszka, H. C. Allen, Jr., and G. Gordon, *J. Chem. Phys.*, *46*:3020 (1967).

552. B. J. Hathaway, I. M. Procter, R. C. Slade, and A. A. G. Tomlinson, *J. Chem. Soc.*, *A*, 2219 (1969).

553. K. T. McGregor and W. E. Hatfield, *J. Chem. Soc. Dalton*, 2448 (1974).

554. R. J. Dudley, B. J. Hathaway, P. G. Hodgson, P. C. Power, and D. J. Loose, *J. Chem. Soc. Dalton*, 1005 (1974).

555. A. Bencini and D. Gatteschi, *Inorg. Chem.*, *16*:1994 (1977).

556. D. E. Billing, R. J. Dudley, B. J. Hathaway, and A. A. G. Tomlinson, *J. Chem. Soc.*, *A*, 691 (1971).

557. D. S. Bieksza and D. N. Hendrickson, *Inorg. Chem.*, *16*:924 (1974).

558. T. R. Felthouse, E. J. Laskowski, and D. N. Hendrickson, *Inorg. Chem.*, *16*:1077 (1977).

559. I. Bertini, D. Gatteschi, and G. Martini, *J. Chem. Soc. Dalton*, 1644 (1973).

560. B. J. Hathaway and P. G. Hodgson, *Spectrochim. Acta*, *30A*:1465 (1974).

561. G. L. Seebach, D. K. Johnson, H. J. Stoklosa, and J. R. Wasson, *Inorg. Nucl. Chem. Lett.*, *9*:295 (1973).

562. S. N. Choi, R. D. Bereman, J. R. Wasson, *J. Inorg. Nucl. Chem.*, *37*:2087 (1975).

563. I. Bertini, D. Gatteschi, and A. Scozzafava, *Gazz. Chim. Ital.*, *104*:1029 (1974).

564. V. K. Voronkova, M. M. Zaripov, Y. V. Yablokov, A. V. Ablov, and M. A. Ablova, *Dokl. Akad. Nauk.*, *220*:623 (1975).

565. F. G. Herring, D. J. Patmore, and A. Storr, *J. Chem. Soc. Dalton*, 711 (1975).

566. H. P. Fritz, B. M. Golla, and H. J. Keller, *Z. Naturforsch.*, *21b*:1015 (1966).

567. H. P. Fritz, I. M. Golla, and H. J. Keller, *Z. Naturforsch.* *23b*:876 (1968).

568. J. R. Wasson, P. J. Cowan, and W. E. Hatfield, *Inorg. Chim. Acta*, *27*:167 (1978).

569. R. B. Wilson, J. R. Wasson, W. E. Hatfield, and D. J. Hodgson, *Inorg. Chem.*, *17*:641 (1973).

570. J. R. Wasson, H. W. Richardson, and W. E. Hatfield, *Z. Naturforsch.*, *32b*:551 (1977).

571. H. Yokoi and A. W. Addison, *Inorg. Chem.*, *16*:1341 (1977).

572. M. Iwasaki, *J. Magnetic Res.*, *16*:417 (1974).

573. J. R. Pilbrow, *Mol. Phys.*, *16*:307 (1969).

574. T. S. Johnston and H. G. Hecht, *J. Mol. Spectrosc.*, *17*:98 (1965).

575. R. Aasa and T. Vanngard, *J. Magnetic Res.*, *19*:308 (1975).

576. R. Lefebvre and J. Maruani, *J. Chem. Phys.*, *42*:1480 (1965).

577. F. K. Kneubuhl, *J. Chem. Phys.*, *33*:1074 (1960).

578. G. G. Belford, R. L. Belford, and J. F. Burkhalter, *J. Magnetic Res.*, *11*:251 (1973).

579. R. L. Belford and G. G. Belford, *J. Chem. Phys.*, *59*:853 (1973).

580. C. N. Bauwell and H. Primas, *Mol. Phys.*, *6*:225 (1963).

581. C. B. Moler and G. W. Stewart, *Joint Report No. STAN-CS-71-232, Computer Science Dept., Stanford University and No. CNA-32, Center for Numerical Analysis, The University of Texas* (1971).

582. L. Kaufman, *Report No. STAN CS-72-76, Computer Science Dept., Stanford University*, 1972

583. K. T. McGregor, R. P. Scaringe, and W. E. Hatfield, *Mol. Phys.* *30*:1925 (1975).

584. R. L. Belford, P. M. Davis, G. G. Belford, and T. M. Lenhardt, *Amer. Chem. Soc. Symp. Ser.*, *5*:40 (1974).

585. C. S. Rudisill and Y. Y. Chu, *AIAAJ*, *13*:834 (1975).

586. A. L. Andrew, *J. Comp. Phys.*, *26*:107 (1978).

587. A. Bencini, D. Gatteschi, C. Zanchini, *J. Amer. Chem. Soc.*, *102*:5234 (1980).

2

3d Metal Complexes with Tripodal Polytertiary Phosphines and Arsines

LUIGI SACCONI and FABRIZIO MANI

University of Florence
Florence, Italy

I. INTRODUCTION

A. General

The coordination chemistry of polydentate ligands containing tertiary
phosphine and arsine entities has been considerably expanded since the
early pioneering work of Chatt, Nyholm, Venanzi, and their collabora-
tors on the metal complexes of the ligands p_3 [1] (I), DAS [2], QAS
[3] (VI), and QP [4] (V). Since that time many other polydentate ligands have been synthesized, both open-chain and tripod-like, and a
large variety of transition metal complexes have been prepared and
described.

McAuliffe [5] has edited a very recent and comprehensive review which covers metal complexes containing P and As as donor
atoms. Other reviews have also appeared on more specific aspects
of this field. The previous author [6] has reviewed complexes containing open-chain tetradentate ligands, whereas Venanzi [7] has
reviewed complexes containing the ligands QP and QAS.

Recently, Mason and Meek [8] have critically discussed the
"versatility" of tertiary phosphine ligands with particular reference to the metal-phosphorus bond. Meek [9] has also reviewed
cobalt(I) and rhodium(I) complexes with polyphosphine ligands.

Despite the existence of these reviews, we considered it
opportune to review the field of 3d metal complexes formed by tri-
and tetradentate tripod-like ligands that have phosphorus or

arsenic as donor atoms. More precisely, this chapter discusses tripod-like ligands containing at least one tertiary phosphine or arsine entity. This type of ligand exhibits sufficiently unusual characteristics, which arise both from the stereochemistry and the electronic properties of the donor atoms, to justify an independent review.

The tripod-like geometry of these ligands gives them their most prominent characteristic feature, since it is the most favourable arrangement for the formation of complexes with trigonal bipyramidal, tetrahedral, and trigonal pyramidal stereochemistries. In fact, tetradentate ligands containing o-phenylene connecting chains give rise almost exclusively to trigonal bipyramidal geometries because of the rigid o-phenylene chains. In contrast, ligands containing the more flexible ethylene or trimethylene connecting chains make other geometries possible: square pyramidal and, more rarely, octahedral.

In terms of the electronic properties of the donor atoms, the polarizability of phosphorus and arsenic atoms, together with the existence of empty d orbitals, which can interact with filled metal d orbitals to form π bonds, favor both spin pairing and the stabilization of low oxidation states in the metal complexes. In this situation the formation of strong bonds between the 3d metals and "soft" ligands that have good π-bonding propensity, such as NO and CO, as well as organic fragments such as CH_3 and C_6H_5, is strongly favored.

It should also be noted that many tripod-like ligands exhibit distinct reducing and basic properties, which may result in the reduction of metal ions and the deprotonation of weakly acidic species such as H_2O and C_2H_5OH. These ligands are also capable of undergoing many other different reactions, such as hydrogenation, phenylation, and acylation. These will be discussed in Sec. IV.

B. Summary of the Tripod-Like Ligands

The ligand p_3 (I), synthesized by Hewertson and Watson [10], contains three -PPh_2 groups, each attached to the same carbon atom of the ethane through a methylene group. Subsequently, the analogous

triarsine, as$_3$ [11] (II), the ethyl-substituted triphosphine Etp$_3$
[12] (III), and the methyl-substituted triarsine, Meas$_3$ [13] (IV),
have been synthesized. All these ligands are potentially tridentate.

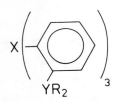

$$\text{CH}_3\text{-C} \begin{cases} \text{CH}_2\text{-ER}_2 \\ \text{CH}_2\text{-ER}_2 \\ \text{CH}_2\text{-ER}_2 \end{cases}$$

E = P; R = Ph, p$_3$ [10]	(I)
E = As; R = Ph, as$_3$ [11]	(II)
E = P; R = Et, Etp$_3$ [12]	(III)
E = As; R = Me, Meas$_3$ [13]	(IV)

The most abundant group of tripod-like ligands can be found
among potentially tetradentate ligands in which the donor atoms are
connected by o-phenyl linkages,

$$X \left(\underset{YR_2}{\bigcirc} \right)_3$$

or by ethylene or trimethylene groups.

$$X \begin{cases} (CH_2)_n\text{-}YR_2 \\ (CH_2)_n\text{-}YR_2 \\ (CH_2)_n\text{-}YR_2 \end{cases} \qquad n = 2, 3$$

The syntheses of many ligands of the first type have been reported by Venanzi and co-workers.

X = Y = P; R = Ph, QP [4]	(V)
X = Y = As; R = Ph, QAS [3]	(VI)
X = Y = As; R = Me, Qas [14]	(VII)
X = As; Y = P; R = Ph, ASTP [15]	(VIII)
X = Sb; Y = P; R = Ph, SBTP [15]	(IX)
X = P; Y = As; R = Ph, PTAS [15]	(X)
X = Sb; Y = As; R = Ph, SBTAS [15]	(XI)
X = P; Y = N; R = Me, TNP (Ptn) [16a]	(XII)

Whereas McAuliffe and co-workers have prepared the ligands (XIII) and (XIV), which contain methylated arsines,

X = Sb, Y = As; R = Me; Sbtas (sbta) [17] (XIII)

X = Bi, Y = As, R = Me; Bitas [18] (XIV)

Dyer and Meek have been interested in ligands containing thiomethyl and selenomethyl groups and have synthesized the ligands

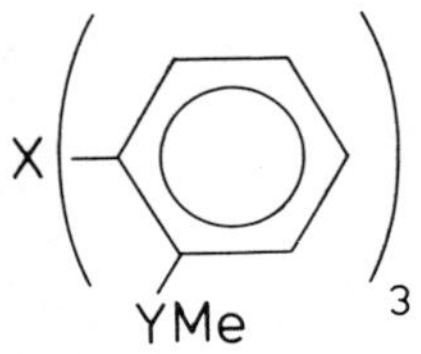

$$X \left(\begin{array}{c} \bigcirc \\ YMe \end{array} \right)_3$$

X = P, Y = S; Pts (TSP) [19] (XV)

X = P, Y = Se; Ptse (TSeP) [20] (XVI)

X = As, Y = S; Asts (TSA) [19]

$$As \begin{array}{l} \nearrow C_6H_4AsPh_2 \\ \longrightarrow C_6H_4AsPh_2 \quad [21] \\ \searrow C_6H_4SMe \end{array}$$

(XVII)

The first example of a ligand containing trimethylene linkages, qas (XVIII), was described by Barcklay and Barnard [22]. Analogous ligands containing phosphorus, arsenic, and antimony donor atoms have been synthesized by Meek and co-workers [23,24].

$$X \begin{array}{l} \nearrow CH_2-CH_2-CH_2-ZMe_2 \\ \longrightarrow CH_2-CH_2-CH_2-ZMe_2 \\ \searrow CH_2-CH_2-CH_2-ZMe_2 \end{array}$$

X = Z = As, qas [22] (XVIII)

X = P; Z = As; tap [23] (XIX)

X = Sb, Z = As, tasb [24] (XX)

The ligand (XXI), containing an NP_3 arrangement of donor atoms joined through ethylenic linkages, has been synthesized by Sacconi and co-workers:

$$N \begin{array}{l} \nearrow CH_2-CH_2-PPh_2 \\ \longrightarrow CH_2-CH_2-PPh_2 \\ \searrow CH_2-CH_2-PPh_2 \end{array}$$

np_3 [25] (XXI)

This ligand has become the archetype of many others having the same triethylamine skeleton with various arrangements of different terminal donors. Some of these ligands that contain at least one tertiary phosphine or arsine group are listed below:

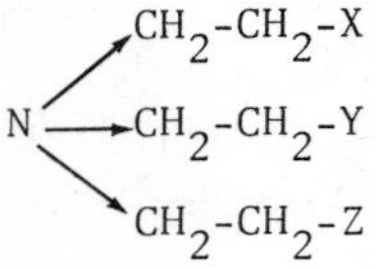

$$
\begin{array}{ll}
N \Bigg\langle \begin{array}{l} CH_2\text{-}CH_2\text{-}X \\ CH_2\text{-}CH_2\text{-}Y \\ CH_2\text{-}CH_2\text{-}Z \end{array} & \\
\end{array}
$$

$X = Y = Z = AsPh_2$; nas_3 [26]	(XXII)
$X = Y = Z = PMe_2$; $Menp_3$ [27]	(XXIII)
$X = Y = Z = AsMe_2$; $Menas_3$ [28]	(XXIV)
$X = Y = PPh_2$; $Z = NEt_2$; n_2p_2 [29]	(XXV)
$X = Y = AsPh_2$; $Z = NEt_2$; n_2as_2 [29]	(XXVI)
$X = Y = NEt_2$; $Z = PPh_2$; n_3p [29]	(XXVII)
$X = Y = NEt_2$; $Z = AsPh_2$; n_3as [29]	(XXVIII)

King and co-workers have synthesized many polytertiary phosphines, among which is the tripod-like ligand (XXIX) containing the donor set P_4:

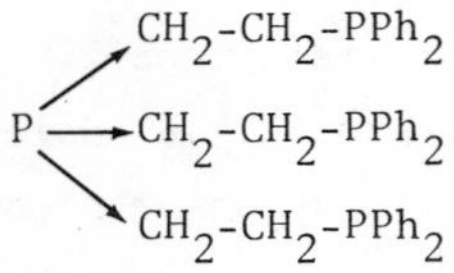

$$
P \Bigg\langle \begin{array}{l} CH_2\text{-}CH_2\text{-}PPh_2 \\ CH_2\text{-}CH_2\text{-}PPh_2 \\ CH_2\text{-}CH_2\text{-}PPh_2 \end{array} \qquad pp_3 \ [30]
$$

(XXIX)

Analogous ligands containing methyl- [31] and neo-pentyl- [32] substituted phosphorus atoms have also been described.

II. TRIDENTATE LIGANDS

A. Complexes with Bipositive Metal Ions

There is a lack of definite data about the stoichiometry and stereochemistry of some of the first complexes reported for nickel-(II) and cobalt(II) with the ligand p_3(I) [33]. Some complexes that do appear to be sufficiently well characterized are the low-spin complexes of cobalt(II) and nickel(II), [Co(NCS)$_2$(p$_3$)] (1) and [Ni(CN)$_2$(p$_3$)], which are accepted as being five-coordinate, and those of formula [NiX$_2$(p$_3$)] (X = Cl, Br, NCS) (2), which are diamagnetic and for which a square planar geometry has been postulated, where p_3

acts as a bidentate ligand [34] (Table 1).

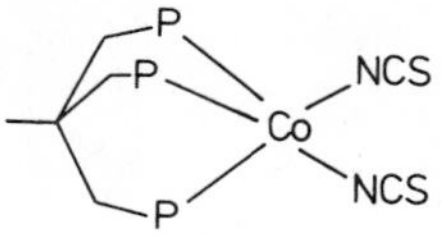

(1) 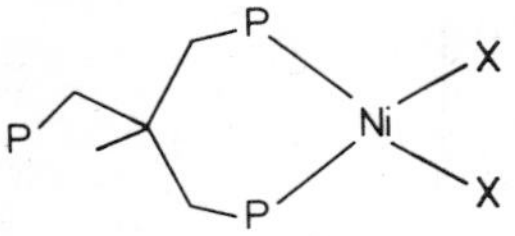   (2)

Both the anhydrous metal halides and the hydrated cobalt(II) ion, in the presence of noncoordinating anions, react with p_3 in a variety of ways, producing five-coordination. The chloride and bromide of cobalt(II) form dimeric complexes $[(p_3)Co(\mu-X)_2Co(p_3)]$-$(BPh_4)_2$ (X = Cl, Br) (a); $Co(H_2O)_6(BF_4)_2$, following a hydrolysis reaction (Table 2), forms the analogous dimeric complex $[(p_3)Co(\mu-OH)_2Co(p_3)](BPh_4)_2$ (b) [35a].

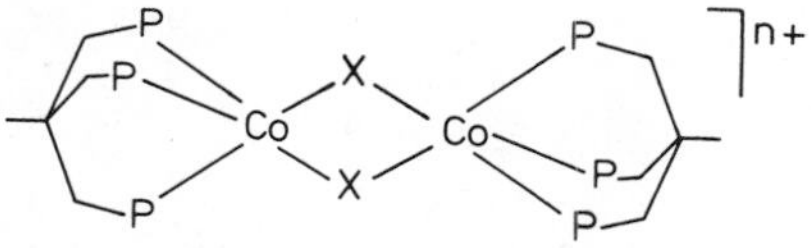

X = Cl, Br, OH, SCH_3; n = 2
X = S; n = 0, 1, 2

(3)

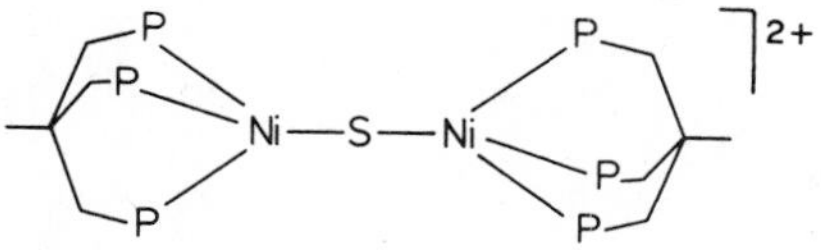

(4)

Similar dimeric complexes containing two SCH_3 or S groups, namely, $[(p_3)Co(\mu-SCH_3)_2Co(p_3)](BPh_4)_2$ (c) and $[(p_3)Co(\mu-S)_2Co(p_3)]BPh_4$ (d) are formed by reaction of $HSCH_3$ and, respectively, H_2S with the di-μ-hydroxo-complex [36]. Compound d has been reduced with sodium borohydride to the species $[(p_3)Co(\mu-S)_2Co(p_3)]$ (e) and oxidized with nitrosylfluoroborate to the species $[(p_3)Co(\mu-S)_2Co(p_3)](BPh_4)_2$ (f). Complexes e and f are diamagnetic; the remaining complexes have magnetic moment values at room temperature in the range 1.48-1.90 μ_B

Table 1 Some Representative Complexes of the Tridentate Ligands p_3, Etp_3, as_3, and $Meas_3$

Complexes	X, Y, M, Z, L	μ_{eff}, μ_B	Coordination Geometry or Number[a]	References
$[Ni(p_3)_2]$		0		1
$[MX(p_3)]$	M = Co; X = halides, SCH_3, BH_4[b]	3.03-3.08	Td	36, 43, 45
	M = Ni; X = Cl, Br, I[b]	1.93-2.20	Td	43, 44
	M = Cu; X = Br, Cl, BH_4	0	Td	43, 45
$[Co(NCS)_2(p_3)]$		2.34	5	34
$[Ni(CN)_2(p_3)]$		0	5	34
$[NiX_2(p_3)]$	X = Cl, Br, NCS	0	SqP1	34
$[M(ZO_4)(p_3)]$	M = Ni; Z = S, Se[b]	3.01-3.12	SqPy	38
	M = Co; Z = S,[b] Se	2.05-2.01	SqPy	38
$[CoX(p_3)]BPh_4 \cdot solv$	X = ac, acac,[b] NO_3; solv = C_3H_6O	2.01-2.08	SqPy-TBPy	35a
$[Co_2X_2(p_3)_2](BPh_4)_2 \cdot solv$	X = Cl, Br, OH,[b] SCH_3	1.48-1.90[c]	SqPy	35a, 36
	solv = C_3H_6O, CH_2Cl_2			

$[Co_2S_2(p_3)_2]BPh_4 \cdot solv^b$	solv = $0.5C_3H_7NO$	1.78	SqPy	36
$[Co_2S_2(p_3)_2](BPh_4)_2 \cdot solv$	solv = $4CH_2Cl_2$	0	SqPy	36
$[Co_2S_2(p_3)_2] \cdot solv^b$	solv = $1.4C_3H_7NO$	0	SqPy	36
$[Ni_2S(p_3)_2](BPh_4)_2 \cdot solv^b$	solv = $1.5C_3H_7NO$	0	SqPy	36
$[M_2H_3(p_3)_2]Y \cdot solv$	M = Fe;^{b} Y = PF_6, BPh_4	0	Oh	40
	M = Co;^{b} Y = BPh_4	3.29	Oh	40
	solv. = CH_2Cl_2, C_4H_8O			
$[Co_2H_3(as_3)_2]BPh_4$		3.17	Oh	40
$[NiX(as_3)]$	Br, I	2.11-2.20	Td	11
$[NiX_2(Meas_3)_2]$	Cl, Br, I	3.02-3.22	Oh	11
$[Ni(Meas_3)_2]Y_2$	ClO_4,BF_4, BPh_4	0	5	11
$[Ni(NCS)_2(Meas_3)]$		0	SqPl	11

---

[a]The limiting geometry the true structure more closely approaches. The coordination number is given for those complexes whose structure either is not known with certainty or cannot be related to any limiting geometry.

[b]Structure determined by X-ray analysis.

[c]Antiferromagnetic behavior.

Table 2   Deprotonation Reactions Assisted by the Ligands p_3

$$2Co(H_2O)_6^{2+} + p_3(\text{excess}) \longrightarrow [Co_2(OH)_2(p_3)_2]^{2+} + 2p_3H^{+a} + H_2O$$

$$2Ni(H_2O)_6^{2+} + p_3(\text{excess}) \xrightarrow{\text{H}_2\text{S}} [Ni_2S(p_3)_2]^{2+} + 2p_3H^{+a} + H_2O$$

---

[a]Here p_3H^+ indicates a possible ionic phosphine species formed by protonation of the triphosphine.

and also exhibit antiferromagnetic behavior with the exception of complex d, which has a temperature-independent paramagnetism. The structures of compounds b, c, d, and e have been determined by X-ray analysis [35a,36]. In these dinuclear complexes each of two cobalt atoms is coordinated to three phosphorus atoms of the p_3 ligand and to two bridging X groups (3). Thus each cobalt atom is five-coordinated in a very distorted square pyramidal geometry.

The main structural difference in the complexes whose crystal structures are known is the different molecular conformations. In complexes b and d the two $Co(p_3)$ units have a configuration intermediate between staggered and eclipsed; on the other hand, the two units are mutually staggered in complexes c and e. Other structural differences can be found in some bond angles and lengths. The Co-Co distances range from 3.067 Å (complex b) to 3.598 Å (complex e), thus ruling out any direct metal-to-metal interaction.

The magnetic behavior of all of the complexes can be rationalized taking into account the valence electrons of each dimeric species and a possible superexchange coupling through the bridging groups. Complex d, with 33 valence electrons, has a magnetic moment corresponding to one unpaired electron, whereas complex f is diamagnetic as a consequence of its 32-electron configuration. The 34-electron complexes a, b, and c exhibit antiferromagnetic behavior arising from a singlet-triplet equilibrium. The value of the exchange coupling constant $|J|$ increases with decreasing electronegativity of the bridging atoms in the order $OH^- < Cl^- < Br^- < SCH_3^-$. The bridging S^{2-} atoms (complex e) can produce complete spin pairing even at room temperature.

The dimeric complexes $[Co_2(\mu-X)_2(p_3)_2]^{2+}$ (X = Cl, Br, SCH_3) react with CO to give paramagnetic (one unpaired electron) five-coordinate species $[CoX(CO)(p_3)]^+$ [35b,36]. Complex <u>c</u>, when reduced with sodiumborohydride, yields the paramagnetic complex $[Co(SCH_3)(p_3)]$ [36], which is isostructural with the other pseudo-tetrahedral cobalt-(I) complexes of the ligand p_3 (Sec. II.B).

Starting with the hydrated nickel(II) ion and using H_2S (Table 2), a diamagnetic complex can be obtained $[(p_3)Ni(\mu-S)Ni(p_3)](BPh_4)_2$ [37] (<u>4</u>). In this dinuclear complex, the nickel atoms are tetracoordinated in a distorted tetrahedral geometry, by three phosphorus atoms and one sulfur atom, which bridges collinearly the two nickel atoms. Both the diamagnetism of this complex and its short Ni-S distance (2.034 Å) can be explained by a dπ-pπ-dπ overlap of two filled p orbitals of bridging sulfur with the appropriate empty d orbitals of each nickel atom. This is the only known case of a tetrahedral complex of nickel(II) that is diamagnetic (see also Sec. IV.B).

If cobalt(II) salts with potentially bidentate anions such as $Co(CH_3COO)_2$, $Co(NO_3)_2$, and $Co(CH_3COCHCOCH_3)_2$ are used, monomeric complexes result with stoichiometry $[CoY(p_3)]BPh_4$ (<u>5</u>) and that are low-spin (μ_{eff} = 2.01-2.08 μ_B) [35a]. Here, the cobalt(II) is five-coordinate, bonding to the anions CH_3COO^-, NO_3^-, and $CH_3COCHCOCH_3^-$, which act as bidentate ligands.

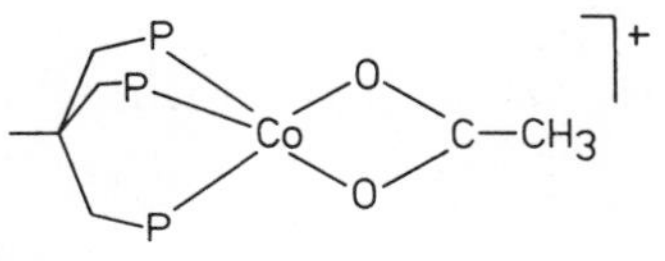

(<u>5</u>)

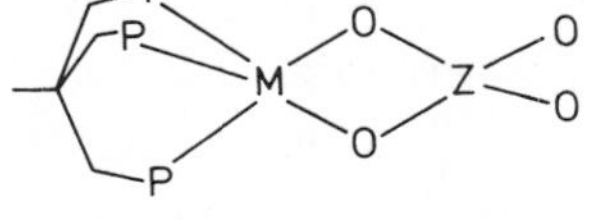

M = Co, Ni; Z = S, Se

(<u>6</u>)

Both the sulfate and selenate derivatives of Ni(II) and Co(II) of stoichiometry $[M(SO_4)(p_3)]$ and $[M(SeO_4)(p_3)]$ (M = Co, Ni) [38] are also five-coordinate, with the anions SO_4^{2-} and SeO_4^{2-} acting as bidentate ligands (6). The coordination geometry of these isomorphous five-coordinate complexes can be considered to be a highly distorted square pyramid. The cobalt complexes are low-spin (μ_{eff} = 2.03-2.01 μ_B), whereas those of nickel are high-spin (3.01-3.12 μ_B). The different spin state of the cobalt and nickel complexes can be correlated with the greater M-P$_{apical}$ to M-P$_{basal}$ bond ratio in cobalt with respect to nickel complexes. It has been generally observed that in square pyramidal 3d metal complexes, the elongation of the apical bond favors the low-spin state [39].

With the ligand Meas$_3$ (IV), the nickel(II) halides form complexes $[NiX_2(Meas_3)_2]$ that are paramagnetic (Table 1) and to which an octahedral geometry has been assigned containing one noncoordinated arsenic atom from each ligand [11]. In solution, the six-coordinate species are in equilibrium with either five-coordinate species (X = Br, I) or planar four-coordinate species (X = Cl). Complexes with weakly coordinating anions, $[Ni(Meas_3)_2]Y_2$ (Y = ClO_4, BF_4, BPh_4), are diamagnetic and presumably have a square pyramidal geometry containing one noncoordinated arsenic atom. The only complex with a metal-to-ligand ratio of 1:1, $[Ni(NCS)_2(Meas_3)]$, is square planar.

1. Hydrido Complexes

Whereas by the action of $NaBH_4$ on the anhydrous halides of cobalt(II) and the ligand p_3, cobalt(I) complexes are formed (Sec. II.B), if the hydrated cobalt(II) ion is used, in the presence of noncoordinating anions, a wide variety of products can be formed, according to the conditions of the reaction (Table 3).

When the reaction is carried out at room temperature in tetrahydrofuran as solvent, a borohydride derivative of cobalt(I) results (Sec. II.B); in contrast, in diethyl ether as solvent, at low temperatures a dinuclear hydrido complex of cobalt(II) is obtained with the formula $[(p_3)Co(\mu-H)_3Co(p_3)]BPh_4$ [40] (7). Analogous complexes are obtained with the ligands as$_3$, and for iron(II) with the ligand p_3

Table 3 Reduction and Hydrogenation Reactions Assisted by the Ligand p_3

$$2Co(H_2O)_6^{2+} + 2p_3 + 3BH_4^- \xrightarrow[\text{ether}]{0°C} [(p_3)CoH_3Co(p_3)]^+ + \frac{3}{2}B_2H_6{}^a + H_2O$$

$$Co(H_2O)_6^{2+} + p_3 + 2BH_4^- \xrightarrow[\text{THF}]{\text{room temp.}} [Co(BH_4)(p_3)] + \frac{1}{2}B_2H_6{}^a + \frac{1}{2}H_2{}^a + H_2O$$

$$2MI_2 + p_3(\text{excess}) \xrightarrow{\text{EtOH}} 2[MI(p_3)] + I_2p_3{}^a \quad (M = Co, Ni)$$

$$MX_2 + p_3 + BH_4^- \xrightarrow{} [MX(p_3)] + X^- + \frac{1}{2}H_2{}^a + \frac{1}{2}B_2H_6{}^a$$

$$(M = Co, Ni; X = Cl, Br)$$

[a] Species that are supposed to be formed in the reactions.

(diamagnetic complex). The magnetism of the cobalt complexes is consistent with two unpaired electrons distributed in each dimeric entity.

D = P, M = Fe, Co; D = As, M = Co

(7)

The structures of the iron(II) complexes with p_3, as well as of the
cobalt(II) complexes with p_3 and as$_3$, are similar and can be thought
of as containing two octahedra, each sharing a common face with three
hydrogen atoms at the corners.

The existence or not of a significant direct metal-metal bond in
binuclear complexes having a structure in which a common face is
shared is a problem that is not yet resolved [41]. Neither on the
basis of the 18-electron rule nor from the metal-metal distance is the
position clear about the metal-metal interaction. For the hydrido
complexes of iron and cobalt one obtains a valence electron number
(VEN) of 15 or 16, respectively. In order to make up the VEN to 18,
one must visualize a triple Fe-Fe bond or a double Co-Co bond in these
complexes. The metal-metal distances in these two complexes are indeed unusually short (2.33 and 2.37 Å), thus suggesting a marked metal-
metal interaction, but are more similar than one would expect for such
a difference in bond order.

The spin state differences and the bonding in these hydrido complexes can perhaps best be rationalized on the basis of a molecular orbital scheme of the type devised by Hoffmann for analogous sandwich compounds [42], without arguing about the extent of the metal-metal interaction.

B. Complexes with Metals in Low Oxidation States

The first metal complexes with the ligand p_3 were those obtained by Chatt and co-workers [1] with nickel(0) and palladium(0), $[M(p_3)_2]$, by the borohydride reduction of $Ni(H_2O)_6(NO_3)_2$ and K_2PdCl_4, respectively. Two easily interconvertible isomers have been described of the diamagnetic nickel complex, but the structures are not known with certainty.

The ligand p_3 itself can reduce the iodides of cobalt(II) and nickel(II), yielding compounds containing monopositive metal ions of formula $[MI(p_3)]$ (M = Co, Ni) [43]. It is recognized that the ligand p_3 acts by accepting iodine atoms, forming iododerivatives of pentavalent phosphorus (Table 3). Likewise, one can obtain the chloro and bromo derivatives of copper(I). To obtain the corresponding chloro and bromo complexes of cobalt(I) and nickel(I), it is necessary to use a specific reducing agent such as $NaBH_4$. In this way one can obtain the complexes of nickel(I) with the ligand as_3 $[NiX(as_3)]$ [11]. A methylthio complex $[Co(SCH_3)(p_3)]$ has been obtained by the reduction with $NaBH_4$ of the dimeric $[Co_2(\mu\text{-}SCH_3)_2(p_3)_2](BPh_4)_2$ [36] (Sec. II.A). All of these metal(I) complexes are isomorphous and have a distorted tetrahedral structure like that of $[NiI(p_3)]$ [44], which has been determined by X-ray analysis [Fig. 1, (8)].

Using instead a hydrated salt of cobalt(II), $Co(H_2O)_6(BF_4)_2$, the borohydride derivative $[Co(BH_4)(p_3)]$ [45] is obtained. In this complex, whose structure was determined by X-ray analysis, the cobalt(I) must be considered to be tetrahedrally coordinated by the three phosphorus atoms, with the BH_4^- anion acting as the fourth ligand [Fig. 1, (9)]. This coordination geometry accounts for the existence of two unpaired electrons as well as the close similarity of the electronic

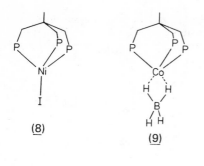

(8)   (9)

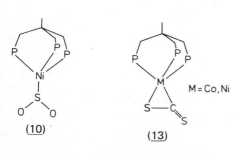

(10)   (13)   M = Co, Ni

Figure 1   Drawings of the X-ray structures of some isomorphous
complexes of cobalt and nickel with the ligand p_3.

spectrum of this complex with that of tetrahedral complexes of
cobalt(I) with p_3, such as $[CoCl(p_3)]$ [43]. An analogous diamagnetic
complex of copper(I) has also been obtained.

The large tetrahedral distortion evident in all these complexes
can be attributed to the rather rigid steric requirements of the ligand p_3.

1.   Complexes with π-Acceptor Coligands and Organometallic Complexes

All the previously mentioned halo complexes of cobalt(I) and
nickel(I) are isomorphous with the analogous diamagnetic and isoelectronic complexes $[Co(NO)(p_3)]$, $[Ni(SO_2)(p_3)]$ [45] [Fig. 1, (10)], and
$[Ni(CO)(p_3)]$ [46], which are formed by reacting $Ni(H_2O)_6(BF_4)_2$ or
$Co(H_2O)_6(BF_4)_2$ with NO, SO_2, or CO under normal conditions of temperature and pressure in the presence of p_3 (Table 4). These reactions
demonstrate the ability of the ligand p_3 to favor the formation of

Table 4   Some Reactions of Neutral Molecules with Aquacations of Nickel(II) and Cobalt(II) in the Presence of the Ligand p_3

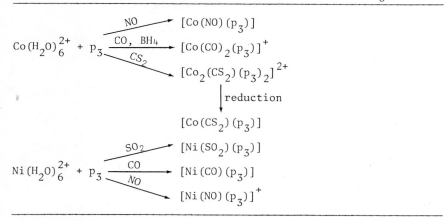

bonds between metals in low oxidation states and "soft" ligands. During the formation of these complexes containing metals in low oxidation state, the potential reducing agents are the molecules NO, CO, and SO_2 and, more probably, the ligand p_3, which is ultimately responsible for stabilizing the low oxidation state of the metal by virtue of its π-acid properties originating in the phosphorus atoms.

It is worth noting that the atoms of nickel, sulfur, and oxygen in complex (10) are all coplanar. The M-S distance (2.013 Å) is smaller than the sum of normal covalent radii of the atoms [47]. Since the forementioned distorted tetrahedral complexes of nickel(0) and cobalt(0) are isomorphous not only with those of nickel(I) and cobalt(I) [(8) and (9)] but also with the pentacoordinate complexes of nickel(II) and cobalt(II) $[M(XO_4)(p_3)]$ (6), it seems clear that their structure and geometry are determined essentially by the steric requirements of the six phenyl groups attached to the ligands p_3. With this in mind, it is interesting to observe that in these complexes the phenyl groups of the two adjacent diphenylphosphine groups are arranged in parallel pairs.

Complexes of formula [Ni(NO)triphosphine]Y (triphosphine = p_3, Etp_3 (III); Y = Cl, Br, I, BF_4, BPh_4) have been prepared by reducing the appropriate nickel(II) salts with KNO_2 in the presence of the triphosphine [12]. They are essentially diamagnetic, being isoelectronic with $[Ni(CO)(p_3)]$, and have tetrahedral geometries containing a substantially linear Ni-N-O bond, as exemplified in the structure of $[Ni(NO)(Etp_3)]BF_4$ [48] (11) which has been determined by X-ray analysis.

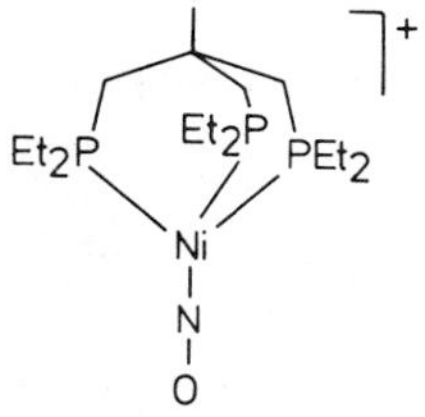

(11)

The complexes obtained from the reaction of carbon disulfide are interesting (Table 4). For instance, from the reaction of $Co(H_2O)_6(BF_4)_2$ with CS_2 and p_3, the binuclear diamagnetic cobalt(I) complex $[(p_3)Co(\mu-CS_2)Co(p_3)](BPh_4)_2$ (12) is obtained [49]. In this complex a bridging CS_2 group holds the two $Co(p_3)$ fragments together; one $Co(p_3)$ is π bonded to one C=S group, whereas the other is bonded by two conventional σ bonds from the two sulfur atoms. When the

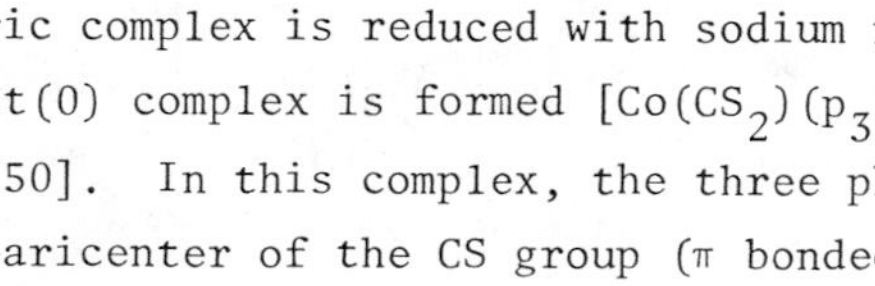

(12)

dimeric complex is reduced with sodium naphthalenide, a monomeric cobalt(0) complex is formed $[Co(CS_2)(p_3)]$ [Fig. 1, (13)] (μ_{eff} = 1.9 μ_B) [50]. In this complex, the three phosphorus atoms together with the baricenter of the CS group (π bonded to the cobalt) form a distorted tetrahedral geometry around the cobalt atom.

The analogous complex of nickel, $[Ni(CS_2)(p_3)]$ [45], has been prepared from CS_2 and $[Ni(p_3)_2]$ and is itself isomorphous with all the other monomeric complexes containing the p_3 ligand. Paramagnetic carbonyl complexes of cobalt(II) containing anionic coligands, $[CoX(CO)(p_3)]BPh_4$ (X = Cl, Br, SCH_3) have been obtained starting from the binuclear compounds $[Co_2(\mu-X)_2(p_3)_2]^{2+}$ and CO (Sec. II.A).

Most of the complexes containing the ligand p_3, as well as other ligands of high π-bonding ability, have been obtained from organometallic complexes containing CO groups, or even from pure metal carbonyls. In each case the starting material always contained the metal in a low oxidation state. In the first case, the ligand p_3 replaces an unsaturated organic group [51], whereas in the second it tends to replace three carbonyl groups [52,53]:

$$[Fe(CO)_3(C_8H_8)] + p_3 \longrightarrow [Fe(CO)_3(p_3)] + C_8H_8$$

$$[Fe(CO)_3(C_8H_8)] + p_3 \longrightarrow [Fe(CO)_2(p_3)] + C_8H_8 + CO$$

$$[Co_2(CO)_4(nor\text{-}C_7H_8)_2] + 2p_3 \longrightarrow [Co_2(CO)_4(p_3)_2] + 2nor\text{-}C_7H_8{}^*$$

$$[Ni(CO)_4] + p_3 \longrightarrow [Ni(CO)(p_3)] + 3CO$$

The two complexes of Fe(0) are assumed to be five-coordinate, the former containing a bidentate p_3 group and the latter containing a tridentate ligand (14). This ligand is also presumed to be bidentate in the dinuclear complex of cobalt(0) (15).

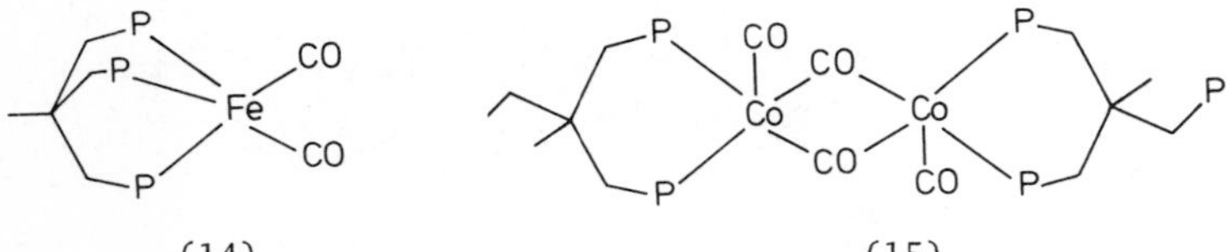

(14)   (15)

Two complexes of cobalt(I) $[Co(CO)_2L]BPh_4$ (L = p_3, as_3) [40b] have also been reported; these are diamagnetic, isoelectronic, and presumably isostructural with the analog of iron(0). However, they were prepared by the reaction of CO on the hydrido complex of cobalt(II) $[Co_2(\mu-H)_3L_2]BPh_4$.

$^*nor\text{-}C_7H_8$ = norbornadiene.

With the methyl-substituted arsine ligand $Meas_3$ (IV), two complexes, one of manganese(I), $[Mn(CO)_3(Meas_3)]ClO_4$, and one of chromium(0), $[Cr(CO)_3(Meas_3)]$, have been reported from which the two low-spin chromium(II) complexes $[CrI(CO)_n(Meas_3)]BPh_4$ (n = 2, 3) have been obtained by oxydation with I_2 [54].

A direct Cu-Mn bond has been reported in the mixed complex $[(Meas_3)Cu-Mn(CO)_5]$, which is diamagnetic [55].

Complexes containing p_3 or as_3 together with fluoro derivatives as coligands have been obtained from the reaction of $[Ni(cdt)]$ (cdt = cyclododeca-1,5,9-triene) with either p_3 or as_3 and various fluoro-carbons (C_2F_4, $CF_2=CFH$, $CF_2=CF-CF_3$) [56]. The same complexes have also been prepared starting from $[Ni(p_3)_2]$ (16, 17).

D = P, As

(16)   (17)

Two isoelectronic complexes, $[Fe(\eta^5-C_5H_5)(p_3)]^+$ and $[Co(\eta^5-C_5H_5)(p_3)]^{2+}$ [57], have been obtained by replacing the carbonyl groups in the complex $[MX_n(\eta^5-C_5H_5)(CO)_2]$ (M = Co, n = 2; M = Fe, n = 1; X = Cl, Br, I) with p_3. The iron complex can be compared with the complex $[Fe(\eta^5-C_5H_5)(CO)_3]^+$ [58], where the three strongly π-accepting carbonyl groups are in the place of the p_3 ligand, which is a relatively weak π acceptor. The corresponding complex with three monodentate phosphine ligands cannot be prepared, demonstrating that the tripod-like structure of the ligand p_3 plays an important role in the formation and stabilization of these complexes.

Cationic cobalt(I) complexes containing both the ligand p_3 and a variety of conjugated polyolefins and acetylenes, $[Co(L)(p_3)]Y$ (L = 1,3-butadiene, isoprene, 1,3-cyclohexadiene, 1,3,5-cycloheptatriene, 1,3,5,7-cyclooctatetraene, phenylacetylene, diphenylacetylene; Y = ClO_4, BPh_4), have been obtained by the reaction of $Co(ClO_4)_2$ with olefins in the presence of p_3 [59]. All the complexes are diamagnetic. The structure of $[Co(\eta^4-C_7H_8)(p_3)]ClO_4$ shows the cobalt atom to be coordinated by three phosphorus atoms of p_3 and by the butadiene fragment of cycloheptatriene molecule acting as a tetra-hapto ligand (18).

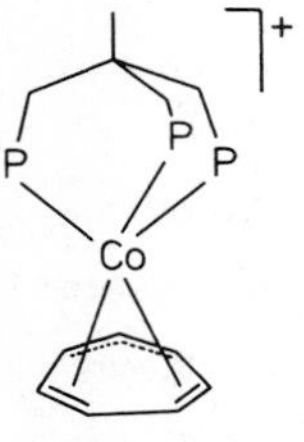

(18)

This compound parallels the isoelectronic one $[Fe(\eta^4-C_7H_8)(CO)_3]$ [60], one p_3 ligand acting like three CO molecules.

By partial substitution of the carbonyl groups in the complex $[Mn(CO)_2(C_5H_5)NO](PF_6)$ with p_3, a dimeric complex (19) is formed, which, on the basis of infrared evidence, appears to contain the p_3 ligand bridging the two manganese atoms [61]. A bridging p_3 group,

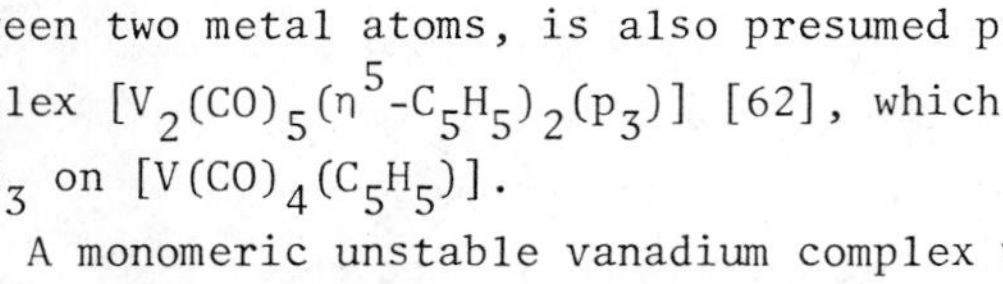

(19)

between two metal atoms, is also presumed present in the vanadium complex $[V_2(CO)_5(\eta^5-C_5H_5)_2(p_3)]$ [62], which is obtained by the action of p_3 on $[V(CO)_4(C_5H_5)]$.

A monomeric unstable vanadium complex with p_3 is also known, $[VI(CO)_3(p_3)]$ [63a], which is probably five-coordinate. Anionic carbonylvanadates, $[Et_4N][V(CO)_n(p_3)]$ (n = 3, 4), have been obtained by irradiation of $[Et_4N][V(CO)_8]$ in the presence of the ligand p_3 [63b].

Finally, hydrido complexes of cobalt containing CO or PPh_3 as coligand, $[CoH(L)(p_3)]$, have been prepared by the action of $NaBH_4$

and L on $[CoCl_2(p_3)]$ [64a]. Recently a hydridocarbonylvanadium
complex, $[VH(CO)_3(p_3)]$, has also been obtained [64b].

III. TETRADENTATE LIGANDS

A. Ligands in Which the Donor Atoms Are Linked by o-Phenylene Groups

1. *Complexes with the Ligands QP, QAS, Qas*

Having first synthesized the ligand QP (V), Venanzi and co-
workers have described most of the complexes so far known containing
this ligand and bi- and tripositive 3d metal ions, viz., chromium
[65,66], iron [67], cobalt [68], and nickel [69a]. The majority of
the metal(II) complexes have the general formula [MX(QP)]Y (X =
halogen or pseudohalogen; Y = halogen, pseudohalogen, BPh_4, ClO_4;
Table 5). The complexes of iron, cobalt, and nickel are low-spin;
those of chromium are high-spin.

An essentially trigonal bipyramidal coordination geometry has
been assigned to these complexes, principally on account of spectral
evidence,* and this has been confirmed by X-ray analysis of the complex $[CoCl(QP)]BPh_4$ [70] (20), in which the coordination polyhedron
is significantly distorted from a trigonal symmetry. This distortion

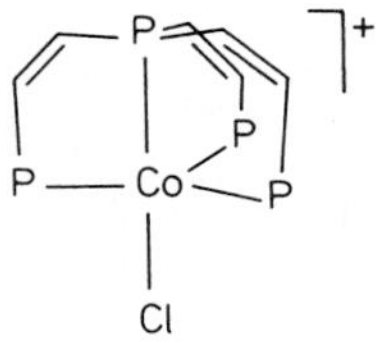

(20)

has been attributed to the Jahn-Teller effect, which removes the
degeneracy of the $d_{x^2-y^2}$ and d_{xy} orbitals in a low-spin d^7 ion with
C_{3v} symmetry (Fig. 2). The spin state of the complexes can be

---

*See Sec. III.C for a discussion of the spectral properties of low-
spin complexes of iron(II), cobalt(II), and nickel(II).

Table 5 Complexes of bi- and tri-Positive Metal Ions with QP, QAS, and Qas Ligands

Complexes	X, Y	μ_{eff}, μ_B	Coordination Geometry[a]	References
[CrX(QP)]BPh$_4$	Cl, Br, I	4.82-4.89	TBPy	65
[CrX$_3$(QP)]	Cl, Br	3.89-3.93	Oh	66
[CrI$_2$(QP)]I$_3$		3.8	Oh	71
[FeX(QP)]BPh$_4$	Cl, Br, I	3.0-3.2	TBPy	67
[FeX$_2$(QP)]	NCS, CN, NO$_2$	0	Oh	67
[CoX(QP)]Y	X = Cl,[b] Br, I, NO$_3$, NCS Y = I, BPh$_4$, ClO$_4$, $\frac{1}{2}$CoX$_4^{2-}$	1.99-2.09	TBPy	68, 70
[CoX$_2$(QP)]BPh$_4$	Cl, Br	0	Oh	68
[NiX(QP)]BPh$_4$	Cl, NO$_3$, ClO$_4$	0	TBPy	69a
[CoCl(QAS)]$_2$CoCl$_4$		2.2-2.5[c]	TBPy	72
[CoCl(QAS)]InCl$_4$		2.2	TBPy	72
[NiX(QAS)]Y	X = Cl, Br, I, NCS, CN, NO$_3$, ClO$_4$ Y = Br, I, ClO$_4$, BPh$_4$	0	TBPy	69a
[NiX(Qas)]Y	X = Cl, Br, I, NCS, NO$_2$, N$_3$, ac Y = I, NCS, NO$_2$, BF$_4$, BPh$_4$	0	TBPy	14

[a]See Table 1.
[b]Structure determined by X-ray analysis.
[c]Average value per metal atom.

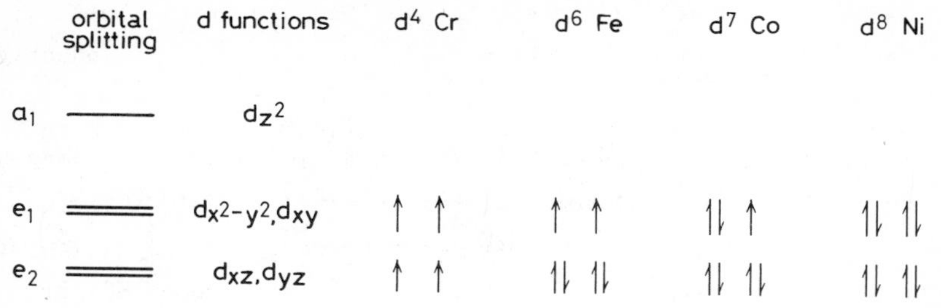

orbital splitting	d functions	d^4 Cr	d^6 Fe	d^7 Co	d^8 Ni
a_1	d_{z^2}				
e_1	$d_{x^2-y^2}, d_{xy}$	↑ ↑	↑ ↑	↑↓ ↑	↑↓ ↑↓
e_2	d_{xz}, d_{yz}	↑ ↑	↑↓ ↑↓	↑↓ ↑↓	↑↓ ↑↓

Figure 2 Schematic crystal field splitting and orbital occupancy in the trigonal bipyramidal complexes $[MX(QP)]^+$.

explained on the basis of the d-orbital splitting shown in Fig. 2 for donor atoms producing a large ligand field.

Although the tripod-like geometry of the ligand QP is the chief factor deciding both the coordination geometry and number, the metal can also play an important role. For instance, with d^6 metals, the ligand QP can arrange itself so that the donor atoms occupy four apices of an octahedron, forming six-coordinate diamagnetic complexes such as those of iron(II), $[FeX_2(QP)]$ (X = CN, NCS) [67] and cobalt-(III), $[CoX_2(QP)]^+$ (X = Cl, Br) [68].

QP does not always act as a tetradentate ligand. For instance, with chromium(III), although in the complex $[CrI_2(QP)]I_3$ [71] it is tetralegate, in the complexes $[CrX_3(QP)]$ (X = Cl, Br) it is assumed to be trilegate [66].

The ligand QAS (VI) has poorer coordinating power than QP, at least with respect to 3d transition metals. In fact, the only complexes of QAS with bipositive 3d metal ions are those of nickel [69a] and cobalt [72] (Table 5), the latter being relatively unstable. The complexes $[NiX(QAS)]Y$ are diamagnetic and have similar trigonal bipyramidal structures to complexes with QP. The only two complexes with cobalt, $[CoCl(QAS)][InCl_4]$ and $[CoCl(QAS)]_2[CoCl_4]$, are assumed to contain the low-spin trigonal bipyramidal species $[CoCl(QAS)]^+$.

The ligand Qas (VII), which differs from QAS by having $-AsMe_2$ as donor instead of $-AsPh_2$, forms with nickel(II) exactly analogous complexes to those above, with identical compositions and stereochemistry [14]. Nevertheless, a greater donor capacity is displayed in

these complexes compared to those with QAS, because of the less
serious steric interactions of the methyl groups as well as inductive
effects.

Carbonyl Complexes of Metals in Low Oxidation States. Diamagnetic, trigonal bipyramidal complexes of cobalt(I), [Co(CO)Q]Y [Q =
QP, QAS, Y = Cl, BPh_4, $Co(CO)_4$], have been prepared by the action of
the tetradentate ligand Q on $[Co_2(CO)_8]$ in the presence of HCl or
$NaBPh_4$ [68]. The ease with which the cobalt(I) complexes can be prepared with the ligand QAS, together with their remarkable stabilities,
both in the solid state and in solution, must be correlated with their
d^8 configuration, which, when forming pentacoordinate complexes, results in an 18-electron system. Likewise, nickel(II), which is also
d^8, readily forms pentacoordinate complexes with QAS. In the case of
the complexes of QP with metals in oxidation states lower than +2,
the tendency to attain a stable electronic configuration appears to
account for the differing bonding modes of this ligand. For example,
in the isoelectronic and isostructural complexes with chromium(0) and
manganese(I), QP is bilegate in $[MnX(CO)_3(QP)]$ [73], trilegate in
$[Cr(CO)_3(QP)]$ [71] and $[Mn(CO)_3(QP)]Y$ [73], and tetralegate in
$[Cr(CO)_2(QP)]$ [71] and $[Mn(CO)_2(QP)]Y$ (Y = Cl, Br, I) [73]. These
d^6 complexes are six-coordinate and diamagnetic. Upon oxidation with
mild halogenating agents ($SbCl_5$, I_2, Br_2), the chromium(0) complexes
give chromium(I) compounds, $[Cr(CO)_2(QP)]Y$ (Y = BPh_4, $SbCl_6$), which
can be further oxidized to Cr(III) complexes [71].

The large number of d^6, diamagnetic, octahedral complexes with
differing metals [chromium(0), manganese(I), iron(II), and cobalt-
(III)] points to the particular stability of the low-spin d^6 configuration in an octahedral field, or, from another point of view, to the
attainment of a VEN of 18.

2. Complexes of Mixed Ligands Containing
N, P, As, Sb, Bi, S, Se Donors

A number of mixed ligands containing either P or As and other
group V or VI donors (Sec. I.B, VIII-XIV [15-18]; XV-XVII [19-21])
have been synthesized, and their coordination behavior toward

nickel(II) and cobalt(II) has been studied. These ligands have the same tripod-like skeleton containing o-phenylene linkages and are potentially tetradentate. The majority of the known complexes have the formula [MXL]Y (X = halide, pseudohalide; Y = uninegative anion of poor coordinating ability such as BPh_4^- or ClO_4^-).

All the nickel complexes [15-21,74] are diamagnetic, pentacoordinate, with an essentially trigonal bipyramidal geometry (with the exception of the complexes with the ligand Bitas, XIV), which has been assigned on the basis of their electronic spectra. The only structures that have been determined by X-ray analysis are those of [NiCl(Pts)]ClO_4 [75] (21) and [NiBr(As$_3$S)]ClO_4 [21] (22); the first has an almost regular trigonal bipyramidal structure, whereas the second is more distorted because of the different equatorial donor atoms. In both complexes the nickel atom is outside the equatorial plane, displaced toward the axial halogen atom. The short Ni-L$_{apical}$ distances (Ni-P, 2.113 Å; Ni-As, 2.218 Å) are significant in both complexes. These short bond distances have been attributed both to

(21)   (22)

steric factors inherent the rigid o-phenylene bridges as well as to electronic factors inherent d^7 or d^8 low-spin configurations containing an empty d_{z^2} orbital in C_{3v} symmetry.

By comparison, the complexes [NiX(Bitas)]BPh_4 (X = halide) seem to possess a square pyramidal coordination geometry, on the basis of spectral evidence, with a planar BiAs$_2$X donor set [18]. On the basis of infrared evidence, the structure (23) has been proposed for the dimeric complex [Ni$_2$(Bitas)$_3$](ClO_4)$_4$.

A square pyramidal geometry has also been assigned to the two complexes [Ni(Sbtas)$_2$]Y$_2$ (Y = ClO_4, BPh_4) [17]. Finally, nickel(II)

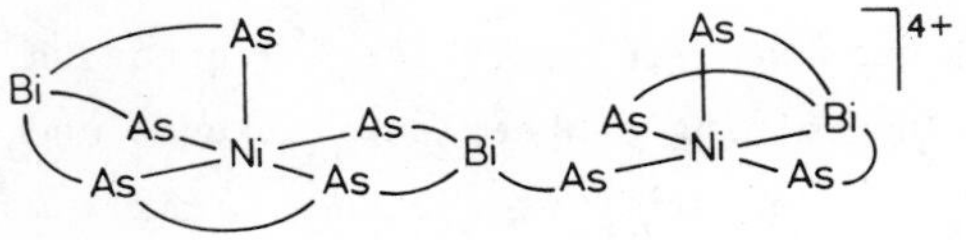

(23)

and cobalt(II) complexes have been obtained with the ligand Ptn
(XII, donor set PN_3) of formula $[MX_2(Ptn)]$. These are tetracoordinate and tetrahedral, with the ligand bound through the apical phos-
phorous atom and only one peripheral nitrogen atom [16b].

The complexes of cobalt with ASTP (VIII), SBTP (IX), and PTAS
(X) are low-spin, and are presumed to be analogous with those of the
tetraphosphine QP [15].

B. Ligands in Which the Donor Atoms Are Linked by Trimethylene Chains

In the ligands qas (XVIII), tap (XIX), and tasb (XX)[22-24], the
terminal $-AsMe_2$ groups are joined to the apical atom (As, P, Sb) by
trimethylene chains. The coordination behavior of these ligands has
been studied for nickel(II) [23,24,76,77] and cobalt(III) [78]. The
nickel complexes have the formula [NiXL]Y (X = halide, pseudohalide,
or other uninegative coordinating anion; Y = BPh_4, ClO_4); they are
diamagnetic, with five-coordinate, essentially trigonal bipyramidal
structures. This assignment is supported by their electronic spectra
and confirmed in the case of the complex [Ni(CN)(tap)]ClO_4 [79] (24),
where the structure has been determined by X-ray analysis and found
to be almost regular trigonal bipyramidal. The nickel atom is

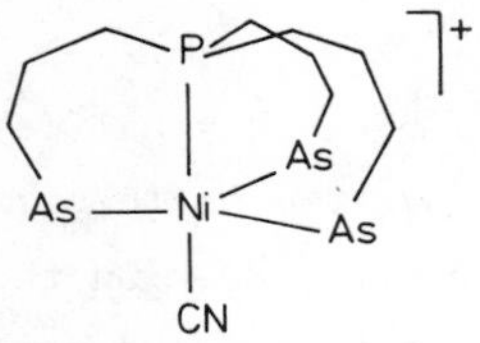

(24)

displaced toward the apical atom of the ligand [P-Ni-As, 94.7° (av);
As-Ni-C, 85.3° (av)], in contrast to low-spin trigonal bipyramidal

complexes of nickel(II) and cobalt(II) with tripod-like ligands containing o-phenylene bridging groups (Sec. III.A.1). This, together with the longer $Ni-P_{apical}$ distance [2.206 Å in (24); 2.113 Å in (21)], is a consequence of the longer trimethylene bridges and their greater flexibility.

Complexes of the type $[NiX(tap)]^+$ are labile towards X coligand substitution in the presence of an excess of a wide range of ligands, including CN^-, NCS^-, $SC_2H_5^-$, NO_2^-, H_2O, OH^-. Some of these complexes have been also isolated in the solid state.

Diamagnetic complexes, whether of cobalt(III), $[CoX_2L]^+$ (L = qas, tap, tasb; X = halide, pseudohalide) [78], or of nickel(II), $[NiX_2(tap)]$, are assumed to be six-coordinate [23].

Other potentially tetradentate ligands containing both olefine and tertiary phosphine groups as donors have been synthesized [80].

$$As \Big\langle \begin{array}{l} (CH_2CH_2CH_2AsMe_2)_{3-n} \\ \\ (CH_2CH_2CH=CH_2)_n \end{array}$$

n = 1, tasol

n = 2, dasdol

n = 3, astol

For the five-coordinate complexes $[NiX(tasol)]Y$ (X = Cl, Br; Y = Cl, Br, ClO_4) only, the infrared spectra demonstrate the nickel-olefin bond. In the other complexes, $[NiI_2(tasol)]$, square planar, and $[NiX(dasdol)_2]Y$, five-coordinate, the ligands are thought to be bound through tertiary arsine only. The ligand astol does not appear to form any complexes with nickel(II).

C. Spectral Properties of Low-Spin, Trigonal Bipyramidal Complexes of Iron(II), Cobalt(II), and Nickel(II)

The main feature of the electronic spectra of low-spin, trigonal bipyramidal complexes of Ni(II), $[NiXL]^+$ (L = tetradentate ligands containing either o-phenylene [14,15,17,19,20,69b] or trimethylene [23,24,76] bridges, where X = uninegative anion) is a very intense band at 1.5-2.0 μm^{-1} (ε_{max} = 2000-5000) accompanied, sometimes, by a less strong band or shoulder at 2.1-2.6 μm^{-1} (ε_{max} = 400-1000) (Figs. 3 and 4). Norgett et al. have considered these spectra in detail and

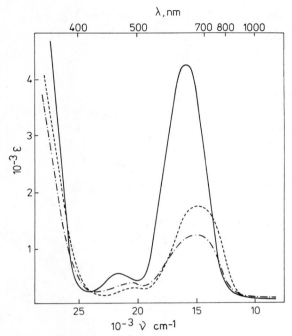

Figure 3 Absorption spectra of the complexes $[NiBr(QAS)]^+$ (——),
$[NiBr(Ptse)]^+$ (---), and $[NiBr(Pts)]^+$ (-·-·-) in dichloromethane
solutions. (Adapted from Ref. 20; reprinted with permission from
Inorg. Chem., *6*, 1967, p. 149. Copyright by the American Chemical
Society.)

and have interpreted them using the crystal field approach [81]. On
the basis of the energy-level diagram for the singlet terms in D_{3h}
symmetry (Fig. 5), the two observed bands have been assigned as follows:

$$(e'')^4(e')^4 \longrightarrow (e'')^4(e')^3 a_1' \qquad (1.5\text{--}1.8 \ \mu m^{-1})$$

and

$$(e'')^4(e')^4 \longrightarrow (e'')^3(e')^4 a_1' \qquad (2.1\text{--}2.6 \ \mu m^{-1})$$

(these transitions correspond to $^1A_1' \longrightarrow {}^1E'$ and $^1A_1' \longrightarrow {}^1E''$ in
weak field notation. In D_{3h} symmetry, only the first transition is

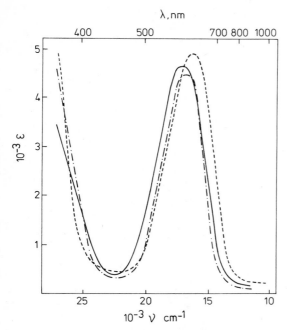

Figure 4   Absorption spectra of the complexes $[NiCl(QP)]^+$ (——),
$[NiCl(ASTP)]^+$ (---), and $[NiCl(SBTP)]^+$ (-•-•-) in dichloromethane
solutions. (Adapted from Ref. 15.)

allowed, and hence one expects that it would be more intense than the
second transition, which is allowed in C_{3v} symmetry.

The intensity of the band has been interpreted in terms of the
delocalization between the $d_{x^2-y^2}$ and d_{xy} orbitals of the metal and
the orbitals of like symmetry on the equatorial donor ligand atoms.
One also must take into consideration the mixing of the d-d transitions with transitions of the type $\pi \longrightarrow \pi^*$ arising from the benzene
rings of the ligands. In fact, the intensity of the bands is re-
duced as one goes from the complexes with QAS to those with Qas and
qas, that is, from ligands containing aromatic groups to those containing aliphatic groups [14]. The frequency of the main transition
is dependent on the particular variation of the donor atoms, both
equatorial and apical. In the former case (the series of complexes

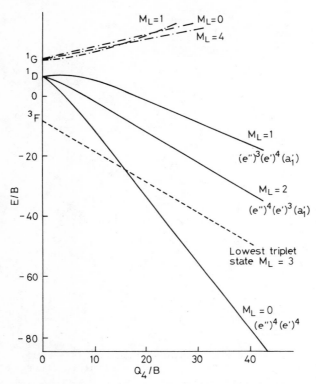

Figure 5   Energy-level diagram for trigonal bipyramidal low-spin nickel(II) complexes.   (Reproduced by permission of the Chemical Society, Ref. 81a.)

with QP, PTAS, Pts, Ptse), the frequency varies with the relative position of the donor atoms in the spectrochemical series [20].   In the latter, (the series of complexes with QP, SBTP, ASTP), however, one anomalous spectrochemical sequence, namely, $P > Sb > As$ [15], has been found.   This has been attributed to the compression of the metal atom by apical donor atom (L_{ap}-M-$L_{eq} < 90°$); this compression increases with increasing size of the donor atom and more than compensates for the decrease in ligand field strength from As to Sb. Complexes with ligands containing trimethylene bridges may or may not exhibit this anomaly, depending on the counterion [24].

The spectra of $[CoX(QP)]^+$ [68] and $[CoX(L)]^+$ [72] (L = QAS, ASTP, SBTP, PTAS) have bands at around 0.9, 1.6, and 2.0 μm^{-1} with a

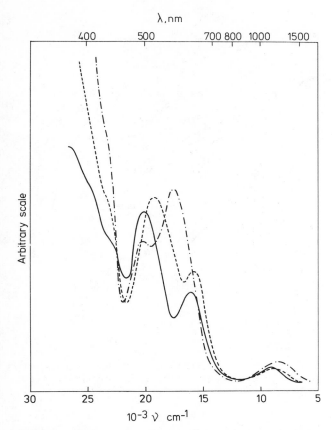

Figure 6   Electronic spectra of complexes $[CoCl(QP)]BPh_4$ (---),
$[CoCl(ASTP)]BPh_4$ (——), and $[CoCl(SBTP)]BPh_4$ (-·-·-) in a 2-methyl-
tetrahydrofuran/CH_2Cl_2 4:1 glass at ca. 100 K.   (Adapted from Ref.
72.)

shoulder at about 2.4 μm^{-1} (Fig. 6).   In view of the significant de-
parture from C_{3v} symmetry in these complexes, the spectra of low-spin
trigonal bipyramidal cobalt(II) complexes have been interpreted on
the basis of the diagram of Fig. 7 [72,82].

The spectrum of $[FeX(QP)]^+$ [67] is shown in Fig. 8.   On the
basis of the energy-level diagram shown in Fig. 9, the bands at ca.
0.9 μm^{-1}, 1.8 μm^{-1}, and the shoulder at ca. 2.5 μm^{-1} have been
assigned as follows [81a]:

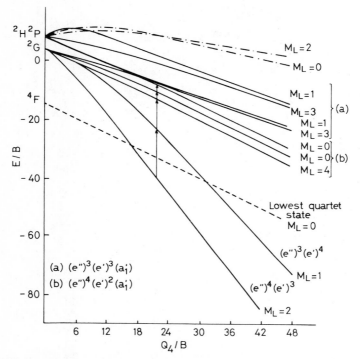

Figure 7 Energy-level diagram for trigonal bipyramidal, low-spin complexes of cobalt(II). (Reproduced by permission of the Chemical Society, Ref. 81a.)

$$(e'')^4(e')^2 \longrightarrow (e'')^3(e')^3$$

$$(e'')^4(e')^2 \longrightarrow (e'')^4(e')a_1'$$

$$(e'')^4(e')^2 \longrightarrow (e'')^3(e')^2a_1'$$

D. Ligands in Which the Donor Atoms Are Linked
    by Ethylene Chains

    1. *Complexes with the Ligand pp$_3$*

    The ligand pp$_3$ (XXIX) [30] contains the same donor set of atoms as the ligand QP, the essential difference being that the three PPh$_2$ groups are linked to the apical phosphorus atom by flexible etylene rather than rigid o-phenylene bridges.

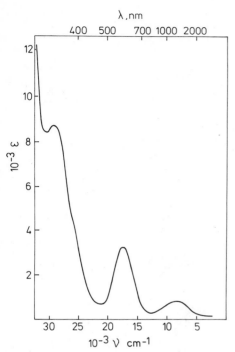

Figure 8 Absorption spectrum of [FeBr(QP)]$^+$ in dichloromethane solution. (Adapted from Ref. 81a by permission of the Chemical Society.)

The complexes of chromium(II) [83], iron(II) [30,84], cobalt(II), and nickel(II) [30] with this ligand have the general formula [MX(pp$_3$)]Y (Table 6). Excluding the complexes of chromium(II) that have μ_{eff} corresponding to four unpaired electrons, the complexes of all the other metals are low-spin, analogous to the corresponding complexes with QP.

The structure has been determined by X-ray analysis for only two complexes, [CoBr(pp$_3$)]PF$_6$ [85] (25) and [FeBr(pp$_3$)]BPh$_4$ [86] (26). The coordination geometry of the cobalt complex can be considered as a distorted square pyramid with a phosphorus atom in an apical position and the other three phosphorus atoms together with the bromine essentially forming the base; the cobalt atom is located 0.37 Å above the base. In the case of the iron(II) complex, the coordination

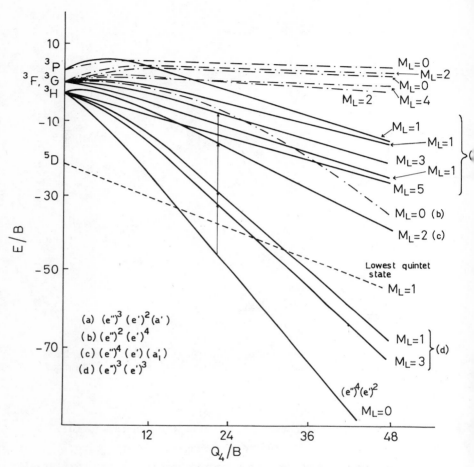

Figure 9   Energy-level diagram for trigonal bipyramidal, low-spin complexes of iron(II).   (Reproduced by permission of the Chemical Society, Ref. 81a.)

Table 6 Complexes of Bipositive Metal Ions with np_3, pp_3, nas_3, and $Menp_3$ Ligands

Complexes	X, Y, L	μ_{eff}, μ_B	Coordination Geometry or Number[a]	References
$[CrX(L)]BPh_4$	X = Cl, Br, I; L = np_3, pp_3	4.55-4.97	TBPy	94, 83
$[FeX(L)]Y$	L = np_3, X = Cl, Br,[b] I; Y = BPh_4, PF_6	5.19-5.29	TBPy-Td	86, 95
	L = pp_3, X = Cl, Br;[b] Y = BPh_4, PF_6	3.0	TBPy	30, 84, 86
$[Fe(NCS)_2(L)]$	pp_3, np_3	0	Oh	84, 95
$[CoX(np_3)]Y$	X = Cl,[b] Br,[b] I;[b] Y = Br, BPh_4, PF_6	4.26-4.36	TBPy-Td	25, 88, 91
$[CoX(np_3)]Y$	X = NCS; Y = BPh_4	2.00	SqPy	25, 88
	X = I;[b] Y = I, BF_4, PF_6	2.02	SqPy	25, 88, 92
$[CoX(pp_3)]PF_6$	Cl, Br[b]	1.7	SqPy	30, 84, 85
$[NiX(np_3)]Y$	X = Y = Cl, Br, I[b]	0	TBPy	25, 88, 89
	X = Cl,[b] Br, I; Y = BPh_4, PF_6	0	TBPy	25, 88, 90
$[NiCl(pp_3)]PF_6$		0	TBPy	30
$[NiX(nas_3)]BPh_4$	Cl, Br	0	TBPy	26
$[Co_3(Menp_3)_4](BF_4)_6 \cdot solv$	solv = $4C_3H_7NO$	1.96[c]	5	27
$[Ni_3(Menp_3)_4](BF_4)_6 \cdot solv$	solv = $4C_3H_7NO$	0	5	27
$[Ni_3(Menp_3)_4]Br_3(BPh_4)_3$		0	5	27

[a] See Table 1.
[b] Structure determined by X-ray analysis.
[c] Average value per metal atom.

geometry is essentially trigonal bipyramidal. In general, a five-
coordinate structure that is somewhere between the two extremes of
square pyramidal and trigonal bipyramidal can be assigned, on the
basis of spectral evidence, to the remaining complexes.

The structure of the cobalt complex, which is greatly distorted
from trigonal bipyramidal geometry despite the trigonal symmetry of
the ligand, appears to support the hypothesis that a Jahn-Teller
mechanism operates in d^7 trigonal bipyramidal low-spin complexes.
The greater distortion in the pp_3 cobalt complex compared to that in
the analogous complex with QP is attributable to the greater flexibility of the ligand containing the ethylene linkages.

A diamagnetic, six-coordinate complex of iron(II) is also known,
$[Fe(NCS)_2(pp_3)]$ [84].

Another ligand, pp_3-neo [32], has been reported, which differs
from pp_3 because it has neopentyl groups substituted at the peripheral phosphorus atoms. The only known complex of this ligand is
$[NiCl(pp_3-neo)]PF_6$, which is diamagnetic and square planar, with the
ligand coordinated by three of its donor atoms [87].

2. Complexes with the Ligands np_3, nas_3, $Menp_3$

Complexes with the ligands np_3, nas_3, and $Menp_3$ possess the
common characteristic of an apical nitrogen atom in place of the
phosphorus, arsenic, and antimony in the ligands already discussed.
The introduction of a "hard" atom, together with the flexibity of
the ethylene linkages, has the effect that the apical nitrogen atom
displays a smaller tendency to bind as compared with the phosphorus
or arsenic atoms. The most important reactions of these ligands are
with the halides or pseudohalides of 3d dipositive metals, possibly
in the presence of $NaBPh_4$. In this way, complexes of the general
formula $[MX(np_3)]Y$ (Table 6) can be obtained with the ligand np_3
(XXI).

All the nickel(II) complexes [25,88] are diamagnetic and have
an essentially trigonal bipyramidal geometry, as exemplified by the
X-ray structural analysis of the two complexes $[NiI(np_3)]I$ [89] (27)
and $[NiCl(np_3)]PF_6$ [90].

[structure diagram showing Ni complex]

(27)

[Co(NCS)(np$_3$)]BPh$_4$ is low-spin. The complexes of cobalt with the formula [CoX(np$_3$)]Y (X = Cl, Br) are high-spin, however, regardless of the nature of the counterion. Complexes of formula [CoI(np$_3$)]Y are high-spin when Y = BPh$_4$ but low-spin when Y = I, PF$_6$, BF$_4$ [25,88]. The high-spin complexes have a trigonal bipyramidal structure that is tetrahedrally distorted [91] (28) whereas the low-spin complexes are square pyramidal [92] (29). The [CoI(np$_3$)]$^+$ cation can thus exist in two isomers, which are magnetic as well as geometric in nature.

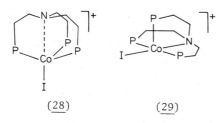

(28)   (29)

The fact that the five-coordinate low-spin complexes of cobalt with np$_3$ have a geometry substantially distorted from trigonal bipyramidal, regardless of the trigonal symmetry of the ligand, can be ascribed, in the same way as for analogous complexes with QP (Sec. III.A.1) and pp$_3$ (Sec. III.D.1), to the presence of the Jahn-Teller effect in a d^7, low-spin configuration with a ligand field of C$_{3v}$ symmetry and to the CFSE which, for the same configuration, is more favorable for a square pyramidal geometry than for a trigonal bipyramidal one [93]. The stability of the configuration e^{4}t$_2^3$ in a tetrahedral ligand field is, on the other hand, a probable cause of the tetrahedral distortion of the high-spin complexes. In the case of the complex cation [CoI(np$_3$)]$^+$, a small difference in the packing forces in the solid state due to the counterion is sufficient to make one geometry more stable than the other. The chromophore $CoNP_3I$ is therefore the crossover point between the low-spin and high-spin configurations. In solution, with polar solvents, the low-spin complexes $[CoI(np_3)]Y$ exhibit a temperature-dependent equilibrium between the low-spin and high-spin forms.

The high-spin complexes of chromium(II), $[CrX(np_3)]BPh_4$ [94], together with those with QP [65] and pp_3 [83], are among the few examples of five-coordinate complexes of this ion.

The complexes of iron(II) with np_3, $[FeX(np_3)]Y$ (Table 6), have a coordination geometry intermediate between tetrahedral and trigonal bipyramidal, and similar to that found in the corresponding high-spin complexes of cobalt(II) [95]. For example, in the complex $[FeBr(np_3)]PF_6$, where the structure has been determined by X-ray analysis [86], the apical nitrogen atom is found at a distance of 2.65 Å from the metal, thus displaying a significant tetrahedral distortion. This intermediate structure between four- and five-coordination explains the quintuplet state in these complexes of iron(II), where the $\sum n^0$ value of the NP_3X donor set [96] exceeds that of the trigonal bipyramidal complexes of iron(II), which show a spin equilibrium between a triplet and quintuplet state [97].

The diamagnetic, six-coordinate complexes of iron(II), $[Fe(NCS)_2(np_3)]$ [95], and cobalt(III), $[CoCl_2(np_3)]BPh_4$ (for which the structure has been determined by X-ray analysis [98]) are the only known examples in which the ligand np_3 is found in octahedral coordination.

From the previous examples, we can conclude that the ligand np_3, like QP and pp_3, has a favorable geometry with respect to the formation of five-coordinate trigonal bipyramidal complexes. Nonetheless, its considerable flexibility also allows it to form square pyramidal, tetrahedrally distorted trigonal bipyramidal or octahedral complexes according to the particular electronic configuration of the metal ion.

The tripod-like ligand nas_3 (XXII) is analogous to np_3, but with tertiary arsine groups as the peripheral donors. Its reactions with nickel halides are interesting. With $NiBr_2$ and NiI_2, complexes with the formula $[NiX(nas_3)]BPh_4$ [26] result, which are trigonal bipyramidal and diamagnetic. With $NiCl_2$, on the contrary, one obtains a five-coordinate compound containing a phenyl group bound to the metal [99] (Sec. IV.E). Cobalt(II) and other bipositive 3d metals do not give complexes with nas_3.

The ligand $Menp_3$ (XXIII), differing from np_3 by having methyl rather than phenyl terminal groups, invariably yields cobalt(II) and nickel(II) complexes of the formula $[M_3(Menp_3)_4](BF_4)_6$ (M = Co, Ni) and $[Ni_3(Menp_3)_4]Br_3(BPh_4)_3$ [27]. The nickel complexes are diamagnetic, whereas the cobalt complexes contain one unpaired electron. The peripheral $-PMe_2$ groups, having less steric hindrance than the $-PPh_2$ groups in np_3, permit the fifth axial donor of the five-coordinate complexes to be the PMe_2 group of another ligand. From this, a radial structure results (30) with the central $Menp_3$ ligand almost completely flattened.

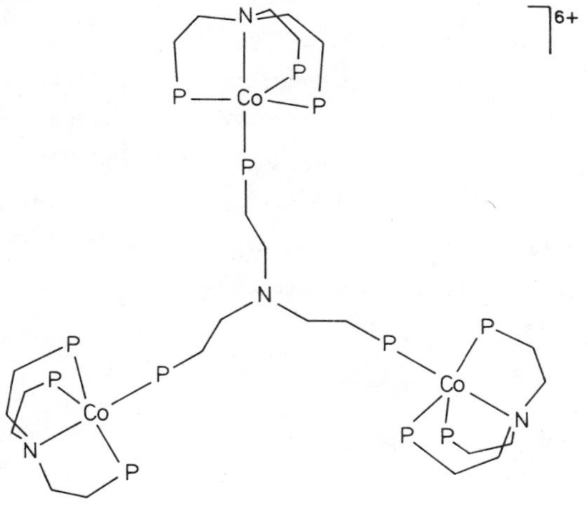

(30)

3. *Complexes with Mixed Ligands* n_3p, n_3as, n_2p_2, n_2as_2

These so-called hybrid ligands, which contain one or two atoms of nitrogen in equatorial positions as well as the one in the apical position, form complexes with nickel(II) and cobalt(II), normally with the formula [MX(ligand)]BPh$_4$ (Table 7), in a similar fashion to np$_3$ and nas$_3$ [29]. However, the properties of the complexes are often different from those of the corresponding complexes with np$_3$ and nas$_3$.

The complexes of cobalt, [CoX(ligand)]$^+$ (ligand = n$_3$p, n$_3$as, n$_2$p$_2$) are all high-spin, five-coordinate with trigonal bipyramidal structures that are distorted toward capped tetrahedra [100,101] (31).

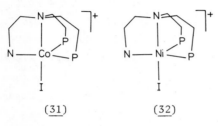

(31)   (32)

Typical spectra of these complexes are shown in Fig. 10. Using the known diagrams for the D$_{3h}$ group [102], the four main bands have been assigned, in order of increasing energy, to spin-allowed transitions from the ground state $^4A_2'$(F) to $^4E''$(F), $^4E'$(F), $^4A_2$(P), and $^4E''$(P). The larger distortion in n$_3$p complexes relative to that in np$_3$ complexes is shown by the uncoupling of the band at ca. 2.0 μm^{-1}, as a consequence of the splitting of the degenerate $^4E''$(P) level.

The dithiocyanate complexes, either of cobalt or nickel, [M(NCS)$_2$L], are again five-coordinate and high-spin, since the ligands n$_3$p and n$_3$as act in tridentate fashion with one -PPh$_2$ or -AsPh$_2$ group uncoordinated [103] (33).

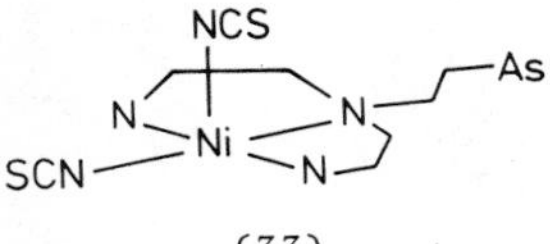

NCS
N―Ni―N―As
SCN―Ni―N

(33)

The nickel complexes [NiXL]BPh$_4$ can be square planar diamagnetic, as with n$_3$as [104], n$_2$as$_2$ or n$_3$p [105] (34), which are trilegate, or five-coordinate high-spin as with n$_3$p or low-spin as with n$_2$p$_2$ [100] (32). The behavior of the ligand n$_3$p is interesting, since it can

Table 7 Complexes of Bipositive Metal Ions with the Ligands n_3p, n_3as, n_2p_2, and n_2as_2

Complexes	X	μ_{eff}, μ_B	Coordination Geometry or Number[a]	References
$[CoX(n_3p)]BPh_4$	Cl,[b] Br, I	4.50-4.56	TBPy-Td	29, 101
$[Co(NCS)_2(n_3p)]$		4.54	5	29
$[CoX(n_2p_2)]BPh_4$	Cl, Br, I,[b] NCS	4.35-4.39	TBPy-Td	29, 100
$[CoX(n_3as)]BPh_4$	Cl, Br, I	4.49-4.69	5	29
$[Co(NCS)_2(n_3as)]$		4.58	5	29
$[NiX(n_3p)]BPh_4$	Cl, Br	3.24-3.29	TBPy	29
$[NiX(n_3p)]BPh_4$	Cl, Br,[b] I, CN	0	SqPl	29, 105
$[Ni(NCS)_2(n_3p)]$		3.34	SqPy	29
$[NiX(n_3as)]BPh_4$	Cl, Br, NCS[b]	0	SqPl	29, 104
$[NiI(n_3as)]BPh_4$		3.25	5	29
$[Ni(NCS)_2(n_3as)]$[b]		3.37	SqPy	29, 103
$[NiX(n_2p_2)]BPh_4$	Cl, Br, I, NCS	0	5	29
$[NiI(n_2p_2)]I$[b]		0	TBPy-SqPy	29, 100
$[NiI(n_2as_2)]BPh_4$		0	SqPl	29
$[Ni(NCS)_2(n_2as_2)]$		3.24	Oh	29

[a]See Table 1.

[b]Structure determined by X-ray analysis.

form paramagnetic trigonal bipyramidal complexes in which it is tetralegate or diamagnetic square planar complexes in which it is trilegate.

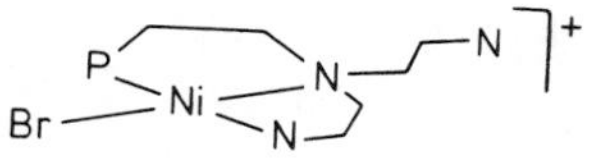

(34)

Thus it can be seen that the ligands that are not always tetradentate in complexes with cobalt(II) or nickel(II) are those that contain more than one nitrogen atom in their donor set, that is,

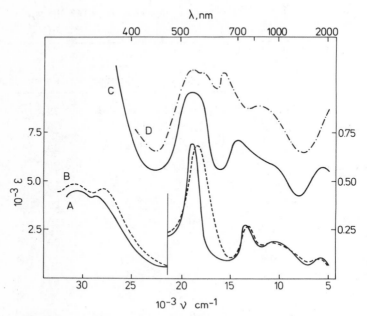

Figure 10   Absorption spectra of $[CoCl(np_3)]^+$ (curve A) and $[CoBr(np_3)]^+$ (curve B) in dichloromethane solutions.   Reflectance spectra of $[CoBr(n_3p)]BPh_4$ (curve C) and $[Co(NCS)_2(n_3p)]$ (curve D).

those ligands that have a low value of overall nucleophilicity, $\sum n^0$ [96].   Factors that may adversely affect the coordination of a donor group include the ability of the thiocyanate group to form strong bonds in preference to the most bulky tertiary phosphine, arsine, and amine, together with the tendency of nickel(II) to form square planar complexes.   The two mixed ligands that contain only one atom of nitrogen, np_3 and nas_3, are, on the other hand, always tetradentate toward bipositive metal ions.

Typical spectra of both low-spin and high-spin nickel(II) complexes with these ligands are shown in Figs. 11 and 12.

The spectra of the diamagnetic square planar complexes, in the solid state, have an asymmetric band in the region $1.4-1.9$ μm^{-1}, which is resolved, in solution, into one band at $1.8-2.0$ μm^{-1} and a shoulder at low energy.   In contrast to this, the spectra of the diamagnetic trigonal bipyramidal complexes have the shoulder at

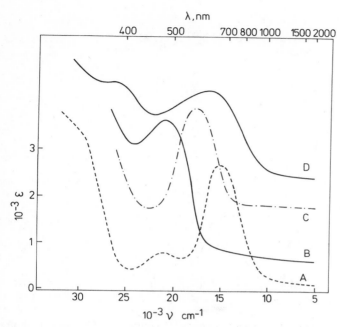

Figure 11   Absorption spectrum of $[NiCl(np_3)]^+$ (curve A) in nitro-
ethane solution; reflectance spectra of red, $[NiBr(n_3p)]BPh_4$ (curve
B); $[NiBr(n_3as)]BPh_4$ (curve C); and $[NiCl(n_2p_2)]BPh_4$ (curve D).

higher frequency than the principal band. Both these types of
spectra are sufficiently different from each other (Fig. 11) to form
a reasonable criterion, in the absence of structural data, for the
assignment of the coordination geometry of low-spin complexes of
nickel(II) formed by polyphosphine and polyarsine ligands.

IV.   REACTIONS ASSISTED BY TETRADENTATE TRIPOD-LIKE PHOSPHINES AND ARSINES

A.   Aquo and Hydroxo Complexes

The hydrated salts of cobalt(II) that contain poorly coordinating
anions, such as $Co(H_2O)_6(BF_4)_2$, react in a variety of different ways
with np_3 according to the nature of the solvent [106]. With THF (or
acetone) and EtOH, different reactions can be proposed as follows:

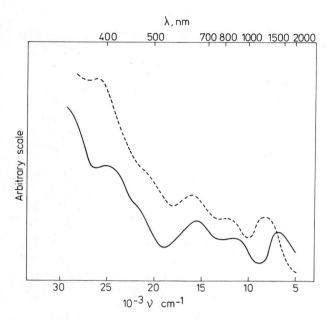

Figure 12   Reflectance spectra of green, [NiBr(n$_3$p)]BPh$_4$ (---); and [Ni(NCS)$_2$(n$_3$p)] (——).

$$Co(H_2O)_6^{2+} + 2Y^- + np_3 \xrightarrow[\text{(acetone)}]{\text{THF}} [Co(H_2O)(np_3)]Y_2 + 5H_2O$$
$$(Y = BF_4)$$

$$\xrightarrow{EtOH} [Co(OH)(np_3)]Y + np_3H^+ + 5H_2O +$$
$$(Y = ClO_4, BF_4)$$

In the first case, the ligand np$_3$, containing four donor atoms, substitutes five molecules of water of solvation, whereas in the second case, the ligand behaves as a base, forming OH$^-$, which acts as the fifth ligand. This last type of reaction also occurs with other tripod-like ligands containing phosphorus donors, such as with the ligand QP and pp$_3$ either with cobalt(II) or nickel(II) [106] and with the tridentate ligand p$_3$ with cobalt(II). In the last case the dimeric di-μ-hydroxo complex [Co$_2$(μ-OH)$_2$(p$_3$)$_2$](BPh$_4$)$_2$ is obtained [Sec. II.A; (3)]. It is significant that hydroxo and aquo complexes with tertiary phosphines can be formed relatively easily. In fact, whereas tertiary phosphines and arsines give rise to many stable metal complexes with ligands such as NO, CO, and CN^-, which have good π-acid characteristics, complexes with "hard" ligands, such as F^-, OH^-, and H_2O, are relatively uncommon [107].

The hydroxo and aquo complexes of cobalt(II) with pp_3 and QP are low-spin (Table 8). Their structures, determined by means of X-ray analysis of the two complexes $[Co(H_2O)(pp_3)]^{2+}$ and $[Co(OH)(pp_3)]^+$ (35), are very similar to that of the complex $[CoBr(pp_3)]^+$ (25), and can be described as distorted square pyramidal [106]. The corresponding cobalt(II) complexes with np_3 are high-spin, and their structure,

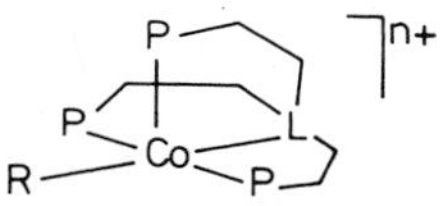

$L = P; \quad R = H_2O, \; n = 2; \quad R = OH, \; n = 1$

$L = N; \quad R = SCH_3, \; n = 1$

(35)

indicated from their electronic spectra, is consistent with a tetrahedrally distorted trigonal bipyramidal arrangement, as found in other five-coordinate high-spin cobalt(II) complexes with the same ligand [Sec. III.D.2, (28)]. In conclusion, the replacement of an atom of phosphorus by an atom of nitrogen in an apical position seems to cause the low-spin square pyramidal structure to go to a high-spin, tetrahedrally distorted trigonal bipyramidal arrangement with an accompanying increase in the distance of "hard" nitrogen atom from the metal.

The aquo and hydroxo complexes of nickel(II) with pp_3 and QP are diamagnetic and have a trigonal bipyramidal geometry. Nickel tetrafluoroborate with pp_3 also gives rise to the complex $[Ni(pp_3)_2$-$(BF_4)_2$, which is diamagnetic and probably square planar [106].

The hydrated nickel(II) ion reacts in a different way with np_3 than with pp_3 as well as differently from $Co(H_2O)_6^{2+}$, forming a variety of products according to the exact reaction conditions [108] (Sec. IV.C).

Table 8 Aquo, Hydroxo, Thio, Mercapto, Methylthio Complexes of the Tetradentate Ligands QP, pp_3, np_3, and n_2p_2

Complexes	L, R, Y, X	μ_{eff}, μ_B	Coordination Geometry[a]	References
$[FeR(pp_3)]BPh_4$	SH[b], SCH_3	3.06-3.31	TBPy	110
$[Co(H_2O)(np_3)](BF_4)_2 \cdot solv$	solv = C_4H_8O	4.56	TBPy-Td	106
$[Co(OH)(np_3)]Y \cdot solv$	ClO_4, BF_4, BPh_4; solv = C_2H_6O	4.21-4.47	TBPy-Td	106
$[Co(H_2O)L](BF_4)_2 \cdot solv$	pp_3,[b] QP; solv = C_4H_8O	2.07-2.10	SqPy	106
$[CoR(L)]Y \cdot solv$	R = OH;[b] L = pp_3,[b] Y = BF_4, BPh_4	1.98-2.10	SqPy	106
	R = SH, SCH_3; L = pp_3, np_3; Y = BPh_4	2.03-2.12	TBPy-SqPy	110
$[Co_2S(np_3)_2]$[b]	solv = C_2H_6O	0	Td	112
$[Ni(H_2O)QP](BF_4)_2 \cdot solv$	solv = C_4H_8O	0	TBPy	106
$[NiR(L)]Y$	R = OH; L = pp_3, QP; Y = BF_4, BPh_4	0	TBPy	106
	R = SH,[b] SCH_3; L = np_3, pp_3;[b] Y = BPh_4	0	TBPy	110
	R = SH,[b] SCH_3; Y = BF_4, BPh_4	0	SqPl	111
$[NiR(n_2p_2H)]Y_2$		0	SqPl	106
$[Ni(pp_3)_2](BF_4)_2$	X = S, R = H, CH_3	1.92	TBPy	112
$[Ni(XR)(np_3)]$	X = Se, R = H	1.92	TBPy	112

[a]See Table 1.

[b]Structure determined by X-ray analysis.

224

B. Thio, Mercapto, and Methylthio Complexes

Complexes containing H_2S, HS^-, or S^{2-} as coligands are uncommon, and it is well established that the action of H_2S on transition metal ions, even if in the presence of ligands containing potentially strong donor atoms, results, in general, in insoluble, binary sulfide polymers (for a relevant review of this field, see Ref. 109). However, in the presence of the ligands np_3 and pp_3 (as well as with the tridentate p_3, Sec. II.A), the hydrated ions of iron(II), nickel-(II), and cobalt(II) readily yield monomeric complexes of the general formula $[M(SR)L]BPh_4$ ($R = H$, CH_3; $L = pp_3$, np_3; Table 8) [110] upon treatment with H_2S or CH_3SH.

The SR^- anions (in an analogous way to OH^- in the hydroxo complexes) are formed by the basic action of the tertiary phosphines np_3 and pp_3 toward H_2S and CH_3SH, which likewise results presumably in the formation of ionic phosphine species of the type np_3H^+ and pp_3H^+. The reaction can be represented in the following way (see also Table 2):

$$M(H_2O)_6^{2+} + 2np_3 + RHS \longrightarrow [M(RS)(np_3)]^+ + np_3H^+ + 6H_2O$$

$$R = H,\ CH_3$$

This hypothesis seems to be confirmed by the nickel complexes $[Ni(SR)(n_2p_2H)]Y_2$ [111] ($R = H$, CH_3; $Y = ClO_4$, BF_4), which can be obtained with the tetradentate ligand n_2p_2 (XXV) and either H_2S or CH_3SH. In these complexes, one nitrogen atom of n_2p_2 has been quaternized by the transfer of a proton from the species RHS and thus the ligand acts as tridentate. The structure, determined by X-ray analysis of the complex $[Ni(HS)(n_2p_2H)](BF_4)_2$ (36), is square planar.

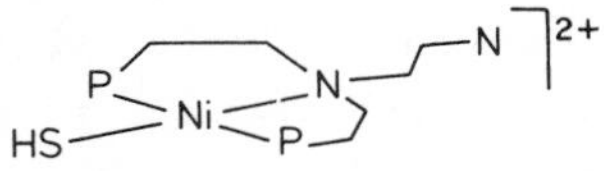

(36)

The mercapto and methylthio complexes of iron(II), cobalt(II), and nickel(II) with np_3 and pp_3 are low-spin and five-coordinate (Table 8): They all have an essentially trigonal bipyramidal coordination geometry (37), which, in the case of the cobalt(II) complexes, is strongly distorted toward a square pyramid (35) [which is the usual arrangement found in low-spin five-coordinate cobalt(II) complexes]. The mercapto and methylthio complexes are thus structurally

M = Fe, Ni

(37)

similar to the hydroxo complexes with the same ligands. Only in the case of cobalt(II) complexes with np_3 is the spin state different, and hence also the details of the five-coordinate structure. This can be correlated with the different value of $\sum n^0$ for the set of donor atoms NP_3S compared with NP_3O [96].

From the preceding structure determinations it can be inferred that the formation of mercapto and methylthio complexes in the presence of np_3 and pp_3 and their poor reactivity is attributable not only to the electronic properties of the donor atoms but also to the characteristic structural properties found in tripod-like ligands. The coordination of HS^- and CH_3S^-, in the presence of np_3 and pp_3, is possible only in the coordination position of the metal that is not occupied by the phosphorus atoms, which compete with sulfur in the formation of strong covalent bonds. Furthermore, the phenyl groups on the phosphorus atoms surround the sulfur, hindering the participation of its unpaired electrons in bond formation with other metal atoms and thus preventing the formation of polymeric sulfides.

A diamagnetic, dinuclear complex of cobalt(I), $[(np_3)Co(\mu\text{-}S)Co(np_3)]$ [112], can be obtained by reducing the complex $[Co(SH)(np_3)]$-BPh_4 with sodium ethoxide. In the dinuclear cobalt(I) complex (d^8) (38), the two metal ions are tetracoordinated in a centrosymmetric

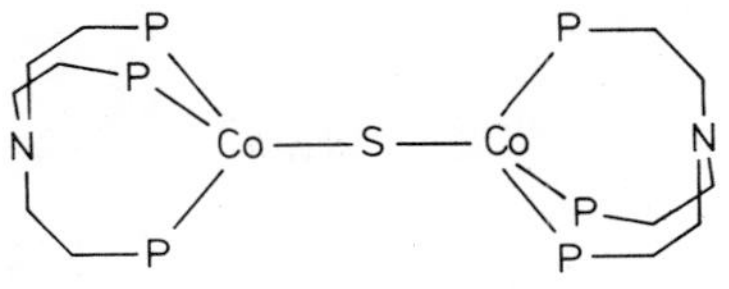

(38)

arrangement by three phosphorus atoms and one of sulfur. The two
metal ions are thus connected by a linear bridge through the sulfur
atom with a bond distance of 2.128(1) Å. The apical nitrogen, on
the other hand, is not bound to the metal. The structure of this
complex is analogous to that of the dinuclear complex of nickel(II),
which is also d^8, with the ligand p_3, $[(p_3)Ni(\mu-S)Ni(p_3)]^{2+}$ (4), np_3
in the cobalt complex behaving as a tridentate ligand like p_3 in the
nickel complex. Both the diamagnetism and short M-S distance in
these two tetrahedral complexes can be explained by a $d_\pi-p_\pi-d_\pi$
interaction between the two d^8 metal ions through the bridging sulfur (see also Sec. II.A). These nickel(II) and cobalt(I) derivatives
are the only two examples of tetrahedral complexes of a d^8 metal that
are diamagnetic. In order to prepare mercapto, methylthio, and
methylselenido complexes of nickel(I), it is necessary that an oxidation addition reaction of H_2S, CH_3SH, or H_2Se on the nickel(0)
complex $[Ni(np_3)]$ occur (Sec. IV.C). The reaction is of the general
type

$$[Ni(np_3)] + HYR \longrightarrow [Ni(YR)(np_3)] + \tfrac{1}{2}H_2$$

$$Y = S, R = H, CH_3; Y = Se, R = H$$

The electronic spectra show that these complexes have a trigonal
bipyramidal coordination geometry analogous to that of the halo or
pseudohalo complexes $[NiX(np_3)]$ (Sec. IV.C).

C. Hydrido Complexes and Complexes of Metals in Low Oxidation States

The ligand np_3, in the presence of $Ni(H_2O)_6(BF_4)_2$, can act as a reducing agent and promote hydrogenation reaction [108]. With regard

to the reaction scheme

$$Ni(H_2O)_6(BF_4)_2 + np_3 \quad \xrightarrow{\substack{(CH_3)_2CO \\ \\ C_2H_5OH \\ \\ BH_4^-}} \quad
\begin{array}{ll}
[Ni(np_3)]BF_4 & \\
[NiH_x(np_3)]BF_4 & x < 0.25 \\
[NiH_x(np_3)]BF_4 & x > 0.25
\end{array}$$

when acetone is used as solvent only the first reaction takes place, whereas in ethanol both reactions occur simultaneously. The reduction and hydrogenation reactions promoted by np_3 may be explained according to the following scheme:

$$2np_3 + H_2O \longrightarrow np_3O + np_3H^+ + \tfrac{1}{2}H_2 + e^-$$

$$np_3 + RCH_2OH \longrightarrow RCHO + np_3H^+ + H^-$$

$[NiH_x(np_3)]BF_4$ complexes with a x value greater than 0.25 have been obtained using the specific reducing agent $NaBH_4$. Complexes with an x value close to unity can be obtained only by the oxidative addition of HBF_4 or $HClO_4$ to the nickel(0) complex $[Ni(np_3)]$.

The presence of hydrogen bound to nickel is indicated by the infrared band at 595 cm^{-1}, which is consistently shifted in the deuterated derivative. The magnetic moments of various $[NiH_x(np_3)]BF_4$ complexes range from 2.08 μ_B when x $\approx$ 0 to 0.64 μ_B when x $\approx$ 0.9, indicating that these nonstoichiometric hydrides are made up from complexes of nickel(I), $[Ni(np_3)]^+$, and diamagnetic hydrido complexes of nickel(II), $[NiH(np_3)]^+$, which are crystallographically identical whether or not they contain an atom of hydrogen. The structure of the cation $[NiH(np_3)]^+$ (39) is trigonal bipyramidal, with N and H in the axial positions [108].

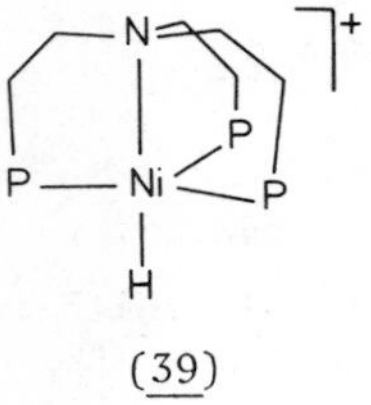

(39)

Nonstoichiometric hydrido complexes $[NiH_x(np_3)]ClO_4$ have been also obtained starting with $Ni(H_2O)_6(ClO_4)_2$, whereas starting from $Ni(H_2O)_6(NO_3)_2$ they form only by the action of BH_4^-; in its absence a nickel(I) complex is formed containing coordinated NO_3^-:

$$Ni(H_2O)_6(NO_3)_2 + np_3
\begin{array}{l}
\nearrow [Ni(NO_3)(np_3)] \\
\xrightarrow[\overset{BH_4^-}{BH_4^- \ exc.}]{} [NiH_x(np_3)]NO_3 \\
\searrow [Ni(np_3)]
\end{array}$$

This seems to indicate that a necessary requisite to the formation of nonstoichiometric hydrido complexes, in the absence of specific hydrogenating reagents, is the absence of a coordinating anion. Moreover, the formation of these nonstoichiometric hydrides seems to be a property peculiar to nickel(II) and the ligand np_3. For instance, hydroxo complexes are formed by both the reactions of $Ni(H_2O)_6^{2+}$ with pp_3 and of $Co(H_2O)_6^{2+}$ with np_3 (Sec. IV.A). On the other hand, with the use of BH_4^-, cobalt(II) forms a high-spin cobalt-(I) complex, $[Co(np_3)]BF_4$ (40) [108].

$$Co(H_2O)_6(BF_4)_2 + np_3
\begin{array}{l}
\nearrow [Co(np_3)]BF_4 \\
\xrightarrow[\overset{BH_4^-}{BH_4^- \ exc.}]{} \\
\searrow [CoH(np_3)]
\end{array}$$

This complex, together with $[Ni(np_3)]BPh_4$ (41) [113], $[Ni(np_3)]$ (42) [114a], and $[Ni(nas_3)]BF_4$ (see later), are the only examples of a trigonal pyramidal geometry. The complex $[Ni(np_3)]$ is formed under

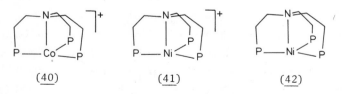

(40)   (41)   (42)

conditions of excess $NaBH_4$ and by prolonged boiling in n-butanol. Using the same reaction conditions but starting with hydrated

cobalt(II), the pure diamagnetic hydrido complex of cobalt(I),
[CoH(np$_3$)] (43), is formed, which is isoelectronic and isostructural
with [Ni(np$_3$)], with which it forms a continuous series of solid
solutions [114b].

(43)

In conclusion, the formation of hydrido complexes with np$_3$ seems
to be confined to the low-spin electronic configuration d^8 as found
in nickel(II) and cobalt(I). This appears to show that the attainment of a VEN of 18 is the necessary condition to obtain hydrido complexes. Confirmation of this can be seen when one forms complexes of
nickel(0), d^{10}, or nickel(I), d^9, by the action of BH$_4^-$ on salts of
nickel(II) in the presence of np$_3$, since the products are not hydrido
complexes but derivatives of the type [Ni(np$_3$)] and [NiX(np$_3$)] (X =
halide) according to the coordinative ability of the anion present in
the salt that was used as starting material.

In the same way as for nickel(II), when one treats cobalt(II)
salts containing a coordinating anion (halide or pseudohalide) with
NaBH$_4$, one ends up with a cobalt(I) derivative, in which the coordinating anion occupies the fifth coordinating position [114b]:

$$MX_2 + np_3 + NaBH_4 \longrightarrow [MX(np_3)] + NaX + \frac{1}{2}B_2H_6 + \frac{1}{2}H_2$$

All the complexes [MX(np$_3$)] are five-coordinate, with a coordination
geometry that can be derived from a trigonal bipyramid. This is exemplified in the complex [NiI(np$_3$)] (44) [115], for which an X-ray analysis has been carried out. It must be remembered that, whereas the
nickel(II) complexes, d^8, [NiX(np$_3$)]$^+$, are all low-spin, those of co-
balt(I), which is also d^8, [CoX(np$_3$)], are low-spin when X = CN or H,
but high-spin with coligands X = Cl, Br, I, and NCS, which have low
nucleophilicity values [96].

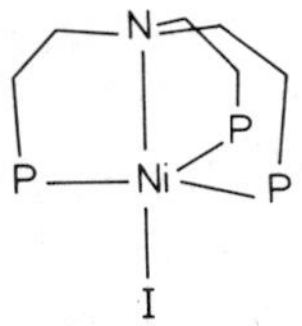

(44)

In the presence of the ligand pp_3, the anhydrous salts of co-
balt(II), together with the hydrated ion, are both reduced to cobalt-
(I) compounds with BH_4^-. However, different complexes are obtained
depending on the nature of the original starting material, on the
solvent, and on the amount of BH_4^- used [116] (Table 9). Nickel salts
containing poorly coordinating anions (I^-, NO_3^-, BF_4^-) with pp_3 and BH_4^-
yield pure hydrido complexes of nickel(II), $[NiH(pp_3)]Y$ [116]. With
pp_3 just as for np_3, hydrido complexes will form only with metals
having a d^8 configuration.

The isoelectronic d^8 complexes of cobalt(I) and nickel(II) with
pp_3 are diamagnetic. The molecular structure of $[CoH(pp_3)]$ [116] is
almost regular trigonal bipyramidal, with the cobalt atom five-coord-
nated by four atoms of phosphorus from the ligand and by one atom of
hydrogen. An analogous trigonal bipyramidal structure has been

Table 9 Reaction Schemes of Cobalt(II) and Nickel(II) with the pp_3
Ligand and BH_4^-

$$CoX_2 + pp_3 + BH_4^- \xrightarrow{C_2H_5OH} [CoX(pp_3)]$$

$$X = Cl, Br, I, NCS$$

$$CoX_2 + pp_3 + BH_4^-(excess) \searrow_{DMF}$$

$$[CoH(pp_3)]$$

$$Co(H_2O)_6^{2+} + pp_3 + BH_4^- \nearrow^{(CH_3)_2CO}$$

$$[Ni(H_2O)_6]Y_2 + pp_3 + BH_4^- \longrightarrow [NiH(pp_3)]Y$$

$$Y = I, NO_3^-, BF_4$$

assigned to the remaining complexes of cobalt(I) and nickel(II) with pp_3.

Reducing either $Ni(H_2O)_6(BF_4)_2$ or the anhydrous halides of co-balt(II) and nickel(II) with $NaBH_4$ in the presence of nas_3, one invariably obtains complexes in which the metal has been reduced to oxidation state +1 [117]. These complexes are either tetracoordinate trigonal pyramidal, $[Ni(nas_3)]BF_4$, or pentacoordinate trigonal bipyramidal $[MX(nas_3)]$ (M = Co, Ni), depending on the coordinating ability of the counteranion that is present in solution.

If a half-equivalent of BH_4^- is used, in the presence of $NaBPh_4$, dinuclear complexes of nickel(I) are obtained with the formula $[(nas_3)Ni(\mu-X)Ni(nas_3)]BPh_4$ (X = Br, I). An X-ray analysis carried out on the iodo derivative (45) shows that there is a linear bridge through the iodine atom connecting the two nickel atoms, which are five-coordinated with a trigonal bipyramidal geometry [118]. The

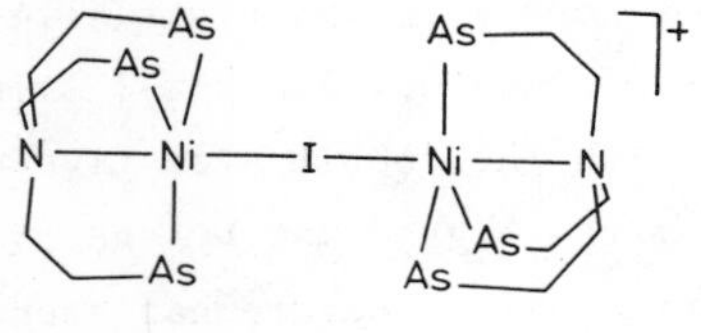

(45)

magnetic moments of the two complexes (lower than 1.0 μ_B at room temperature) indicate a strong antiferromagnetic interaction between the two nickel atoms. The stoichiometry of the reaction demonstrates that the reduction of the nickel is due partly to the BH_4^- ion and partly to the BPh_4^- ion. The reducing action of the latter ion can be represented schematically as

$$2BPh_4^- \longrightarrow 2BPh_3 + Ph_2 + 2e^- \quad (Ph_2 = diphenyl)$$

The behavior of the BPh_4^- parallels that of the BH_4^- ion:

$$2BH_4^- \longrightarrow B_2H_6 + H_2 + 2e^-$$

Hydrido complexes of iron(II) with pp_3, $[FeH(pp_3)]BF_4$, are formed by

the action of $NaBH_4$ on the halo complex $[FeX(pp_3)]BPh_4$. If the
reaction is carried out in an atmosphere of nitrogen, the hydrido
dinitrogen complex, $[FeH(N_2)(pp_3)]BPh_4$, is formed (the same reaction
also takes place with np_3). The molecule of dinitrogen can be substituted by other neutral molecules such as CO, CH_3CN, or C_6H_5N. All
these complexes are diamagnetic and presumably have a coordination
number of 5 or 6 according to their individual stoichiometries [95].

The easy addition of neutral molecules such as CO or N_2 to the
hydrido complexes of iron(II) [on the contrary, in the hydrido complexes of cobalt(I) and nickel(II), the same neutral molecules
replace the hydrogen atoms; Sec. IV.D] can be attributed to the
attainment of the 18-electron configuration in the six-coordinate
complexes of iron(II).

D. Complexes with NO, CO, SO_2 as Coligands

Numerous nitrosyl and carbonyl complexes with the ligands np_3 and pp_3
have been reported. The greater part of these have been prepared by
one of the following methods (Table 10): (1) by the reaction of CO
or NO with hydrido complexes and metal complexes in low oxidation
states [114b,119]; or (2) by the reaction of tripod-like ligands with
metal carbonyl complexes [30,120]. In the latter case, the ligand is
either tridentate or bidentate, substituting respectively either
three or two carbonyl groups. Only in the case of the iron complex
$[Fe(NO)(np_3)]BPh_4$ does the reaction start from the divalent metal compound $Fe(H_2O)_6(BF_4)_2$ [119].

The fact that these reactions generally take place at normal
temperature and under atmospheric pressure clearly demonstrates the
ability of the ligand np_3 to stabilize strong bonds between the 3d
metals and "soft" coligands that have good π-bonding ability.

The reaction of $[Mn(CH_3)(CO)_5]$ with pp_3 is interesting, since in
THF the acetyl derivative is formed with insertion of CO into the CH_3
-Mn bond. In the complex $[\{Fe_2(CO)_2(C_5H_5)_2\}_2(pp_3)]$ the pp_3 ligand is
bound to four metal atoms, having substituted the terminal carbonyl
groups in the dinuclear starting material [30].

Table 10   Reaction Schemes for Nitrosyl and Carbonyl Complexes of the Ligands np_3 and pp_3

$$[NiH_x(np_3)]^+ \quad \xrightarrow{CO} \quad [Ni(CO)(np_3)]^+$$
$$\xrightarrow{NO} \quad [Ni(NO)(np_3)]^+$$

$$[Ni(np_3)] \xrightarrow{CO} [Ni(CO)(np_3)]$$

$$[Co(np_3)]^+ \xrightarrow{CO} [Co(CO)(np_3)]^+$$

$$[CoH(np_3)] \xrightarrow{NO} [Co(NO)(np_3)] \xrightarrow{BPh_4^-} [Co(NO)(np_3)]BPh_4$$

$$[Cr(CO)_6] \xrightarrow{L} [Cr(CO)_3L] \qquad (L = pp_3, np_3, n_2p_2)$$

$$[Mn(CH_3)(CO)_5] \quad \xrightarrow{pp_3, THF} \quad [Mn(CH_3CO)(CO)_2(pp_3)]$$
$$\xrightarrow[o-xylene]{pp_3} \quad [Mn(CH_3)(CO)_2(pp_3)]$$

$$[MnBr(CO)_5] \xrightarrow{pp_3} [MnBr(CO)(pp_3)]$$

$$[Mn(CO)_3(\eta-C_5H_5)] \xrightarrow{pp_3} [Mn(CO)(\eta-C_5H_5)(pp_3)]$$

$$[Mn(CO)_2(\eta-C_5H_5)(NO)]PF_6 \xrightarrow{pp_3} [Mn(\eta-C_5H_5)(NO)(pp_3)]PF_6$$

$$[FeI(CO)_2(\eta-C_5H_5)] \xrightarrow{pp_3} [Fe(CO)(\eta-C_5H_5)(pp_3)]I$$

$$[Fe(CO)_2(\eta-C_5H_5)]_2 \xrightarrow{pp_3} [\{Fe_2(CO)_2(\eta-C_5H_5)_2\}_2(pp_3)]$$

The structures of the most important nitrosyl and carbonyl complexes that have been studied by means of X-ray analysis suggest that the attainment of the 18-electron configuration is the main factor in deciding the coordination number and hence the coordination geometry of these complexes.

The two diamagnetic, isoelectronic complexes [Fe(NO)(np_3)]BPh_4 (46) [119] and [Co(CO)(np_3)]BPh_4 [121] have almost regular trigonal

(46)   (47)

bipyramidal structures, with the ligand np_3 forming four bonds with
the metal, which attains the VEN of 18. On the other hand, in the
two similarly diamagnetic and isoelectronic complexes $[Ni(NO)(np_3)]$-
BPh_4 (47) [119] and $[Ni(CO)(np_3]$ [121], the apical nitrogen is not
bound to the metal atom, which is now tetracoordinated in a distorted
tetrahedral arrangement. In this way, the nickel achieves a VEN of
18 in these two complexes. The two isoelectronic complexes $[Ni(CO)$-
$(np_3)]BPh_4$ [114b] and $[Co(NO)(np_3)]BPh_4$ [119] have substantially
similar structures but have a VEN of 17 and hence an unpaired electron. The M-N-O angles in the three complexes $[M(NO)(np_3)]^+$ (M = Fe,
Co, Ni) are 164(7), 165(7), and 167.7(2.1)°, respectively; hence each
MNO group can be considered to be essentially linear [122].

The magnetic and spectral properties of these complexes have
been interpreted using a simple molecular orbital scheme, which
can be adapted to either the carbonyl or nitrosyl complexes. In
fact, the magnetic and spectral properties of a wide range of isoelectronic complexes are essentially the same, and are independent
of the nature of the metal or whether the coligand is NO or CO. The
formation of the molecular orbitals is represented schematically in
Fig. 13, from the interaction of the metal d orbitals with the σ lone
pairs on either NO or CO and with the degenerate antibonding π* orbitals on NO or CO [119]. By assigning all the metal d electrons
(possibly also an electron from the π* orbital of NO) to the levels
of Fig. 13b, both the electronic configuration and the spin state of
the complexes can be deduced. The electronic spectra exhibit a band
in the region 1.6-2.1 μm^{-1} that has been assigned, on the basis of
Fig. 13b, to the transition to the lowest nonoccupied energy level.

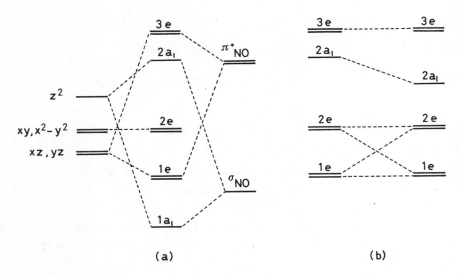

(a)   **(b)**

Figure 13   Schematic molecular orbital diagrams.   (a) Interaction
between NO orbitals and metal orbitals in C_{3v} symmetry.   (b) The
effect of tetrahedral distortion on the orbital energies.   [Re-
printed with permission from C. A. Ghilardi, A. Sabatini, and L.
Sacconi, *Inorg. Chem.*, 15:2763.   Copyright (1976) American Chemical
Society.]

The spectra of the cobalt complexes have another less intense band
at around 0.7 μm^{-1}, which can be assigned to the transition
$[2a_1^2 2e^3] \longleftarrow [2a_1^1 2e^4]$.

   Carbonyl and nitrosyl complexes have also been obtained with
the ligands nas_3 [117,119] and are similar in every way to the
corresponding complexes with np_3 and pp_3 [123].

   A diamagnetic nickel(0) complex that is analogous to the complexes described above but that contains SO_2, $[Ni(SO_2)(np_3)]$ (48),
has been prepared by the reduction of the nickel(II) complex
$[NiBr(np_3)]BPh_4$ with BH_4^- in the presence of SO_2 [124].   This complex
has a tetrahedrally distorted trigonal bipyramidal structure with
Ni-N$_{apical}$ bond distance of 2.315 Å.   The SO_2 group that is bonded
to the metal is bent.   It should be noticed that in the complex
$[Ni(SO_2)(p_3)]$ (Sec. II.B, 10), the SO_2 group is coplanar with the
nickel atom, which is thus able to act as an electron acceptor toward

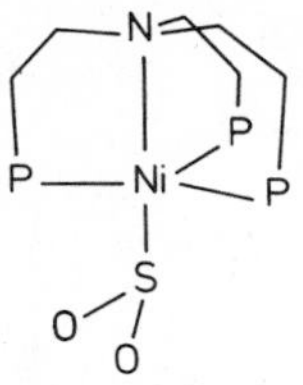

(48)

the SO_2. On the contrary, in the complex with np_3 it is the SO_2 that formally acts as an electron acceptor, gaining electrons from the nickel(0) and assuming a tetrahedral hybridization with a lone pair of electrons. The different bonding mode of SO_2 can be explained, in the valence bond model, considering the 16-electron configuration of the (p_3)Ni moiety, which can act as a Lewis acid, whereas the (np_3)Ni moiety, with an 18-electron configuration, behaves as a Lewis base. One can attain the same results on the basis of the MO scheme used for nitrosyl and carbonyl complexes (Fig. 13).

E. Alkyl, Aryl, and Acyl Complexes

Whereas the ligand nas_3 yields the complexes $[NiX(nas_3)]BPh_4$ [26] (Sec. III.D.2) with $NiBr_2$ and NiI_2, upon continued boiling with $NiCl_2$ in butanol it gives the diamagnetic complex $[Ni(C_6H_5)(nas_3)]-BPh_4$, in which the phenyl residue is σ bonded to the nickel atom [99,117]. The structure of this five-coordinate complex is trigonal bipyramidal (49).

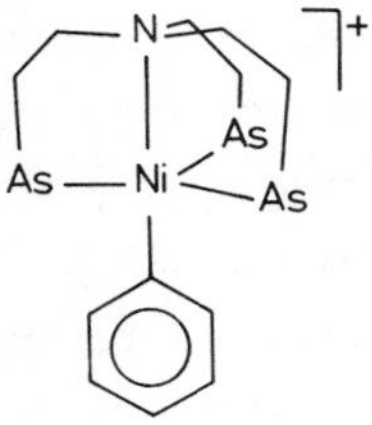

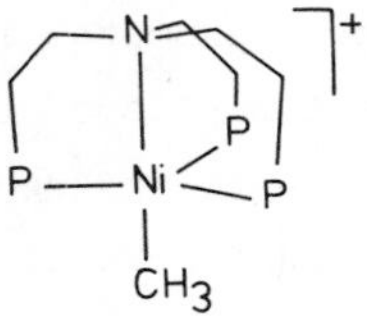

(49)   (50)

Analogous complexes containing a σ-bonded phenyl group have been obtained, not only with other phenyl-substituted tertiary arsines,

QAS and DAS, but also with the methyl-substituted tetraarsine Meas$_3$. This indicates that the substituted phenyl group comes from the BPh$_4^-$ ion according to the reaction scheme

$$NiCl_2 + L + 2NaBPh_4 \longrightarrow [Ni(C_6H_5)L]BPh_4 + BPh_3 + 2NaCl$$

The products of this reaction have been completely identified. The phenylation reaction of BPh$_4^-$, which has, as yet, been verified only in the presence of tertiary polyarsines, is thus equivalent to the corresponding hydrogenation reaction of BH$_4^-$:

$$BPh_4^- \longrightarrow Ph^- + BPh_3$$
$$2BH_4^- \longrightarrow 2H^- + B_2H_6$$

Apart from this exceptional type of reaction, it is possible to obtain nickel [125] and cobalt [126] complexes with alkyl groups σ-bonded to the metal of general formula [M(R)L]BPh$_4$ (Table 11) by means of Grignard reactions with the halo complexes of bipositive metals, [MX(L)]BPh$_4$. In the same way, cobalt(I) complexes with pp$_3$ and np$_3$ and σ bonded alkyl or aryl residues, [Co(R)L], have been obtained from the halo complexes of cobalt(I), [CoX(L)] (L = np$_3$, pp$_3$) and the appropriate organolitium compounds [126]. An essentially trigonal bipyramidal geometry can be assigned to the diamagnetic, isoelectronic complexes of both cobalt(I) and nickel(II), as well as to the paramagnetic cobalt(II) complexes. The structure of [Ni(CH$_3$)-(np$_3$)]BPh$_4$ (50) [125] has been determined by X-ray analysis.

The stability of all these organometallic complexes has been attributed to the special characteristic of the tripod-like ligands np$_3$, nas$_3$, and pp$_3$, which are able to arrange themselves, with their six phenylene groups, right around the metal atom, thereby restoring on the complex a degree of kinetic inertness that is not present, for example, in the organometallic complexes of nickel with monodentate phosphines. These latter complexes are consequently much less stable [127].

Carbon monoxide reacts rapidly with the above organometallic complexes of nickel(II) at room temperature and atmospheric pressure,

giving acyl derivatives [126]. The compounds that are first isolated·
are nevertheless different depending on whether the ligand is np_3 or
nas_3. With nas_3 the acyl derivative can be isolated directly:

$$[Ni(R)(nas_3)]^+ + CO \longrightarrow [Ni(COR)(nas_3)]^+ \qquad R = CH_3, C_2H_5,$$
$$CH_2C_6H_5$$

With np_3, at first, a solid solution of the carbonyl derivative of
nickel(I), d^9, and the acyl derivative of nickel(II) (d^8, diamagnetic)
in the ratio 1:1 is formed:

$$2[Ni(R)(np_3)]^+ + 2CO \longrightarrow [Ni(CO)(np_3)]^+ \cdot [Ni(COR)(np_3)]^+ + R$$

When the solid solution is dissolved in THF and C_2H_5OH, the pure acyl
compounds $[Ni(COR)(np_3)]BPh_4$ separate.

The structures of both the acetyl and the carbonyl derivatives,
which have been determined by X-ray analysis, are essentially trigonal bipyramidal (51,52) [126,128]. The solid solution is assumed to

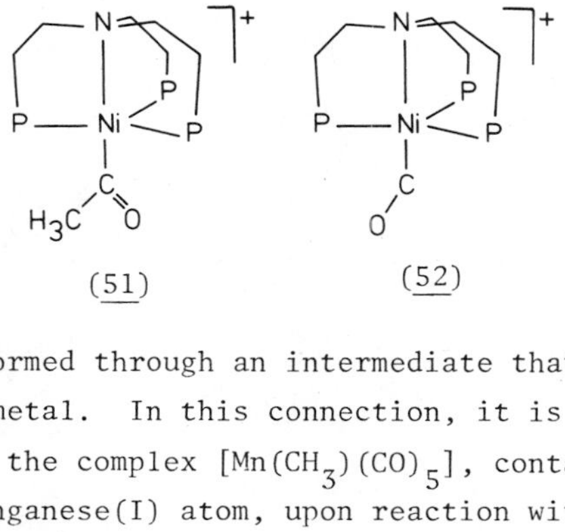

$$(51) \qquad\qquad (52)$$

be formed through an intermediate that contains CO and CH_3 bound to
the metal. In this connection, it is perhaps interesting to remember
that the complex $[Mn(CH_3)(CO)_5]$, containing both CH_3 and CO bound to
a manganese(I) atom, upon reaction with the tetraphosphine pp_3, gives
an acetyl derivative [30] (Sec. IV.D). In a similar way to the nickel complex, $[Co(CH_3)(np_3)]BPh_4$ reacts with CO, giving the acetyl
derivative [126].

These insertion reactions of CO into organometallic compounds of
nickel and cobalt with np_3 and nas_3 are important. It is well established that the insertion of CO into metal-carbon σ bonds to give

acyl derivatives usually occurs by the action of tertiary phosphines, CO, amines, etc. [30,129] on metal complexes that already contain both CO and alkyl residues. In some of these reactions it has been shown that the CO does not directly enter in the metal-carbon bond, but instead, it first replaces a carbonyl group already present in the complex, which, in turn, now forms the acyl derivative. This supports the hypothesis that the acyl derivatives with np_3 and nas_3 are formed through an intermediate that contains both the alkyl residue and CO coordinated to the metal.

F. Alkylsulfito Complexes

SO_2 reacts with $Ni(H_2O)_6(BF_4)_2$ and np_3 (or pp_3), forming different complexes depending on whether the reaction is carried out in THF (or acetone) [130] or in ethanol (or methanol) [131], hydrogenosulfito

$$Ni(H_2O)_6^{2+} + 2np_3 + SO_2 \xrightarrow[ROH]{THF} \begin{array}{l} [Ni\{SO_2(OH)\}(np_3)]^+ + np_3H^+ + 5H_2O \\[2em] [Ni\{SO_2(OR)\}(np_3)]^+ + np_3H^+ + 6H_2O \end{array}$$

$$R = CH_3, C_2H_5$$

and alkylsulfito complexes, respectively, being formed (Table 11). An analogous reaction also takes place with $Co(H_2O)_6^{2+}$ in an alcoholic solution [131]. The structures of the alkylsulfito complexes of cobalt [132] and nickel [131], determined by X-ray analysis, are essentially trigonal bipyramidal (53). The Ni-S bond distance (2.130 Å)

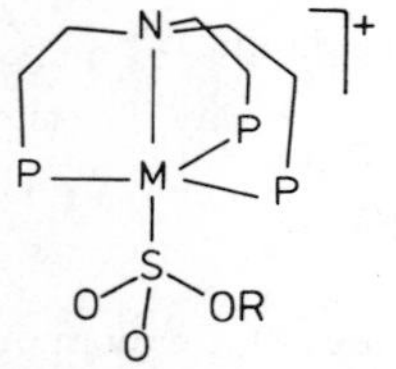

M = Co, Ni; R = C_2H_5

(53)

Table 11 Some Representative Organometallic Complexes with the Ligands pp_3, np_3, nas_3, $Menas_3$, and QAS

Complexes	R, L	μ_{eff}, μ_B	Coordination Geometry[a]	References
[CoR(L)]	R = CH_3, C_6H_5; L = np_3, pp_3	0	TBPy	126
[CoR(np_3)]BPh$_4$·solv	CH_3, $CH_2C_6H_5$, $COCH_3$; solv = C_4H_8O	2.03–2.18	TBPy	126
[Co{SO$_2$(OC$_2$H$_5$)}(L)]BPh$_4$	np_3,[b] pp_3	2.39–2.25	TBPy	131, 132
[Ni(C$_6$H$_5$)(L)]BPh$_4$	nas_3,[b] $Menas_3$, QAS	0	TBPy	99, 117a
[NiR(L)]BPh$_4$	R = CH_3,[b] C_2H_5, $CH_2C_6H_5$ L = np_3,[b] nas_3, pp_3	0	TBPy	125
[Ni(COR)(L)]BPh$_4$·solv	R = CH_3,[b] C_2H_5, $CH_2C_6H_5$ L = np_3,[b] nas_3; solv = C_4H_8O	0	TBPy	126, 128
[Ni(COR)(np$_3$)]BPh$_4$·[Ni(CO)(np$_3$)]BPh$_4$·solv	CH_3,[b] C_2H_5, $CH_2C_6H_5$; solv = C_4H_8O	2.1	TBPy	126
[Ni{SO$_2$(OR)}(np$_3$)]BF$_4$·solv	H, CH_3,[b] C_2H_5; solv = $\frac{1}{2}CH_3OH·\frac{1}{2}H_2O$	0	TBPy	130, 131
[Ni{SO$_2$(OR)}(pp$_3$)]BPh$_4$	CH_3, C_2H_5	0	TBPy	131

[a] See Table 1.
[b] Structure determined by X-ray analysis.

is shorter than the sum of the covalent radii of the two atoms, in
agreement with that found in other complexes containing M-S bonds
[37,112].

The formation of these alkylsulfito complexes can be rationalized
by considering the unsaturation of the metal atom that is bound to np_3
in solution, such that it is able to behave as an electron acceptor
from SO_2. This results in a shift in the electron density that makes
it easier for the ester to form. The basic nature of the phosphine
in this case also aids the formation of the ester.

V.   COMPLEXES WITH CYCLO-TRIPHOSPHORUS
AND CYCLO-TRIARSENIC AS COLIGANDS

In the presence of hydrated nickel(II) and cobalt(II) ions, the ligands p_3 and np_3 are able to react with both white phosphorus and
yellow arsenic, breaking up the structure of tetraatomic P_4 and As_4
molecules in such a way as to form triatomic triangular cyclo-triphosphorus and cyclo-triarsenic species. These groups act as trihapto
ligands, yielding sandwich or double-sandwich complexes.

Thus the two diamagnetic complexes $[Co(\eta^3\text{-}P_3)(p_3)]$ [133] (54)
and $[Co(\eta^3\text{-}P_3)(np_3)]$ [134] (55) are obtained from the reaction between cobalt(II) tetrafluoroborate and a solution of P_4 in THF under
very mild conditions (20-50°C); these complexes are stable either in
the solid state or in solution. They have sandwich structures in

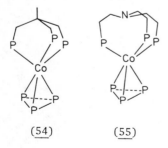

(54)   (55)

which the cobalt is six-coordinated by three phosphorus atoms of the
cyclo-triphosphorus and by three other phosphorus atoms from either
the ligand p_3 or np_3, which acts as a tridentate ligand in which the

apical nitrogen atom is not coordinated to the metal. In this way
the cyclo-triphosphorus acts as a three-electron donor, and the cobalt atom attains a VEN of 18 in agreement with the observed diamagnetism in these complexes.

Under analogous reaction conditions to the above, and using solutions of P_4 or As_4 in THF, the dinuclear complexes of general formula
$[(p_3)M-\mu-(\eta^3-D_3)M(p_3)]Y_2$ (M = Co, Ni; D = P, As; Y = BF_4, BPh_4) have
also been prepared. Their magnetic moments correspond to one unpaired
electron for each dinuclear species [135]. Upon reduction with $NaBH_4$,
they give the corresponding complexes $[(p_3)M-\mu-(\eta^3-D_3)M(p_3)]Y$ (M = Co,
two unpaired electrons; M = Ni, diamagnetic), whereas upon electro-
chemical oxidation at the anode they form the triply charged diamagnetic cation $[(p_3)Co-\mu-(\eta^3-P_3)Co(p_3)]^{3+}$ [136]. Finally, the mixed
complex having two unpaired electrons $[(p_3)Co-\mu-(\eta^3-P_3)Ni(p_3)](BPh_4)_2$
is formed by allowing $[Co(\eta^3-P_3)(p_3)]$ to react with $Ni(H_2O)_6(BF_4)_2$.

X-ray studies on some of these dinuclear complexes have shown
that they have a double-sandwich type of structure (56) in which the
trihapto-P_3 and trihapto-As_3 groups form a bridge between the two
$M(p_3)$ residues. Each metal is six-coordinated by three phosphorus
atoms from the ligand p_3 and by another three phosphorus or arsenic
atoms from the cyclo-triphosphorus or cyclo-triarsenic, which act as
three-donor ligands.

$M_1 = M_2 = Co;$ D = P, As; n = 1, 2
$M_1 = M_2 = Ni;$ D = P, As; n = 1, 2
$M_1 = Ni;$ $M_2 = Co;$ D = P; n = 2

(56)

The average P-P bond-length in the trihapto-P_3 group (2.164 Å)
and As-As bond length in the trihapto-As_3 group (2.21 Å) are both only
slightly less than the distances found in the tetrahedral P_4 and As_4

molecules, respectively. The structures of these dinuclear complexes can be compared to that found in the tris-μ-hydrido complexes of cobalt and iron with the same ligand p_3 [Sec. II.B, (7)]. As in these complexes, the bonding and the spin state in the trihapto-P_3 and trihapto-As_3 complexes can be rationalized on the basis of a semiquantitative molecular orbital treatment derived from the one proposed by Hoffmann for other analogous double-sandwich complexes [42].

ACKNOWLEDGMENTS

The authors are grateful to Dr. S. Midollini for many helpful comments and suggestions in preparing this chapter. Thanks are due also to Mr. F. Cecconi for his skillful preparation of the drawings.

ABBREVIATIONS

ac: acetate

acac: acetylacetonate

as_3:   1,1,1-tris(2-diphenylarsinomethyl)ethane (II)

As_3S: bis(o-diphenylarsinophenyl)-o-methylthiophenylarsine (XVII)

astol: tris(3-butenyl)arsine

ASTP: tris(o-diphenylphosphinophenyl)arsine (VIII)

Asts(TSA):   tris(o-methylthiophenyl)arsine

Bitas: tris(o-diphenylarsinophenyl)bismutine (XIV)

CFSE: crystal field stabilization energy

DAS: o-phenylenebisdimethylarsino

dasdol: bis(3-butenyl)-3'-dimethylarsinopropylarsine

DMF: N,N'-dimethylformamide

Et: ethyl

Etp_3:   1,1,1-tris(2-diethylphosphinomethyl)ethane (III)

IR: infrared

Me: methyl

$Meas_3$:   1,1,1-tris(2-dimethylarsinomethyl)ethane (IV)

$Menas_3$:   tris(2-dimethylarsinoethyl)amine (XXIV)

$Menp_3$:   tris(2-dimethylphosphinoethyl)amine (XXIII)

MO: molecular orbital

nas_3: tris(2-diphenylarsinoethyl)amine (XXII)

n_2as_2: N,N-bis(2-diphenylarsinoethyl)-2'-diethylaminoethylamine (XXVI)

n_3as: N,N-bis(2-diethylaminoethyl)-2'-diphenylarsinoethyl amine (XXVIII)

np_3: tris(2-diphenylphosphinoethyl)amine (XXI)

n_2p_2: N,N-bis(2-diphenylphosphinoethyl)-2'-diethylaminoethylamine (XXV)

n_3p: N,N-bis(2-diethylaminoethyl)-2'-diphenylphosphinoethylamine (XXVII)

Oh: octahedral

P_3: 1,1,1-tris(2-diphenylphosphinomethyl)ethane (I)

Ph: phenyl

pp_3: tris(2-diphenylphosphinoethyl)phosphine (XXIX)

pp_3-neo: tris(2-dineopentylphosphinoethyl)phosphine

PTAS: tris(o-diphenylarsinophenyl)phosphine (X)

Ptn(TNP): tris(o-dimethylaminophenyl)phosphine (XII)

Ptse(TSeP): tris(o-methylselenophenyl)phosphine (XVI)

Pts(TSP): tris(o-methylthiophenyl)phosphine (XV)

qas: tris(3-dimethylarsinopropyl)arsine (XVIII)

Qas: tris(o-dimethylarsinophenyl)arsine (VII)

QAS: tris(o-diphenylarsinophenyl)arsine (VI)

QP: tris(o-diphenylphosphinophenyl)phosphine (V)

sbta(Sbtas): tris(o-dimethylarsinophenyl)stibine (XIII)

SBTAS: tris(o-diphenylarsinophenyl)stibine (XI)

SBTP: tris(o-diphenylphosphinophenyl)stibine (IX)

solv: solvent of crystallization

SqPl: square planar

SqPy: square pyramidal

tap(ptas): tris(3-dimethylarsinopropyl)phosphine (XIX)

tasb(sbtas): tris(3-dimethylarsinopropyl)stibine (XX)

tasol: bis(3-dimethylarsinopropyl)-3-butenylarsine

TBPy: trigonal bipyramidal

Td: tetrahedral

THF: tetrahydrofuran

TPy: trigonal pyramidal

VEN: valence electron number

REFERENCES

1. J. Chatt, F. A. Hart, and H. R. Watson, *J. Chem. Soc.*, 2537 (1962).

2. R. S. Nyholm, *J. Chem. Soc.*, 2061 (1950); J. Chatt and F. G. Mann, *J. Chem. Soc.*, 610 (1939).

3. T. E. W. Howell, S. A. J. Pratt, and L. M. Venanzi, *J. Chem. Soc.*, 3167 (1961).

4. J. G. Hartley, L. M. Venanzi, and D. G. Goodall, *J. Chem. Soc.*, 3930 (1963).

5. B. Chiswell in *Transition Metal Complexes of Phosphorus, Arsenic and Antimony Ligands* (C. A. McAuliffe, ed.), Macmillan, London, 1973, p. 271.

6. C. A. McAuliffe, in *Advances in Inorganic Chemistry and Radiochemistry* (H. J. Emeleus and A. G. Sharpe, eds.), Vol. 17, Academic Press, New York, 1975, p. 165.

7. L. M. Venanzi, *Angew. Chem. Int. Ed. Engl.*, *3*:453 (1964).

8. R. Mason and D. W. Meek, *Angew. Chem. Int. Ed. Engl.*, *17*:183 (1978).

9. D. W. Meek, D. L. DuBois, and J. Thiethof, in *Inorganic Compounds with Unusual Properties* (R. B. King, ed.), American Chemical Society, Washington, D.C., 1976, p. 335.

10. W. Hewertson and H. R. Watson, *J. Chem. Soc.*, 1490 (1962); J. Chatt and F. H. Hart, *J. Chem. Soc.*, 1378 (1960).

11. S. Midollini and F. Cecconi, *J. Chem. Soc.*, 681 (1973).

12. D. Berglund and D. W. Meek, *Inorg. Chem.*, *11*:1493 (1972).

13. A. S. Kasenally, R. S. Nyholm, and M. H. B. Stiddard, *J. Amer. Chem. Soc.*, *86*:1884 (1964); J. R. Phillips and J. H. Vis, *Can. J. Chem.*, *45*:675 (1967).

14. O. St. C. Headly, R. S. Nyholm, C. A. McAuliffe, L. Sindellari, M. L. Tobe, and L. M. Venanzi, *Inorg. Chim. Acta*, *4*:93 (1970).

15. B. R. Higginson, C. A. McAuliffe, and L. M. Venanzi, *Inorg. Chim. Acta*, *5*:37 (1971).

16. (a) H. P. Fritz, I. R. Gordon, K. E. Schwarzhans, and L. M. Venanzi, *J. Chem. Soc.*, 5210 (1965); (b) R. E. Christopher, I. R. Gordon, and L. M. Venanzi, *J. Chem. Soc.*, 205 (1967).

17. L. Baracco, M. T. Halfpenny, and C. A. McAuliffe, *J. Chem. Soc.*, 1945 (1973).

18. W. Levason, C. A. McAuliffe, and S. G. Murray, *J. Chem. Soc.*, *Chem. Commun.*, 164 (1975).

19. G. Dyer and D. W. Meek, *Inorg. Chem.*, *4*:1398 (1965).

20. G. Dyer and D. W. Meek, *Inorg. Chem.*, *6*:149 (1967).

21. M. Mathew, G. J. Palenik, G. Dyer, and D. W. Meek, *J. Chem. Soc.*, *Chem. Commun.*, 379 (1972).

22. G. A. Barclay and A. K. Barnard, *J. Chem. Soc.*, 4269 (1961).

23. G. S. Benner, W. E. Hatfield, and D. W. Meek, *Inorg. Chem.*, *3*: 1544 (1964).

24. C. A. McAuliffe and D. W. Meek, *Inorg. Chim. Acta*, *5*:270 (1971).

25. L. Sacconi and I. Bertini, *J. Amer. Chem. Soc.*, *90*:5443 (1968).

26. L. Sacconi, I. Bertini, and F. Mani, *Inorg. Chem.*, *7*:1417 (1968).

27. C. Bianchini, C. Mealli, S. Midollini, and L. Sacconi, *Inorg. Chim. Acta*, *31*:L433 (1978).

28. L. Sacconi, P. Dapporto, and P. Stoppioni, *Inorg. Chem.*, *15*: 325 (1976).

29. L. Sacconi and R. Morassi, *J. Chem. Soc.*, 2904 (1969).

30. R. B. King, R. N. Kapoor, M. S. Saran, and P. N. Kapoor, *Inorg. Chem.*, *10*:1851 (1971); R. B. King and J. C. Cloyd, Jr., *Inorg. Chem.*, *14*:1550 (1975).

31. R. B. King and J. C. Cloyd, Jr., *J. Amer. Chem. Soc.*, *97*:53 (1975).

32. R. B. King, J. C. Cloyd, Jr., and R. H. Reimann, *J. Org. Chem.*, *41*:972 (1976).

33. R. Davis and J. E. Fergusson, *Inorg. Chim. Acta*, *4*:23 (1970).

34. D. Berglund, Ph.D. dissertation, Ohio State University, 1969.

35. (a) C. Mealli, S. Midollini, and L. Sacconi, *Inorg. Chem.*, *14*: 2513 (1975); (b) C. A. Ghilardi, S. Midollini, and L. Sacconi, *J. Organometal. Chem.*, *186*:279 (1980).

36. C. A. Ghilardi, C. Mealli, S. Midollini, V. I. Nefedov, A. Orlandini, and L. Sacconi, *Inorg. Chem.*, *19*:2454 (1980).

37. C. Mealli, S. Midollini, and L. Sacconi, *J. Chem. Soc.*, *Chem. Commun.* 765 (1975).

38. C. Benelli, M. Di Vaira, G. Noccioli, and L. Sacconi, *Inorg. Chem.*, *16*:182 (1977).

39. P. L. Orioli, *Coord. Chem. Rev.*, *6*:285 (1971); L. Sacconi, *Coord. Chem. Rev.*, *8*:351 (1972); R. Morassi, I. Bertini, and L. Sacconi, *Coord. Chem. Rev.*, *11*:343 (1973).

40. (a) P. Dapporto, G. Fallani, S. Midollini, and L. Sacconi, *J. Amer. Chem. Soc.*, *95*:2021 (1973); (b) P. Dapporto, S. Midollini, and L. Sacconi, *Inorg. Chem.*, *14*:1643 (1975).

41. M. J. Bennet, J. V. Brencic, and F. A. Cotton, *Inorg. Chem.*, *8*: 1060 (1969); F. A. Cotton and D. A. Ucko, *Inorg. Chim. Acta*, *6*: 161 (1972); S. I. Murahashi, T. Mizoguchi, T. Hasohawa, I. Moritani, Y. Kai, M. Kohara, N. Yasuaka, and N. Kasai, *J. Chem. Soc., Chem Commun.*, 563 (1974); F. A. Cotton and S. A. Koch, *J. Amer. Chem. Soc.*, *99*:7371 (1977); H. Vahrenkamp, *Angew. Chem. Int. Ed. Engl.*, *17*:379 (1978); R. H. Summerville and R. Hoffmann, *J. Amer. Chem. Soc.*, *101*:3821 (1979).

42. J. W. Lauher, M. Elian, R. H. Summerville, and R. Hoffmann, *J. Amer. Chem. Soc.*, *98*:3219 (1976).

43. L. Sacconi and S. Midollini, *J. Chem. Soc. Dalton*, 1213 (1972); A. Bencini, C. Benelli, D. Gatteschi and L. Sacconi, *Inorg. Chim. Acta*, *37*:195 (1979).

44. P. Dapporto, F. Fallani, S. Midollini, and L. Sacconi, *J. Chem. Soc., Chem. Commun.*, 1160 (1972); P. Dapporto, G. Fallani, and L. Sacconi, *Inorg. Chem.*, *13*:2847 (1974).

45. P. Dapporto, S. Midollini, A. Orlandini, and L. Sacconi, *Inorg. Chem.*, *15*:2768 (1976).

46. Unpublished results from this Laboratory.

47. L. H. Vogt, Jr., J. L. Kats, and S. E. Wiberly, *Inorg. Chem.*, *4*:1157 (1965); C. Berbeau and R. J. Dubey, *Can. J. Chem.*, *51*: 3684 (1973); R. R. Ryan, P. G. Eller, and G. J. Kubas, *Inorg. Chem.*, *15*:797 (1976).

48. P. G. Eller, Ph.D. dissertation, Ohio State University, 1971.

49. C. Bianchini, C. Mealli, A. Meli, A. Orlandini, and L. Sacconi, *Angew. Chem. Int. Ed. Engl.*, *18*:673 (1979); C. Bianchini, A. Meli, A. Orlandini, and L. Sacconi, *Inorg. Chim. Acta*, *35*:L375 (1979).

50. C. Bianchini, C. Mealli, A. Meli, A. Orlandini, and L. Sacconi, *Inorg. Chem.*, *19*:2968 (1980).

51. H. Behrens, H. D. Feilner, and E. Lindner, *Z. Anorg. Allg. Chem.*, *385*:321 (1971); H. Behrens and W. Aquila, *Z. Anorg. Allg. Chem.*, *356*:8 (1967).

52. J. Chatt and R. H. Watson, *J. Chem. Soc.*, 4980 (1961); J. Chatt and F. A. Hart, *J. Chem. Soc.*, 812 (1965).

53. H. Behrens and K. Lutz, *Z. Anorg. Allg. Chem.*, *356*:225 (1967); J. Ellerman and H. A. Lindner, *Z. Naturforsch.*, *31b*:1350 (1976).

54. R. S. Nyholm, M. R. Snow, and M. H. B. Stiddard, *J. Chem. Soc.*, 6564 (1965).

55. A. S. Kasenally, R. S. Nyholm, and M. H. B. Stiddard, *J. Chem. Soc.*, 5343 (1965).

56. P. K. Maples, M. Green, and F. G. A. Stone, *J. Chem. Soc. Dalton*, 388 (1973); J. Browing and B. R. Penfold, *J. Chem. Soc., Chem. Commun.*, 198 (1978).

57. R. B. King. L. W. Hauk, and K. H. Pannell, *Inorg. Chem.*, *8*:1042 (1969).

58. E. O. Fischer and K. Fichtel, *Chem. Ber.*, *94*:1200 (1961); A. Davidson, M. L. H. Green, and G. Wilkinson, *J. Chem. Soc.*, 3172 (1961); R. B. King, *Inorg. Chem.*, *1*:964 (1962).

59. C. Bianchini, P. Dapporto, A. Meli, and L. Sacconi, *J. Organometal. Chem.*, *193*:117 (1980).

60. R. Burton, L. Pratt, and G. Wilkinson, *J. Chem. Soc.*, 594 (1961).

61. R. B. King and A. Efraty, *Inorg. Chem.*, *8*:2374 (1969).

62. H. Behrens and H. Brandl, *Z. Naturforsch.*, *B22*:1353 (1967).

63. (a) W. S. Tsang, D. W. Meek, and W. Wojcicki, *Inorg. Chem.*, 7: 1264 (1968); (b) D. Rehder and U. Puttfarcken, *J. Organometal. Chem.*, *184*:343 (1980).

64. (a) D. L. Dubois and D. W. Meek, *Inorg. Chim. Acta*, *19*:L29 (1976); (b) U. Puttfarcken and D. Rehder, *J. Organometal. Chem.*, *185*:219 (1980).

65. F. Mani and P. Stoppioni, *Inorg. Chim. Acta*, *16*:177 (1976).

66. I. V. Howell, L. M. Venanzi, and D. C. Goodall, *J. Chem. Soc.*, 395 (1967).

67. M. T. Halfpenny, J. G. Hartley, and L. M. Venanzi, *J. Chem. Soc.*, 627 (1967).

68. J. G. Hartley, D. G. E. Kerfoot, and L. M. Venanzi, *Inorg. Chim. Acta*, *1*:145 (1967).

69. (a) G. Dyer, J. G. Hartley, and L. M. Venanzi, *J. Chem. Soc.*, 1293 (1965); (b) G. Dyer and L. M. Venanzi, *J. Chem. Soc.*, 2771 (1965).

70. T. L. Blundell, H. M. Powell, and L. M. Venanzi, *J. Chem. Soc., Chem. Commun.*, 763 (1967); T. L. Blundell and H. M. Powell, *Acta Crystallogr.*, *B27*:2304 (1971).

71. I. V. Howell and L. M. Venanzi, *J. Chem. Soc.*, 1007 (1967).

72. B. R. Higginson, C. A. McAuliffe, and L. M. Venanzi, *Helv. Chim. Acta*, *58*:1261 (1975).

73. B. Chiswell and L. M. Venanzi, *J. Chem. Soc.*, 417 (1966).

74. J. W. Dawson and L. M. Venanzi, *J. Amer. Chem. Soc.*, *90*:7229 (1968); J. W. Dawson, B. C. Lane, R. J. Mynott, and L. M. Venanzi, *Inorg. Chim. Acta*, *5*:25 (1971).

75. L. P. Haugen and R. Eisenberg, *Inorg. Chem.*, *8*:1072 (1969).

76. G. S. Benner and D. W. Meek, *Inorg. Chem.*, *6*:1399 (1967).

77. J. R. Ferraro, *Inorg. Nucl. Chem. Lett.*, *6*:823 (1970).

78. G. Kordosky, G. S. Benner, and D. W. Meek, *Inorg. Chim. Acta*, *7*:605 (1973).

79. D. L. Stevenson and L. F. Dahl, *J. Amer. Chem. Soc.*, *89*:3424 (1967).

80. C. A. McAuliffe and D. G. Watson, *J. Organometal. Chem.*, *78*:C51 (1974).

81. (a) M. J. Norgett, J. H. M. Thornley, and L. M. Venanzi, *J. Chem. Soc.*, 540 (1967); M. J. Norgett, J. H. M. Thornley, and L. M. Venanzi, *Coord. Chem. Rev.*, *2*:99 (1967); (b) J. W. Dawson, H. B. Gray, J. E. Hix, J. R. Preer, and L. M. Venanzi, *J. Amer. Chem. Soc.*, *94*:2979 (1972).

82. M. J. Norgett and L. M. Venanzi, *Inorg. Chim. Acta*, *2*:107 (1968).

83. F. Mani, P. Stoppioni, and L. Sacconi, *J. Chem. Soc. Dalton*, 461 (1975).

84. M. Bacci, S. Midollini, P. Stoppioni, and L. Sacconi, *Inorg. Chem.*, *12*:1801 (1973).

85. M. Di Vaira, *J. Chem. Soc. Dalton*, 2360 (1975).

86. L. Sacconi and M. Di Vaira, *Inorg. Chem.*, *17*:810 (1978).

87. R. B. King, J. C. Cloyd, Jr., and R. H. Reimann, *Inorg. Chem.*, *15*:449 (1976).

88. L. Sacconi and I. Bertini, *J. Amer. Chem. Soc.*, *89*:2235 (1967).

89. P. Dapporto and L. Sacconi, *J. Chem. Soc., Chem. Commun.*, 1091 (1969); P. Dapporto and L. Sacconi, *J. Chem. Soc.*, 1804 (1970).

90. M. Di Vaira and L. Sacconi, *J. Chem. Soc. Dalton*, 493 (1975).

91. L. Sacconi, M. Di Vaira, and A. Bianchi, *J. Amer. Chem. Soc.*, *92*:4465 (1970); M. Di Vaira and A. Bianchi Orlandini, *Inorg. Chem.*, *12*:1292 (1973); M. Di Vaira, *J. Chem. Soc. Dalton*, 1575 (1975).

92. P. L. Orioli and L. Sacconi, *J. Chem. Soc., Chem. Commun.*, 1012 (1969); C. Mealli, P. L. Orioli, and L. Sacconi, *J. Chem. Soc.*, 2691 (1971).

93. L. Sacconi, *Pure Appl. Chem.*, *17*:95 (1968).

94. F. Mani and I Sacconi, *Inorg. Chim. Acta*, *4*:365 (1970).

95. P. Stoppioni, . Mani, and L. Sacconi, *Inorg. Chim. Acta*, *11*:227 (1974).

96. L. Sacconi, *J. Chem. Soc.*, 248 (1970); L. Sacconi, *Coord. Chem. Rev.*, *8*:351 (1972).

97. W. S. J. Kelly, G. H. Ford, and S. M. Nelson, *J. Chem. Soc.*, 388 (1971).

98. M. Di Vaira, A. Meli, and L. Sacconi, *Crystal. Struct. Commun.*, 6:727 (1977).

99. P. Dapporto and L. Sacconi, *Inorg. Chim. Acta*, 9:L2 (1974).

100. A. Bianchi, P. Dapporto, G. Fallani, C. A. Ghilardi, and L. Sacconi, *J. Chem. Soc. Dalton*, 641 (1973).

101. P. Dapporto and G. Fallani, *J. Chem. Soc. Dalton*, 1498 (1972).

102. M. Ciampolini, N. Nardi, and G. P. Speroni, *Coord. Chem. Rev.*, 1:222 (1966); M. Ciampolini and I. Bertini, *J. Chem. Soc.*, 2241 (1968).

103. M. Di Vaira, and L. Sacconi, *J. Chem. Soc.*, *Chem. Commun.*, 10 (1969); M. Di Vaira, *J. Chem. Soc.*, 148 (1971).

104. M. Di Vaira and A. Bianchi Orlandini, *J. Chem. Soc. Dalton*, 1704 (1972).

105. I. Bertini, P. Dapporto, G. Fallani, and L. Sacconi, *Inorg. Chem.*, 10:1703 (1971).

106. A. Orlandini and L. Sacconi, *Inorg. Chem.*, 15:78 (1976).

107. P. M. Treichel, W. K. Dean, and J. C. Calabrese, *Inorg. Chem.* 12:2908 (1973); L. Vaska and J. Peone, Jr., *J. Chem. Soc.*, *Chem. Commun.*, 418 (1971); C. A. Reed and W. R. Roper, *J. Chem. Soc.*, *Chem. Commun.*, 1459 (1969).

108. L. Sacconi, A. Orlandini, and S. Midollini, *Inorg. Chem.*, 13:2850 (1974).

109. H. Vahrenkamp, *Angew. Chem. Int. Ed. Engl.*, 14:322 (1975).

110. M. Di Vaira, S. Midollini, and L. Sacconi, *Inorg. Chem.*, 16:1518 (1977).

111. M. Di Vaira, S. Midollini, and L. Sacconi, *Inorg. Chem.*, 17:816 (1978).

112. C. Mealli, S. Midollini, and L. Sacconi, *Inorg. Chem.*, 17:632 (1978).

113. Unpublished results from this Laboratory.

114. (a) C. Mealli, and L. Sacconi, *J. Chem. Soc.*, *Chem. Commun.*, 886 (1973); (b) L. Sacconi, C. A. Ghilardi, C. Mealli, and F. Zanobini, *Inorg. Chem.*, 14:1380 (1975).

115. P. Dapporto and L. Sacconi, *Inorg. Chim. Acta*, 39:61 (1980).

116. C. A. Ghilardi, S. Midollini, and L. Sacconi, *Inorg. Chem.*, 14:1790 (1975).

117. L. Sacconi, P. Dapporto, and P. Stoppioni, *Inorg. Chem.*, 15:325 (1976).

118. L. Sacconi, P. Dapporto, and P. Stoppioni, *J. Amer. Chem. Soc.*, 97:5595 (1975); L. Sacconi, P. Dapporto, and P. Stoppioni, *Inorg. Chem.*, 16:224 (1977).

119. M. Di Vaira, C. A. Ghilardi, and L. Sacconi, *Inorg. Chem.*, *15*: 1555 (1976).

120. M. Bacci and S. Midollini, *Inorg. Chim. Acta*, *5*:220 (1971).

121. C. A. Ghilardi, A. Sabatini, and L. Sacconi, *Inorg. Chem.*, *15*: 2763 (1976).

122. J. H. Enemark and R. D. Feltham, *Coord. Chem. Rev.*, *13*:339 (1974); R. Hoffmann, M. M. L. Chen, M. Elian, A. R. Rossi, and D. M. P. Mingos, *Inorg. Chem.*, *13*:2666 (1974) and references therein.

123. M. Di Vaira, A. Tarli, P. Stoppioni, and L. Sacconi, *Crystal. Struct. Commun.*, *4*:653 (1975).

124. C. Mealli, A. Orlandini, L. Sacconi, and P. Stoppioni, *Inorg. Chem.*, *17*:3020 (1978).

125. L. Sacconi, P. Dapporto, P. Stoppioni, P. Innocenti, and C. Benelli, *Inorg. Chem.*, *16*:1669 (1977); P. Dapporto, L. Sacconi, P. Stoppioni, *Atti Acc. Naz. Lincei Rend. Sc. Fis. Mat. e Nat.*, *60*:641 (1976).

126. P. Stoppioni, P. Dapporto, and L. Sacconi, *Inorg. Chem.*, *17*: 718 (1978).

127. H. F. Klein and H. H. Karsch, *Chem. Ber.*, *105*:2628 (1972).

128. L. Sacconi, P. Dapporto, and P. Stoppioni, *J. Organometal. Chem.*, *116*:C33 (1976).

129. T. H. Coffield, R. D. Closson, and J. Kozikowski, *J. Org. Chem.*, *22*:598 (1957); F. Calderazzo and F. A. Cotton, *Inorg. Chem.*, *1*:30 (1962); R. J. Mawby, F. Basolo, and R. G. Pearson, *J. Amer. Chem. Soc.*, *86*:3994 (1964); K. Noack and F. Calderazzo, *J. Organometal. Chem.*, *10*:101 (1967); K. Noack, M. Ruch, and F. Calderazzo, *Inorg. Chem.*, *7*:345 (1968).

130. Unpublished results from this Laboratory.

131. C. A. Ghilardi, S. Midollini, and L. Sacconi, *Inorg. Chem.*, *16*:2377 (1977).

132. P. Stoppioni and P. Dapporto, *Crystal. Struct. Commun.*, *7*:375 (1978).

133. M. Di Vaira, C. A. Ghilardi, S. Midollini, and L. Sacconi, *J. Amer. Chem. Soc.*, *100*:2550 (1978).

134. F. Cecconi, P. Dapporto, S. Midollini, and L. Sacconi, *Inorg. Chem.*, *17*:3292 (1978).

135. M. Di Vaira, S. Midollini, L. Sacconi, and F. Zanobini, *Angew. Chem. Int. Ed. Engl.*, *17*:676 (1978); M. Di Vaira, S. Midollini, and L. Sacconi, *J. Amer. Chem. Soc.*, *101*:1757 (1979).

136. L. Fabbrizzi and L. Sacconi, *Inorg. Chim. Acta*, *36*:L407 (1979).

3

Stereochemistry of α-Hydroxycarboxylate Complexes

ROBERT E. TAPSCOTT

University of New Mexico
Albuquerque, New Mexico

I. INTRODUCTION

α-Hydroxycarboxylate complexes have numerous analytical, industrial,
and medical applications. Their formation is important in many ion
exchange separations worked out for lanthanides and actinides [1].
Although they have been extensively studied, no general analysis of
their geometries has appeared in the literature. This chapter pre-
sents a review and a systematic analysis of stereochemical studies
reported for these complexes. The review is not exhaustive, although
the survey of X-ray diffraction studies through early 1980 is hopeful-
ly complete. Solution studies have been deliberately slighted in
favor of solid-state investigations whose results are usually less
ambiguously interpretable. Since α-hydroxycarboxylate complexes of
nontransition elements often show structural features similar to
those of the transition metals, discussions of the former have been
included.

   This survey covers complexes of ligands containing α-hydroxy-,
α-alkoxy-, or α-aryloxycarboxylate groups (1) or their sulfur analogs.

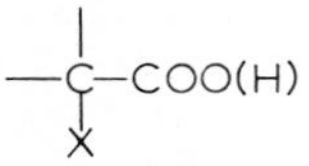

$$-\overset{\displaystyle |}{\underset{\displaystyle X}{C}}-COO(H)$$

X = O(H, R, Ar)

(1)

Brief references to complexes of structurally related ligands such as
the isoelectronic α-aminoacidates and (non-α)hydroxycarboxylates are
included for comparison. Earlier reviews on α-hydroxycarboxylate
complexes [2-5] have emphasized specific systems, and with one exception [4] have not been structurally oriented. Some information on
α-hydroxycarboxylates has also been included in earlier general reviews on carboxylate complexes [6,7] and on calcium binding [8,9].

II. α-HYDROXYCARBOXYLIC ACIDS

A. Sources, Nomenclature, and Structure

The syntheses and reactions of α-hydroxycarboxylic acids have been
recently reviewed [10]. A number of these compounds occur naturally
as phytochemicals in molds [11], algae [12], and higher plants [13],

and as important biochemicals in animals [14]. In particular, they play important roles in the tricarboxylic acid cycle, in catabolism of carbohydrates, and in syntheses of amino acids. Hydroxyacids have been identified in a carbonaceous chondrite [15] and in products of prebiotic discharge experiments [16].

Structures of some of the more common α-hydroxycarboxylates are shown in Fig. 1. The more common enantiomers with their absolute configuration designations [17] are shown for asymmetric naturally occurring compounds [18,19]. After much consideration, it has been decided to employ trivial names for most organic compounds discussed in this chapter.

Of the simple α-hydroxycarboxylic acids, the best known are probably glycolic, lactic, and mandelic acids (Fig. 1). Although mandelic acid has only minor biological roles, glycolic acid, which

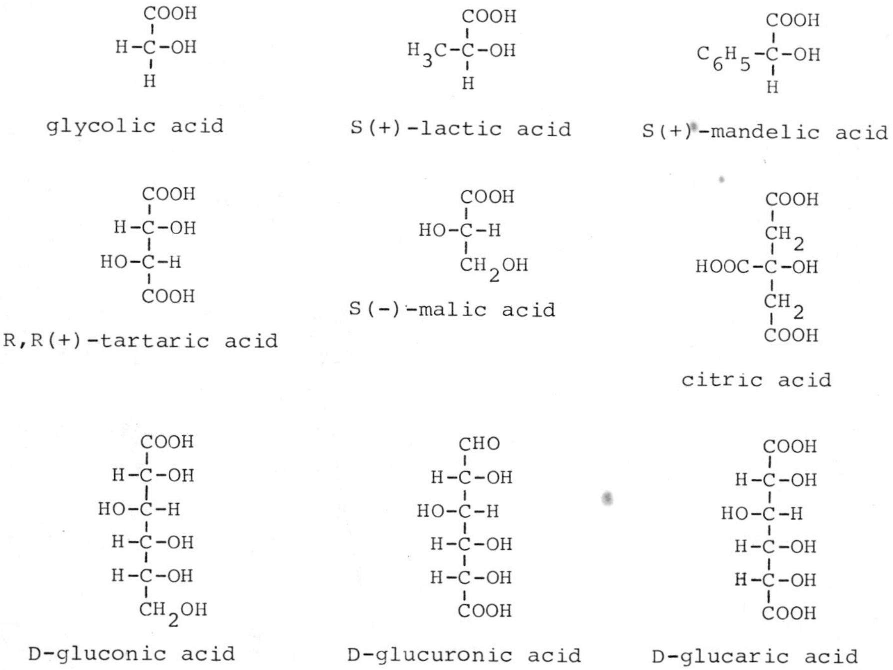

Figure 1 Fischer projections of some common α-hydroxycarboxylic acids.

is involved in plant photorespiration, and lactic acid, which is found in a large number of metabolic and catabolic sequences [20], are important biological materials. Unlike glycolic acid, lactic acid is optically active. Although both enantiomers are found naturally, the S(+) isomer predominates [21].

The biologically important acids tartaric, malic, and citric are perhaps the best known of the multifunctional α-hydroxycarboxylic acids. Tartaric acid is a relatively abundant phytochemical [22], whereas malic and citric acids are exceedingly important and ubiquitous biochemicals best known for their participation in the tricarboxylic acid cycle [23]. A large group of multifunctional acids are the sugar acids (which can include tartaric and malic acids) [24]. Three important families of these acids are the aldonic, uronic, and saccharic acids--the oxidized aldoses. Examples are D-gluconic, D-glucuronic, and D-glucaric acid--all oxidized forms of D-glucose (Fig. 1). The sugar acids are involved in a number of biological processes.

B. Chemical and Physical Properties

All of the α-hydroxycarboxylic acids show a tendency, in some cases marked, to form intermolecular and, when δ- and/or γ-hydroxyl groups are present (as in the sugar acids), intramolecular esters. In general, the solution equilibria between the free acids and esters are not considered here.

Although the presence of esters in aqueous solutions of many α-hydroxycarboxylic acids makes the determination of accurate ionization constants difficult [25], reported thermodynamic data leave no doubt that the presence of an α-hydroxyl group significantly decreases carboxyl group pK_a. Some selected thermodynamic values are given in Table 1. The effect of an α-hydroxyl group can also be seen in the preferential ionization of a carboxylate adjacent to such a group in polyprotic carboxylic acids [36,37].

Although internal hydrogen bonding between hydroxyl and carboxyl groups could lower the carboxylate pK_a's relative to the corresponding

Table 1 Ionization Thermodynamic Quantities for Selected Carboxylic Acids in Water at 25°C

Acid	pK_a	ΔH, kcal mol^{-1}	ΔS, e.u.	References
Acetic acid, CH_3COOH	4.77	-0.02	-22	26, 27
Glycolic acid, $CH_2(OH)COOH$	3.83	+0.11	-17	26, 28
Methoxyacetic acid, $CH_2(OCH_3)COOH$	3.57	-0.96	-20	29
Propionic acid, CH_3CH_2COOH	4.87	-0.14	-23	26, 30
Lactic acid, $CH_3CH(OH)COOH$	3.86	-0.06	-18	31
Succinic acid, $HOOCCH_2CH_2COOH$	4.21	+0.80	-17	26, 32
	5.64	+0.06	-26	26, 33
Malic acid, $HOOCCH(OH)CH_2COOH$	3.46	+0.71	-14	34
	5.10	-0.28	-24	34
(+)-Tartaric acid, $HOOCCH(OH)CH(OH)COOH$	3.04	+0.74	-11	35
	4.31	+0.24	-19	35

fatty acids [38,39], such hydrogen bonding is very unlikely in polar
solvents such as water [39,40]. The effect on carboxylate acidity
of an α-hydroxyl group is primarily an entropy effect, which is, in
fact, partially offset by adverse enthalpy changes owing to enhanced
solvation of α-hydroxycarboxylic acids [29]. The low pK_a's of alkoxy-
acids, on the other hand, are ascribed to a decreased solvation for
such compounds [29].

Table 2 gives pK_a values for the alcohol groups of mandelate(1-)
and tartrate(2-) and for the mercapto group in mercaptoacetate(1-).
These are the only free α-hydroxy- or α-mercaptocarboxylates for
which such data have been reported. Values for related compounds are
also presented. A negative charge in the α position should increase
the pK_a of a hydroxyl or mercapto group by about 2.3 (estimated from
the maximum observed ΔpK_a of 2.9 for dicarboxylic acids after correct-
ing for a symmetry contribution of $\ln 4$) [47]. After an allowance for
this charge effect has been made, a comparison of the hydroxyl group
pK_a values for α-hydroxycarboxylates with those of related alcohols
(Table 2) shows the acidity-increasing effect of a neighboring carboxyl group. Mercapto groups are particularly acidic [48].

C. Geometry

Geometric parameters of uncoordinated α-hydroxycarboxylate groupings
as determined by X-ray or neutron diffraction are given in Table 3.*
Where more than one structure determination has been reported for a
given compound, the results of that determination judged most precise
are given. Groups containing carboxyl or hydroxyl oxygen atoms within bonding distance of any metal ions have been omitted from this
table. Their geometries will be discussed later. In the several
cases where protonated and unprotonated groupings are present in the
same structure, results for that structure are found in two different
parts of Table 3. In those few cases where both coordinated and
uncoordinated groups are found in the same structure, parameters for

---

*Note: Tables 3 through 9 appear on pp. 353-411.

Table 2 Acid Dissociation Constants for Hydroxyl and Mercapto Groups in Water (At 25°C Unless Otherwise Noted)

Compound	pK_a	References
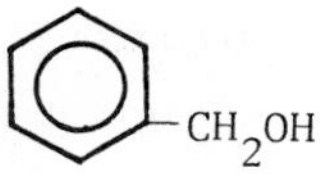 $\bigcirc$—CH_2OH	15.4	41
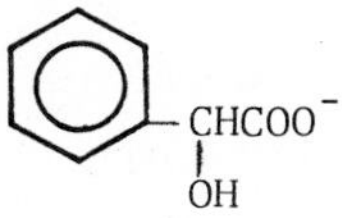 $\bigcirc$—$\underset{OH}{CHCOO^-}$	16.4[a]	42
$\underset{OH}{CH_2}\underset{OH}{CH_2}$	14.8	43
$^-OOC\underset{OH}{CH}{-}\underset{OH}{CH}COO^-$	15.0,[b] 14.2,[c] 15.5[d]	44, 45
$^-OOC\underset{OH}{CH}{-}\underset{O^-}{CH}COO^-$	16.5,[b] 15.7,[c] 17.5[d]	44, 45
$\underset{OH}{CH_2}CH_3$	15.9	43
$\underset{SH}{CH_2}CH_2$	10.6	46
$\underset{SH}{CO_2}COO^-$	10.7	46

[a] 20°C.

[b] In NaOH at 22°C.

[c] In NaOH.

[d] In KOH at 22°C.

the latter are reported here. Preliminary reports have been given
on fentanyl(2+) citrate(2-) [108] and (R)- and (S)-2-hydroxypropyl-
1-ammonium (+)-tartrate(1-) [109]. These structures are not given
in Table 3.

      Table 3 also gives average values of the geometric parameters
for protonated, ionized, and esterified α-hydroxycarboxylate groups
and average absolute deviations. These values are reproduced in Fig.
2. No attempt was made to use weighted averages, since the bonding
parameters are expected to vary with the carbinol carbon substituents
and environment. Although it is known that intermolecular forces
such as hydrogen bonding can affect bonding parameters of carboxyl
groups [64], the relatively large number of values for which the

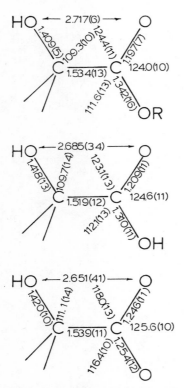

Figure 2 Average dimensions determined for esterified (top), protonated (middle), and ionized (bottom) α-hydroxycarboxylate groups.

averages were obtained should help offset variability due to such interactions. The averages indicate that the bonding parameters of carboxyl groups in α-hydroxyacids differ little from those in alkanoic acids [110,111], although the O-C=O angle and C=O bond length seem to be slightly larger and smaller, respectively, in the former compounds.

The changes in the average bond parameters upon going from esters to acids to ionized carboxylates are consistent with a decreasing double bond character and, therefore, a decreasing steric bulk [112-114] for the carbonyl group. As the lengths of the carboxylate C-O bonds proximal and distal to the hydroxyl (the latter are alkylated or protonated in most esters and acids) increase and decrease, respectively, the C-C-O angles involving these two bonds show respective decreases and increases (Fig. 2). It is interesting that these changes are consistent with a greater covalency in the O-R bond of esters than in the O-H bond of acids--a result not unexpected considering the relative bond polarities and the hydrogen bonding tendencies (weakening the O-H bond) of acids in the solid state. Similar trends in metal ion-complexed α-hydrocarboxylate group geometries may permit the evaluation of relative covalency and degree of metal ion interaction (O-M bond strength?) in metal ion complexes.

An almost generic feature of α-hydroxycarboxylates is the planarity of the five-atom α-hydroxycarboxylate group--a structural feature early observed in crystal structure determinations [115]. Planar groupings are also found in the isoelectronic α-aminocarboxylates [116]. It is interesting that in the absence of α-oxygen or α-nitrogen atoms, a C_α-C_β bond may lie near the plane of the carboxyl group [117,118]. Similarly, in bromoacetic [119] and fluoromalic [77] acids, halogens lie near the α-hydroxycarboxylate planes.

The O-C-C-O dihedral angles between hydroxyl oxygens and the *nearest* carboxyl oxygens, across C-C bonds, are one measure of α-hydroxycarboxylate planarity, and these are listed in Table 3. Positive and negative angles denote clockwise and counterclockwise rotations, respectively (Fig. 3). Most, but by no means all, of the dihedral angles are near the value of 0° required for a planar

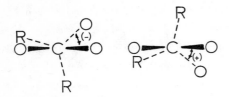

Figure 3 Dihedral angle convention employed here for α-hydroxycarboxylate groups.

grouping--whether the carboxylate is protonated, ionized, or esterified. In some cases, nonplanarity is attributable to specific factors.
The cyclic ether N-acetylneuramic acid derivatives contain two oxygen
atoms attached to the α-carbon, and the carboxyl groups appear to move
toward intermediate positions since they cannot eclipse both (though
the hydroxyl oxygen atoms lie much more nearly in the carboxyl planes
than do the ring oxygens) [81,100]. In the substituted 1-hydroxy-
cyclohexane-1-carboxylic acid [82] and tartronic acid [58], intramolecular contacts between the carboxyl oxygen atoms and other groups (a
methylene hydrogen in the former compound and another carboxyl oxygen
in the latter) force nonplanarity of the α-hydroxycarboxylate groupings. Intermolecular forces also undoubtedly contribute to nonplanarity in these highly hydrogen bonded systems.

Pauling's bent-bond description of the carbonyl double bond [117]
and intramolecular interactions between hydroxyl and carboxyl oxygen
atoms [115] have both been proposed to explain α-hydroxycarboxylate
planarity. Intramolecular hydrogen bonding does not appear to be an
important factor. Although several structures containing an intramolecular hydrogen bond between the hydroxyl and nearest carboxyl oxygen atoms have been reported [51,84,88,90,97], and although theoretical
calculations show that the lowest energy conformation corresponds to
an internally hydrogen bonded system [120], in most cases no such bond
is found in the solid state owing to the possibility for stronger
intermolecular hydrogen bonding. Similarly, although intramolecular
hydrogen bonding between a carboxyl oxygen and an α-hydroxyl group has
been observed in nonpolar solvents [121], such bonding is very unlikely in polar solvents [39,40].

Both molecular orbital [120,122] and molecular mechanics [123] calculations predict planar conformations for the α-hydroxycarboxylate grouping. Moreover, the former calculations predict two minima corresponding to the protonated carboxyl oxygen atom proximal and distal to the hydroxyl oxygen, with the latter conformer lying lower in energy. In agreement with this, most crystal structure determinations on protonated α-hydroxycarboxylates, with few exceptions [56, 61,67,82], have shown the protonated carboxyl oxygen to be distal to the hydroxyl oxygen atom. Similarly, with only two exceptions [105, 107], both of which show highly nonplanar α-hydroxycarboxylate groups, X-ray structure determinations have shown the alkylated oxygen of α-hydroxycarboxylate esters to be distal to the hydroxyl group.

Since C-C=O angles in carboxylate groupings are larger than C-C-O angles (Table 3), the proximal conformers are expected to have smaller values for the hydroxyl oxygen-to-nearest carboxyl oxygen nonbonded distance--a distance that is already very small. This prediction is confirmed by a comparison of geometries for proximal conformer structures (structures where footnote c is cited in Table 3) with those for distal conformers. Thus the distal conformer may be stabilized by a decrease in the hydroxyl oxygen to nearest carboxyl oxygen nonbonded interaction. The presence of such a (repulsive) interaction is indicated by the observation that of the two C-C-O angles in an ionized carboxylate, that involving the carboxyl oxygen nearer a hydroxyl group is generally larger (Table 3).

The distribution of positive and negative dihedral angles, corresponding to enantiomeric conformers (Fig. 3), is skewed for dissymetric α-hydroxycarboxylate groups. When the carbinol carbon has a hydrogen atom and one other group attached in addition to the hydroxyl and carboxyl groups, there is a significant tendency for the α-hydroxycarboxylate group to adopt a conformation such that the dihedral angle is positive (a (+)-synperiplanar, + sp, conformation [124]) when the absolute configuration is R (priority [17]: hydroxyl oxygen > carboxyl oxygen > substituent > hydrogen) (Fig. 4). This stereoselective effect may be of importance in explaining

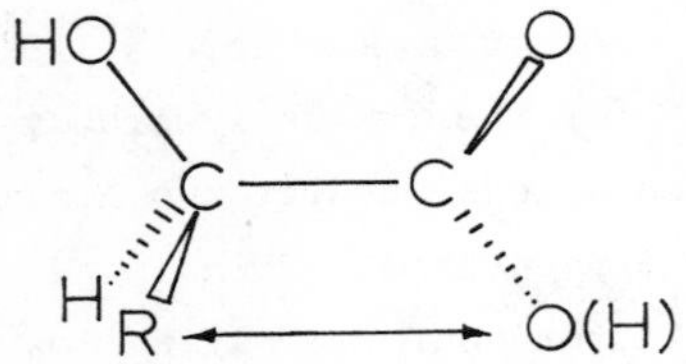

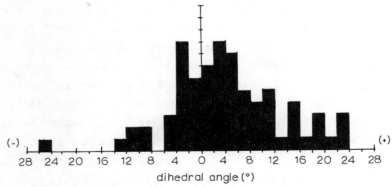

Figure 4 α-Hydroxycarboxylate group conformation most commonly
observed for carbinol carbon atoms of R chirality.

correlations of CD spectra with absolute configuration in α-
hydroxycarboxylates [125] and their complexes [126,127]. Although CD
spectra of lactate chelates have been interpreted assuming a planar
ring [128], in fact, crystallographic data reviewed here indicate
that this is often not strictly true.

A graph of the distribution of dihedral angles for all crystallographically nonequivalent, free, asymmetric α-hydroxycarboxylate groups
for which structural data have been reported is shown in Fig. 5. For
this figure and Fig. 6, coordinates have been inverted where necessary
to give carbinol carbons having R chirality. The distribution of dihedral angles for chiral α-hydroxycarboxylate groups is even more
skewed when tartrate structures are omitted (Fig. 6). Both tartaric
acids [but not tartrate(2-)!] and the ionized portion of bitartrates
show no bias toward a +sp conformation for R chirality.

Figure 5 Distribution of dihedral angles observed for α-hydroxycarboxylate groups bearing a carbinol (bonded to a single hydrogen atom) of
R chirality by X-ray diffraction.

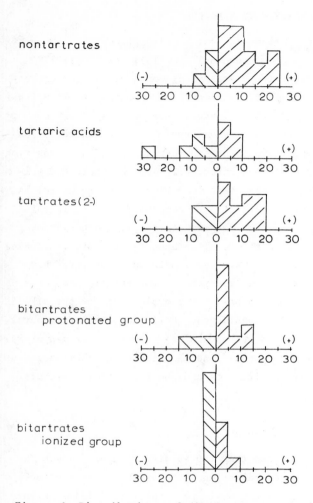

Figure 6 Distributions of dihedral angles by compound type. All results are for α-hydroxycarboxylate carbinols of R chirality.

The tendency toward a particular enantiomeric conformation may be due to steric interactions between the carbinol substituent and the nearer carboxyl oxygen atom (Fig. 4). The isoelectronic amino acids of predominantly L (S) absolute configuration appear to show a skewed distribution of conformations with a -sc ((-)-synclinal [124]) bias [129]. This observation is consistent with the α-hydroxycarboxylate results.

III.   COMPLEXES OF α-HYDROXYCARBOXYLATES

Since even alkali metal cations exhibit a definite coordination
chemistry [130] and have been shown by ^{23}Na [131] and ^{1}H [132] NMR
to complex with α-hydroxycarboxylates in aqueous solution, differen-
tion between simple metal salts and coordination complexes is neither
straightforward nor (usually) meaningful. Moreover, although our
primary interest may be the stereochemistry of transition metal complexes of α-hydroxycarboxylates, the steric features of nontransition
metal α-hydroxycarboxylates are sufficiently similar where compari-
sons can be made that their geometric characteristics are useful in
elucidating the structures of the former compounds. Thus, in this
chapter, all structures in which α-hydroxycarboxylate donor atoms
reside within coordinating distance of an acceptor species (including
a few metalloids and even nonmetals) are included. However, where
alkali metal ions are present as counterions for anionic complexes
formed by other, stronger acceptors, the coordinations of the alkali
metal ions are not generally considered. Following a brief, general
overview of some physical and stereochemical properties of α-hydroxycarboxylate coordination compounds, the detailed stereochemistries as
determined by (primarily) diffraction and spectral studies are discussed by ligand type.

A.   Proton Dissociation and Chelation

A number of studies have shown that the acidity of hydroxyl groups
is greatly increased upon coordination. Chelation is often associated with even larger increases in ligand acidity [133,134]. With
α-hydroxycarboxylates, the five-membered chelate ring (2) is obtained.
Even in aqueous solution, however, hydroxy groups can coordinate without proton loss--at least when other strong coordinating groups are
present (as in the α-hydroxycarboxylates) [135]. A large number of
solid-state structures of simple hydroxyl chelates have been determined with both neutral and anionic ligands--although no attempt will
be made to review the literature in this area--and the metallo

$$(H)O \diagdown \underset{\diagup}{\overset{M}{\diagup}} \diagdown O$$

$$\underset{|}{C} - C \overset{\diagup}{\underset{\diagdown}{O}}$$

(2)

chemistry of alkoxides is extensive [136]. In this chapter the term "deprotonated hydroxyl group" rather than "alkoxide" is used to avoid confusion with α-alkoxycarboxylates, whose coordination chemistry is also reviewed herein.

EPR studies have shown the presence of both 1:1 and 1:2 vanadyl-(IV) glycolates and lactates in aqueous solution [137]. The 1:2 species exhibit first and second proton dissociation values of 3.5 and 5.7 for the glycolate and 3.5 and 5.0 for the lactate corresponding to ionization of one or both coordinated hydroxyl groups. Several pK_a values have been reported for α-hydroxycarboxylates chelated to cobalt (I)--3.3 ± 0.1 for $[Co(en)_2(glycolate)]^{2+}$ [134], 3.4 ± 0.1 for $[Co(en)_2(lactate)]^{2+}$ [134], 2.45 ± 0.1 for $[Co(en)_2(tartrate)(en)_2Co]^{3+}$ [138], and 3.51 for $[Co(phen)_2(lactate)]^{2+}$ [139]. The pK_a values of 2.3 for $[Co(NH_3)_4(tartrate)]^+$ [140] and 3.72 to 4.08 for diastereomers of $[Co(phen)_2(tartrate)]^+$ and $[Co(bipy)_2(tartrate)]^+$ [139] may involve deprotonation of the dangling carboxyl group rather than the coordinated hydroxyl. Proton dissociations from α-hydroxy groups in a series of chromium(III) tartrate complexes occur with pK_a values of 3.8 to 4.0 [141]. Trinuclear uranyl(VI) α-hydroxycarboxylates lose their alcoholic hydrogen atoms below pH7 [142]. Chelation and hydroxyl group ionization in the pH region of 1 to 2 are indicated in equilibrium studies of the reaction of iron(III) with α-hydroxycarboxylates [143].

A number of studies have indicated chelation by α-hydroxycarboxylates in solution. That the formation constants for the 1:1 and 1:2 copper(II) complexes are larger for glycolate than for acetate evidences chelation in the glycolate system [144]. A similar conclusion has been reached for actinide [145-147] and lanthanide [148] glycolates based on their larger-than-expected stability constants.

("Expected" values are based on ligand basicity and acetate complex
stability constants.) Increased volume changes (compared with those
obtained with aliphatic acids) upon coordination of zinc(II) with α-
hydroxycarboxylates in aqueous solution are probably due to release
of electrorestricted water from the hydroxyl groups upon coordination [149].

Differential proton relaxation studies indicate that α-hydroxy-,
α-alkoxy-, and β-hydroxycarboxylates are predominantly bidentate toward copper(II) in aqueous solution, whereas β-alkoxy- and γ-hydroxy-
carboxylates give unidentate complexes [150,151]. Interestingly,
formation constants for actinides with α-, β-, and γ-hydroxycarboxy-
lates indicate that only α-hydroxy derivatives are chelated [152,153].
That ^{23}Na NMR spectra indicate Na$^+$ complexation with hydroxycarboxylates in aqueous solution but no complexation with acetate, formate,
or benzoate under the same conditions [131] indicates that the
hydroxyl group participates, presumably in chelate formation.

Owing to the presence of multiple donor atoms and the possibility
for two or more states of ionization, α-hydroxycarboxylates show a
pronounced tendency to form complexes of varying charge [127] and
degree of aggregation [142-154], even within one system. This tendency to give polynuclear species is notable in the formation of a
number of mixed-metal α-hydroxycarboxylates [155-161]. We have seen,
however, that one feature often found in α-hydroxycarboxylates is a
five-membered chelate ring. This structural feature will be emphasized in our stereochemical analyses.

B. Diffraction and Spectroscopic Studies

Table 4 contains bonding parameters determined for α-hydroxycarboxylate groups within bonding distance of metal ions in solid-state
structures of complexes and metal salts as determined by X-ray and
neutron diffraction. Structures where the carboxylate group has
been incorporated into a ring, as in hydroxy lactones, are not included and will not be discussed in this chapter. (On the other
hand, α-carboxyl cyclic ethers, where the oxygen α to a carboxyl
group is incorporated in a ring, are discussed briefly with other
alkoxides.)

1. Simple α-Hydroxycarboxylates

In this section, the structures of metallo compounds of α-hydroxycarboxylate ligands containing only a single hydroxyl and a single carboxyl group are examined.

Glycolates. Neutron [49] and X-ray [162] diffraction studies have been reported for glycolic (hydroxyacetic) acid, $HOCH_2COOH$, the simplest α-hydroxycarboxylic acid.

The structures of four alkali metal salts of glycolate(1-) have been determined. The anhydrous [163] and monohydrated [164] forms of lithium glycolate are polymeric structures that include chelation to lithium. The two structures exhibit strikingly similar distorted trigonal bipyramidal coordination geometries except that two of the equatorial sites in the coordination sphere of the anhydrous compound (3) contain glycolate oxygen atoms in place of the water molecules found in the monohydrate compound's coordination polyhedron (4). The latter structure [164] contains a probable error in the c lattice parameter (5.79 Å rather than 5.97 Å) and in one reported axial bond length [our calculations using the reported coordinates give 2.01 Å rather than 1.95 Å (4)]. The isomorphous potassium [165] and rubidium [166] hydrogen bisglycolates contain a polymeric network with the metal ions in a coordination geometry best described as trigonal prismatic, with capping of the rectangular faces and with crystallographically equivalent triangular faces (5). The un-ionized glycolate molecule acts as a bidentate ligand through the α-hydroxycarboxylate,

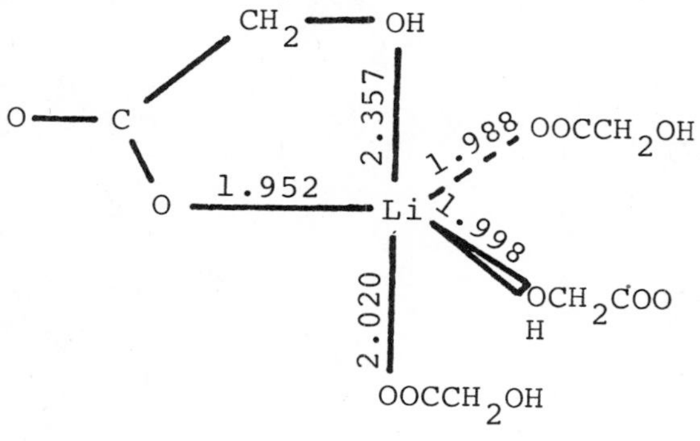

(3)

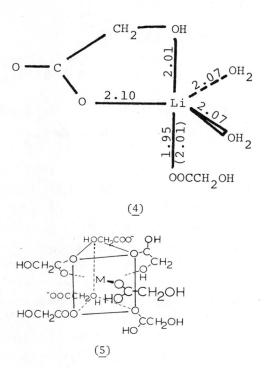

CH2
OH
O
C
O
2.01
2.10
Li
2.07
OH2
2.07
OH2
1.95
(2.01)
OOCCH2OH

($\underline{4}$)

HOCH2COO⁻
OH
HOCH2C
O
CH2
M
CH2OH
⁻OOCCH2O
HO
HOCH2COO
O
CCH2OH
HO

($\underline{5}$)

whereas the ionized glycolate is bidentate through the carboxylate group only. The spread in coordination bond lengths of the potassium structure ($\underline{6}$) is slightly less than that observed for the less-regular rubidium compound ($\underline{7}$).

Examination of the alkali metal structures just discussed shows a characteristic feature of α-hydroxycarboxylates--the tendency of the oxygen donor atoms to bind to more than one metal ion. This

3.22
3.83
3.042
3.97
2.925
2.807
2.778
3.060
K
2.853
2.806
2.874
2.829

($\underline{6}$)

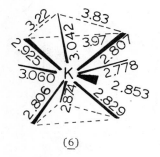

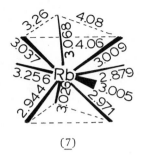

(7)

tendency often causes the formation of polymeric materials. Thus, in the lithium compounds discussed above, there are instances of a "coordination number" of 2 for each type of oxygen atom--a carboxyl in a chelate ring, a carboxyl external to a chelate ring, and a hydroxyl oxygen. In the potassium and rubidium hydrogen bisglycolates, oxygen atoms binding to one, two, and three metal ions are found. Thus, of the 18 crystallographically nonequivalent oxygen atoms in the four alkali metal glycolate structures discussed, oxygen coordination numbers of 0 (three cases), 1 (six cases), 2 (seven cases), and 3 (two cases) are observed.

Two other characteristics of α-hydroxycarboxylate complexes, neither of which is unexpected, can be inferred from an examination of the alkali metal glycolates and will be found to hold for most other compounds discussed here. The first is that polymerization in solid-state structures tends to decrease as the degree of hydration increases. Compare the hydrated and anhydrous forms of lithium glycolate(1-). The second observation is that polymerization tends to decrease as the coordination number of the metal ion decreases. Compare the potassium and lithium compounds. Both of these trends obviously depend on the metal ion achieving its desired coordination number.

The structures of two complexes of glycolate with d-block transition elements have been reported (excluding complexes of lanthanum and scandium). The bis(glycolato(1-))copper(II) complex [167] is a tetragonal bis chelate with two very long axial bonds to carbonyl oxygens of glycolate ligands in neighboring molecules (8). This

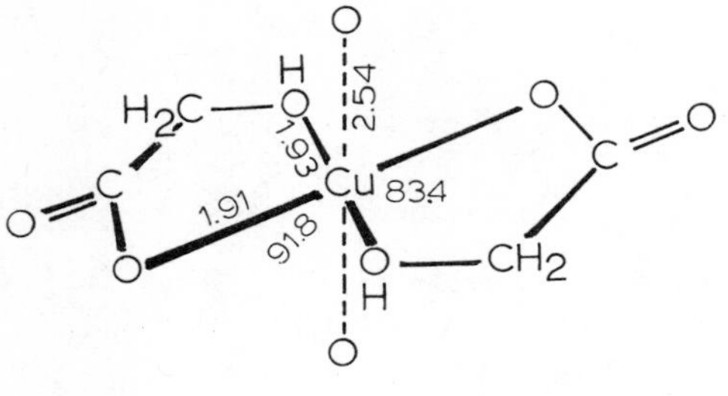

(8)

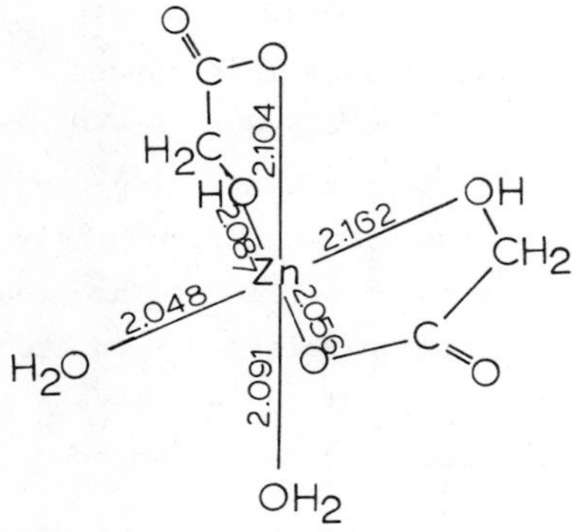

(9)

trans isomer has idealized C_{2h} symmetry with nearly planar chelate
rings and lies on a crystallographic inversion center. Diaquobis-
(glycolato(1-))zinc(II) (9), isomorphous with the Mn(II) and Co(II)
compounds, contains two chelated glycolate ligands and two coordinated water molecules [168,169]. An inspection of molecular models
indicates that there may be two types of significant hydrogen atom
nonbonded interactions present (depending on the rotation of the
coordinated water molecules): those between coordinated water molecules and methylene groups (weak) and those between water molecules
and glycolate hydroxyl groups (possibly strong). Models also indicate that whereas the former interactions are approximately constant
for the three nonenantiomeric isomers possible for a cis-aquo system,
the hydroxyl/water interactions become significant for hydroxyl
groups cis to coordinated water molecules (Fig. 7). Thus the

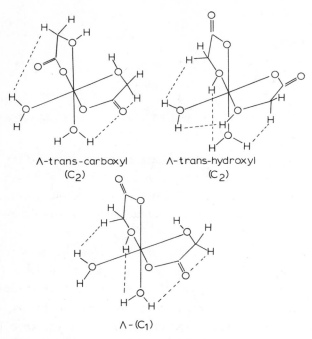

Λ-trans-carboxyl
(C₂)

Λ-trans-hydroxyl
(C₂)

Λ-(C₁)

Figure 7 Nonenantiomeric isomers of cis-diaquobis(glycolato(1-))zinc-(II) with significant nonbonded interactions indicated.

trans-hydroxyl isomer of C_2 symmetry is expected to be the least stable and the C_2 trans-carboxyl isomer, the most stable. That the C_1 isomer, expected to have intermediate steric interactions, is the one observed for the zinc complex may be due to a higher abundance for this species in the solution from which crystals were obtained owing to an entropy contribution of R ℓn 2 resulting from the reduced symmetry [170]. Molecular models also indicate that for a Λ [171] absolute configuration for the complex (as shown in 9), the relatively weak nonbonded interactions between hydrogen atoms of water molecules and methylene groups will be reduced if the chelate rings pucker to give λ [171] conformations (negative ligand dihedral angle, Fig. 8) for chelate rings with hydroxyl groups trans to water molecules and δ conformations for rings with carboxyl groups trans to water molecules. In fact, the crystal structure of diaquobis(glycolato(1-))zinc(II), which shows the

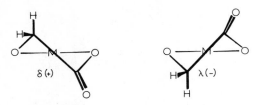

Figure 8   Positive dihedral angle (δ) and negative dihedral angle
(λ) conformations of .chelate rings.

presence of an equal mixture of Δ and Λ enantiomers, shows a slight
δ puckering for the ring predicted to be δ in the Λ isomer although
the ring predicted to be λ is planar.   (Note that Fig. 8 shows a
puckered conformation rather than the asymmetric envelope conformation more commonly seen; see Sec. IV.)

   Six structures of glycolate with lanthanides and scandium have
been reported, and in several cases these have been shown to be
isostructural with other rare earth glycolates for which full structure determinations have not been carried out.   In the structures,
which include both polymeric networks and discrete complexes, coordination numbers of 8 [172-174] and 9 [175-177] are observed.   As is
often the case with high coordination number [178], the actual
coordination polyhedra may be described in terms of more than one
idealized polyhedron.

   In the solid state, erbium(III) triglycolate(1-) dihydrate [173]
and the isomorphous scandium(III) compound [174] contain discrete
$[M(HOCH_2COO)_4]^-$ anions (10) and $[M(HOCH_2COO)_2(H_2O)_4]^+$ cations (11).
There are three crystallographically nonequivalent glycolate ligands

(10)

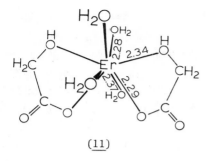

(<u>11</u>)

and two nonequivalent metal ions. The idealized D_{2d} dodecahedron used to describe the coordination polyhedra contains two nonequivalent sets of four coordination sites each (A and B) and two types of edges of interest here (a and m), where the designations of Hoard and Silverton [179] are used (Fig. 9). This polyhedron is composed of two perpendicular trapezoids, which intersect at a 90° angle. Both the anionic and cationic complexes in the triglycolate(1-) dihydrate structures [173,174] have C_2 crystallographic symmetry. In the former complexes, glycolate ligands span the four m edges such that carboxyl and hydroxyl oxygen atoms are located at both A and B sites (<u>10</u>). In the cationic complexes, glycolates span two m edges of one trapezoid with carboxyl oxygens at A sites and hydroxyl oxygens at B sites. Four water molecules at the two A and two B sites of the other trapezoid complete the coordination sphere (<u>11</u>). Although the large chelate "bite" (distance between the donor atoms relative to the donor atom to metal ion distance) of about 1.10 for these complexes indicates a rather shallow minimum in the potential

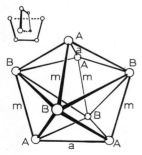

Figure 9 D_{2d} dodecahedron (labels taken from Hoard and Silverton [179]).

energy surface encompassing both the D_{2d} dodecahedron and the
distorted D_2 square antiprism, as indicated by ligand repulsion
calculations [180], and despite the deviation from planarity of the
four atoms that should constitute the trapezoidal planes [173], the
assignment of the coordination polyhedra as distorted dodecahedra
seems satisfactory. The angle of 86° between the best trapezoidal
planes calculated for both the cationic and anionic erbium(III) complexes is reasonably close to the value of 90° required for a D_{2d}
dodecahedron. Lippard and Russ have proposed this angle as a criterion for a choice of coordination polyhedra [181]. The observed m
edge spanning is not unexpected. Most D_{2d} dodecahedral complexes
containing bidentate ligands are chelated along m edges [182].

Aquoglycolato(1-)glycolato(2-)erbium(III) monohydrate [172] is
a network solid containing two crystallographically nonequivalent
chelated glycolate ligands, one of which is deprotonated at the
hydroxyl position. The erbium(III) coordination geometry can be
described as idealized D_{2d} dodecahedral with the glycolate(1-) ligand
spanning the m edge of one trapezoid and the glycolate(2-) ligand
spanning the a edge of the other (12). A coordinated water molecule
and oxygen atoms from glycolates of neighboring coordination polyhedra
complete the coordination. The erbium-ionized hydroxyl oxygen bond
lengths of 2.240 and 2.261 Å are about 0.2 Å shorter than the bond

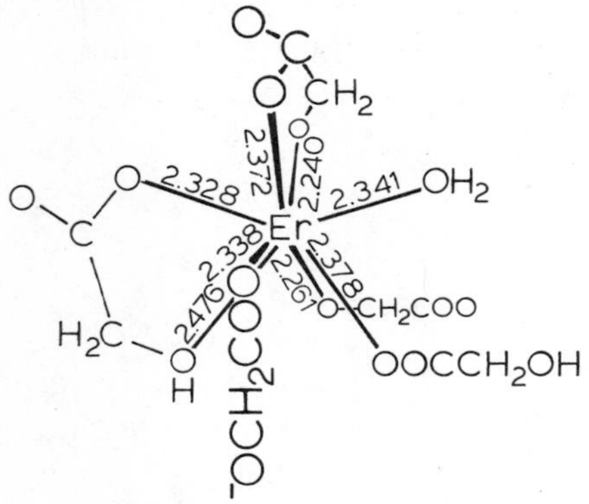

(12)

lengths involving un-ionized hydroxyl oxygen atoms in this and other
[173] erbium(III) glycolate complexes. This large decrease in hydroxyl oxygen-to-metal distances upon proton loss is a common feature of
α-hydroxycarboxylate stereochemistry.

The coordination polyhedra found in the anhydrous gadolinium(III)
[175,177] (13), lanthanum(III) [177] (13), and europium(III) [176]
(14) triglycolates are described as distorted rectangular-face-capped
trigonal prisms. The polyhedron of the gadolinium compound has also
been described as intermediate between this geometry and a capped C_{4v}
tetragonal antiprism [183]. The polyhedra lack any crystallographic
symmetry and are each chelated by three nonequivalent glycolate(1-)
ligands. The other three positions are filled by nonchelating

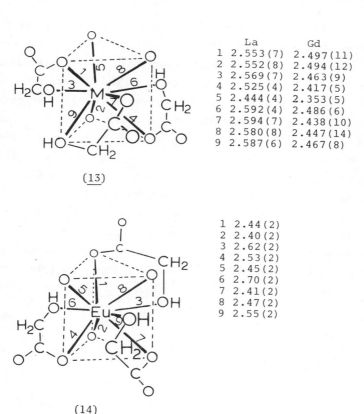

	La	Gd
1	2.553(7)	2.497(11)
2	2.552(8)	2.494(12)
3	2.569(7)	2.463(9)
4	2.525(4)	2.417(5)
5	2.444(4)	2.353(5)
6	2.592(4)	2.486(6)
7	2.594(7)	2.438(10)
8	2.580(8)	2.447(14)
9	2.587(6)	2.467(8)

(13)

1	2.44(2)
2	2.40(2)
3	2.62(2)
4	2.53(2)
5	2.45(2)
6	2.70(2)
7	2.41(2)
8	2.47(2)
9	2.55(2)

(14)

carboxylate oxygen atoms of glycolates of neighboring polyhedra to give a lattice network. Each glycolate oxygen atom in these structures is bonded to exactly one metal ion. As in the potassium and rubidium hydrogen bisglycolates, all bidentate ligands span only prism corners to face-centered vertices. The edges of these trigonal prisms are too long to permit edge spanning by glycolate. The large nonplanarity of the ligands in the isostructural gadolinium and lanthanum compounds probably results from steric restrictions involved in bridging between polyhedra rather than in steric factors within the polyhedra themselves. Note that the conformation changes from one enantiomer to the other for all three ligands of a particular coordination polyhedron upon going from gadolinium to lanthanum (Table 4).

Grenthe [177] has proposed that the significant decrease in the distance between donor atoms within chelate rings upon going from nine-coordinate to eight-coordinate compounds (Table 4) is due to an increased metal ion-donor atom attraction in the latter. Another important factor must, however, be considered. In the nine-coordinate complexes, where all carboxyl oxygen atoms are bound to a metal ion, the two carbon-oxygen bonds of a carboxyl group may be considered to have equivalent bond orders and therefore roughly equal angles between them and the carbon-carbon bond (15). On the other hand, in the eight-coordinate erbium and scandium triglycolates [173, 174], where the carboxyl oxygen atoms external to the chelate rings are not coordinated to metal ions, the external carbon-oxygen bond is expected to have a larger bond order and the internal carbon-oxygen bond a smaller bond order (2). Simple VSEPR arguments [112-114] predict a decrease in the carbon-carbon-carboxyl oxygen angle interior

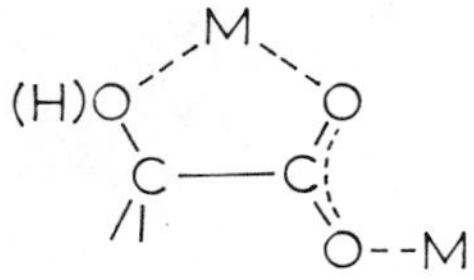

(15)

to the chelate ring (decreasing the donor atom-donor atom distance) and an increase in the corresponding exterior angle. A comparison of carbon-carbon-carboxyl angles for eight- and nine-coordinate rare earth glycolates (Table 4) shows that changes in the angles accompany changes in the donor atom-donor atom distance. Note also that in the free ligands, the C-C-O angle of α-hydroxycarboxylic acids involving the protonated (equivalent to metallated) carboxyl oxygen atoms are on an average significantly larger than those involving the nonprotonated carboxyl oxygen atoms (Table 3). In general, then, we can say that a decreased donor atom-donor atom distance and carbon-carbon-carboxyl oxygen angle within a chelate ring is related to an increased donor atom-metal bonding as long as there is no change in the metal ion bonding involving the carboxyl oxygen exterior to the chelate ring. An inspection of the geometrical parameters for the non-rare earth glycolates will convince the reader that the trends predicted by this simple model hold for the general case.

A preliminary report indicates a pentagonal bipyramidal coordination geometry about the metal ion in uranyl(VI) glycolate(1-), but details of the bonding have not been given [184].

This detailed analysis of the stereochemistry of glycolates permits the following conclusions for these and presumably for related compounds.

1. Polymerization in the solid state *tends* to:
    a. Decrease as the coordination number decreases
    b. Decrease as the degree of hydration increases
    c. Affect ligand conformation
2. Metal-deprotonated hydroxyl oxygen distances are significantly shorter than those involving protonated hydroxyl oxygens.
3. In general, a decrease in donor atom-donor atom distance is associated with increased metal bonding by the carboxyl oxygen atom in a chelate ring and with decreased metal bonding by the exterior carboxyl oxygen.

Lactates [3]. We have seen that there is a marked tendency for α-hydroxycarboxylate groups containing a single nonhydrogen

substituent (other than the hydroxyl and carboxyl groups) at the
carbinol carbon to adopt enantiomeric conformations with positive and
negative dihedral angles for carbinol carbon atoms of R and S chiralities, respectively (Figs. 5 and 6). In a chelate ring formed by an
α-hydroxycarboxylate group there may, however, be an opposing tendency, since the conformational dissymmetries will force a more axial
position for the nonhydrogen substituent, although there may be off-
setting contributions from other nonplanar features of the chelate
ring (see Sec. IV). Axiality of nonhydrogen ring substituents is
known to be unfavorable in other chelate systems [185]. Thus for L-
lactate, (S)-CH$_3$CH(OH)COOH (2-hydroxy-1-propanoic acid), the more
commonly occurring isomer, the adoption of a puckered conformation of
negative dihedral angle (λ) forces the methyl group to be axial (Fig.
10). We may expect, therefore, to see a more randomized distribution
of dihedral angles for chelated dissymmetric α-hydroxycarboxylates
than for the free ligands.

       Larsen and Olsen [186] have proposed that the presence of a
Cotton effect in the d-d region of the circular dichroism spectrum of
copper(II) lactate indicates both chelation and ring puckering in
solution. Others have shown that chelation by α-hydroxycarboxylates
enhances the intensities of optical rotatory dispersion spectra [187].
Infrared and visible spectra indicate that lactate "always" acts as a
chelating ligand through one oxygen of the carboxylate group and the
hydroxyl group when coordinating to copper(II), nickel(II), cobalt-
(II), and zinc(II) [188].

       Of the six structures reported for lactate complexes [91,167,
189-192], four contain copper(II). Bis(L-lactato(1-))(N,N,N',N'-
tetramethylethylenediamine)copper(II) [189] and -nickel(II) [190]

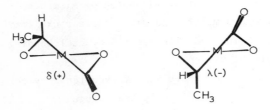

Figure 10 Puckered conformations of (S)-lactate chelate rings.

exhibit similar coordination polyhedra (16) except for an obvious tetragonal distortion in the copper complex. The trans hydroxyl coordination maximizes the interligand distance between lactate methyl groups, whereas the Δ absolute configuration that is formed stereoselectively permits a maximum separation between methyl groups on lactate and diamine. A λ conformation is adopted by the diamine ligand in order to reduce nonbonded interactions between N-methyl groups and hydroxyl protons (see Fig. 7). The lactate chelate rings are nearly planar.

A Δ absolute configuration tris chelate is also found in bis(L-lactato(1-))(N,N,N',N'-tetraethylethylenediamine)copper(II) hemihydrate [191]; however, in this complex, two nonequivalent lactate ligands--one bidentate through carboxyl oxygens only--are present (17). Two crystallographically nonequivalent complexes having the same structure are present. The coordination polyhedron is tetragonally distorted with one very long (and presumably very weak) copper-carboxylate oxygen interaction. Although only Δ tris chelates are found in these three bis(L-lactato(1-)) structures [189-191], both Δ and Λ isomers of $[Co(en)_2(HOCH(CH_3)COO)]^{2+}$ are formed in the reaction of $[Co(en)_2(CO_3)]^{1+}$ with optically pure lactic acid [193], even though the Δ-(R)-lactate and Λ-(S)-lactate diastereomers are expected to have a steric interaction between the lactate methyl group and an ethylenediamine proton. In this case, however, the product isomer distribution may be controlled kinetically.

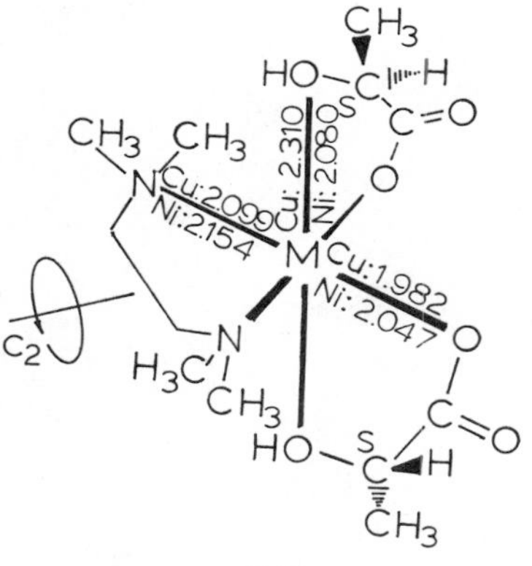

(16)

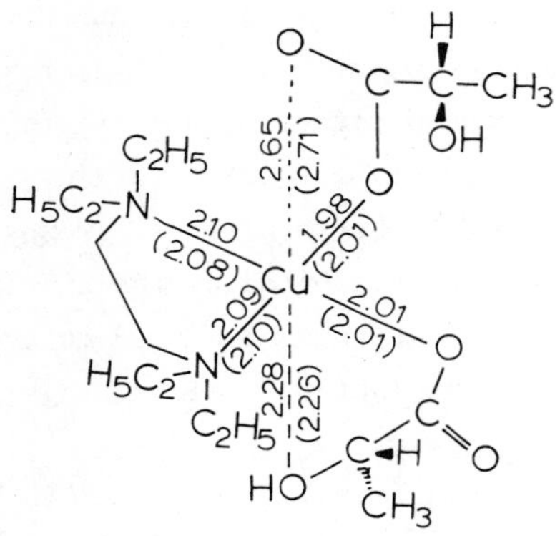

(17)

Both Δ and Λ enantiomers of the tris chelate are present in (D-lactato(1-))(L-lactato(1-))diaquozinc(II) monohydrate [192] (18). The complex, which lacks any symmetry, is strikingly similar to the bis(glycolato(1-))diaquozinc(II) structure [168-169] (9). Both enantiomeric forms of lactate are present in the complex. Somewhat surprisingly, the structure adopted puts the lactate methyl groups on the side of the chelate ring near the coordinated water molecules. Moreover, the chelate ring deformations from nonplanarity are in a direction to reduce the water-methyl group distance.

Aquobis(DL-lactato(1-))copper(II) [167] has a tetragonally distorted coordination polyhedron containing two bidentate ligands of

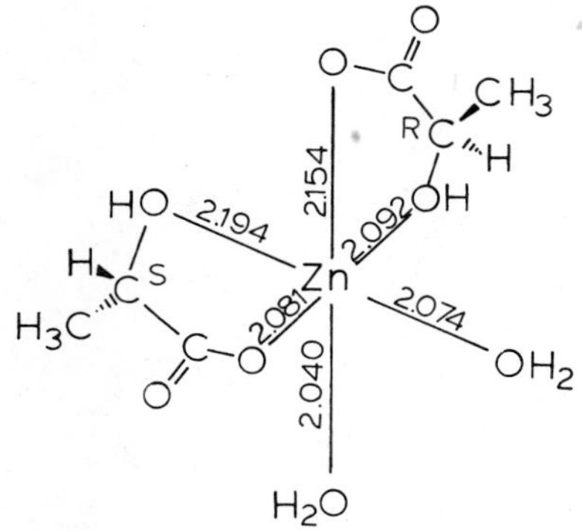

(18)

opposite chirality (19). The trans coordination is similar to that found for the bis(glycolate(1-)) complex [167] (8). A carboxyl oxygen from a lactate on a neighboring complex makes a very long contact to copper. The two chelate rings, which are strikingly similar in their bonding parameters, are bent in the same direction to minimize the water-methyl group interactions.

A monodentate lactate ligand is present in L-lactato(1-)bis(N-isopropyl-2-methylpropane-1,2-diamine)copper(II) L-lactate(1-) 1-hydrate [91] (20). A free lactate molecule is also present. The copper ion lies 0.33 Å out of the plane of the four nitrogen atoms in the square pyramidal coordination polyhedron. Unexpectedly, the nitrogen atoms adopt R absolute configurations, which bring the isopropyl groups near the axially coordinated lactate.

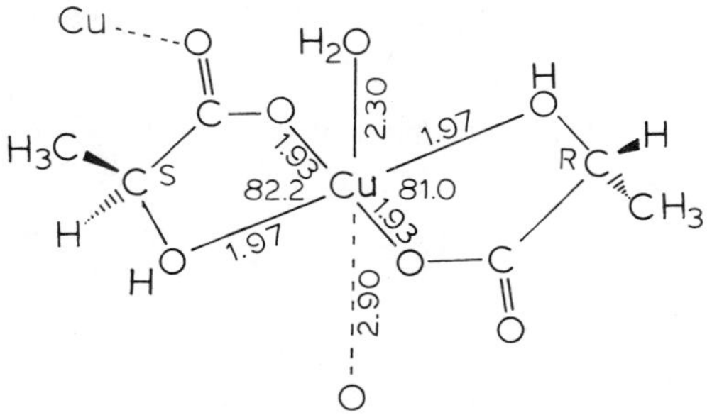

(19)

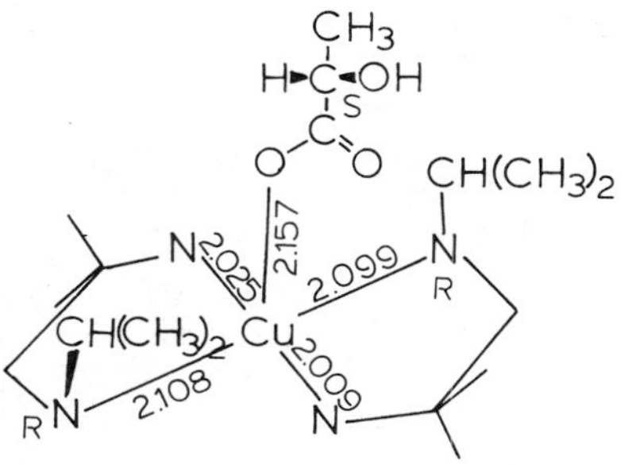

(20)

A review of the lactate structures indicates that:

1. In octahedral complexes containing two (S)-lactate groups, there may be a tendency to give a Δ absolute configuration. This may extend to complexes of other α-hydroxycarboxylates.

2. The trend in conformational dissymmetry observed earlier for free asymmetric α-hydroxycarboxylates seems to hold also for the coordinated ligands, although this forces an axial position for the nonhydrogen atom if the chelate rings have a puckered conformation. Thus assumptions that chelated α-hydroxycarboxylates adopt a conformation such that the alkyl substituent is equatorial [194] may, in fact, not hold.

Other Simple α-Hydroxycarboxylates. The structures of bis(2-hydroxy-2-methylbutyrato(2-))oxochromate(V) (21) (as the potassium salt) [195] and bis(benzilato(2-))oxovanadate(IV) (22) (as the mixed sodium tetraethylammonium salt) [196] are strikingly similar. The metal ions, whose coordination geometry is best described as intermediate between square pyramidal and trigonal bipyramidal (23) ("axial" and "equatorial" angles of 151.6° and 132.9° for vanadium and 155.6°

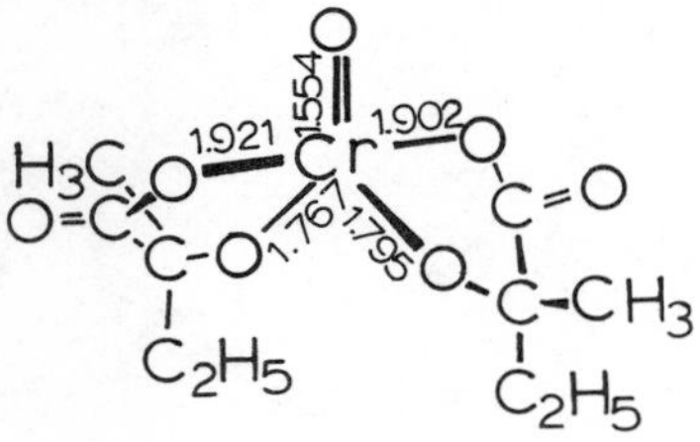

(21)

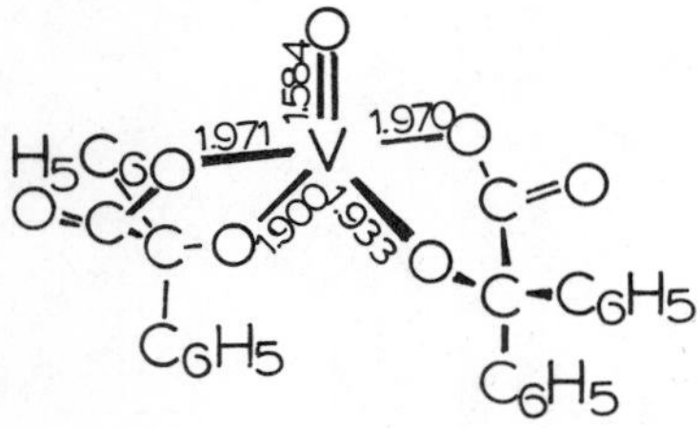

(22)

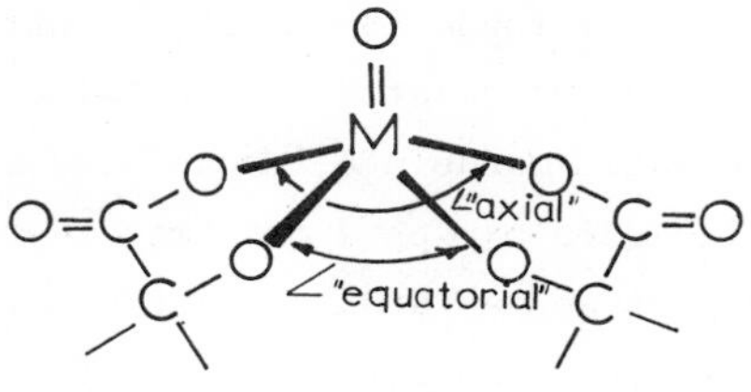

O
‖
O=C—O—M—O—C=O
with structure labels "axial", "equatorial"

(23)

and 132.3° for chromium), are chelated by two completely ionized α-hydroxycarboxylates. As observed for the erbium glycolate complex [172], the metal-to-ionized hydroxyl oxygen distances are very short. Both the chromium and vanadium compounds in the solid state contain a racemic mixture of enantiomers whose coordination geometries can be assigned [197] as Λ (21, 22) and Δ; however, in the chromium compound there is stereoselective association of S ligands with Λ complexes (21) and R ligands with Δ complexes. This stereoselectivity permits the more bulky ethyl groups to point away from the oxo ligand. No stereoselectivity is possible for the vanadium compound containing achiral benzilates. Irregular coordinations of seven and six oxygen atoms from solvent and carboxyl groups are found around the potassium and sodium ions, respectively, in the two compounds.

The tetragonal geometry found for the diaquobis(2-hydroxy-2-methylpropionato(1-))copper(II) complex [167] is very similar to that shown for the bis(glycolate(1-)) compound (8) [167]. Water molecules lie at the axial sites in the coordination polyhedron, which has C_{2h} crystallographic symmetry. The copper ion, however, may be disordered. In this compound, as in other α-hydroxycarboxylates where cis and trans planar or square pyramidal structures are possible, the trans geometry is the only one found by diffraction techniques. On the other hand, EPR studies of several vanadyl(IV) α-hydroxycarboxylates [137,198] have shown both cis and trans isomers in solution, although the amount of cis isomer decreases with bulky alkyl substituents on the carbinol carbon [198]. Structure determinations have shown the formation of cis complexes for bis(glycinato(1-))copper(II) monohydrate [199] and bis(1-alaninato(1-))copper(II) [200].

Eight-coordinate calcium ions are found in discrete centrosymmetric binuclear complexes in calcium mandelate(1-) hydroxide 3-acetic
acid [201]. In addition to a chelated mandelate, each coordination
sphere [shown in part in (24)] includes oxygen atoms from acetate,
hydroxide, and the carboxyl group of an α-hydroxycarboxylate ligand
bidentate to the adjacent calcium ion. There is some question concerning the location of protons in this compound.

A discrete mandelate(2-) complex (25) is found in the crystal
structure of diaquobis((±)-mandelato(2-))germanium(IV) dihydrate [202].
The centrosymmetric coordination polyhedron is approximately octahedral with a trans coordination by two water molecules and chelation
by two mandelate(2-) ligands of opposite chirality.

Infrared spectra indicate that in the solid state, the complexes
M(mandelate)$_3 \cdot 3H_2O$ [M = Ti(III), Cr(III), Fe(III), Al(III), In(III)]

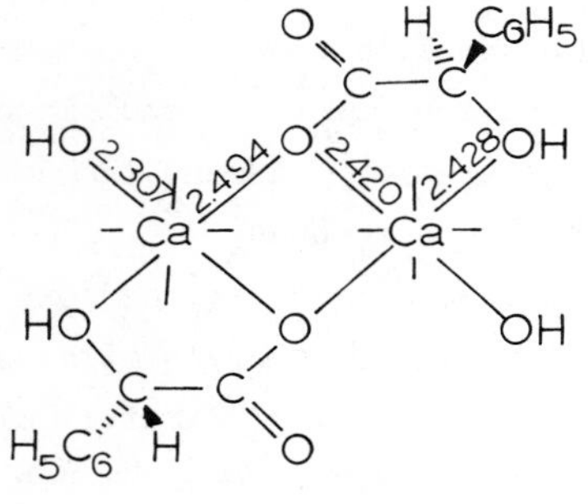

(24)

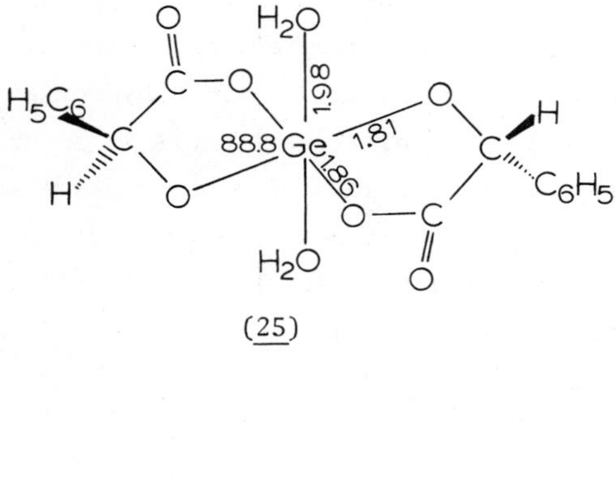

(25)

are tris chelates with coordination through both the un-ionized
hydroxyl group and a carboxylate oxygen [203].

A highly irregular fivefold coordination is found around the
potassium ions in the network lattice of potassium benzilate(1-)
[204]. Although the coordination sphere does not include an α-
hydroxycarboxylate bidentate through α-hydroxyl and carboxyl oxygens,
it does include a benzilate chelated through the two carboxyl oxygens
only. The α-hydroxycarboxylate group is distinctly nonplanar (dihedral angle of 49.0°).

2. Complexes of α-Hydroxycarboxylate Ligands with Multiple Carboxyl Functions

Tartronates. Tartronic acid, HOOCCH(OH)COOH (2-hydroxy-1,3-
propanedioic acid), has two stereochemical features of interest.
First, because of the close intramolecular contact between the carboxyl oxygens, the two α-hydroxycarboxylate groups cannot be
simultaneously planar. For the free acid, dihedral angles of -17.5°
and +14.0° are observed in the solid (Table 3) [58]. Second, the
attempt at α-hydroxycarboxylate planarity forces the separate carboxylate groups into close contact (closest O-O approach, involving the
protonated oxygens, is 2.902 Å [58]), which should enhance the ability of the ligand to be bidentate through carboxylates only (as in
some malonate complexes [205]).

Potassium hydrogen tartronate contains potassium ions in an
irregular eightfold coordination in two different crystal forms [206].
Both the α form (26), where the two carboxylates are equally protonated owing to the proton being located on symmetry elements, and the

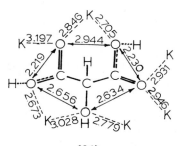

(26)

β form (27), where the two carboxylates differ in ionization, have
chelation involving the two carboxylates as well as one of the α-
hydroxycarboxylate groups. There are a number of short oxygen-oxygen

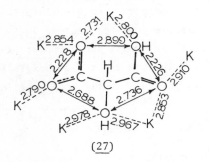

(27)

contacts in these ions. Some of the potassium-oxygen contacts are
rather long in the α form, and one of the "half-protonated" carboxyl
oxygens of this structure lies proximal to the α-hydroxyl group--an
unusual feature in α-hydroxycarboxylic acids (vide supra). Unlike
the conformation of the free acid, where the two α-hydroxycarboxylate
groupings are about equally nonplanar, in the potassium hydrogen
tartronates, one of the α-hydroxycarboxylate portions of the molecular ions is distinctly more nonplanar than the other (Table 4).

The dish-shaped trinuclear anion (28) in rubidium tri-μ-tartron-
ato(3-)tricuprate(II) 6-hydrate contains tartronate ligands, each of

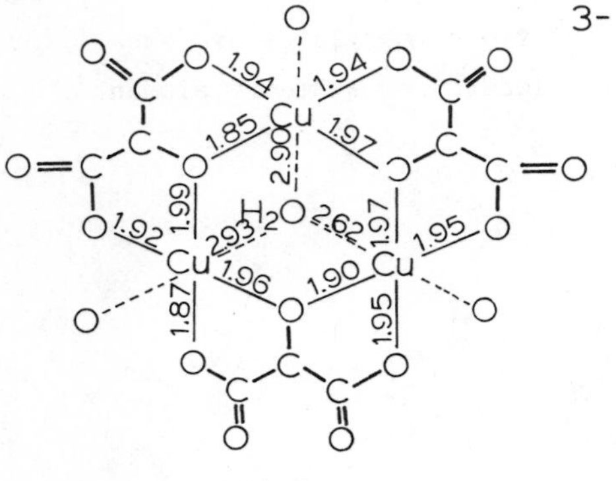

(28)

which bind two copper ions in five-membered chelate rings [208].
One water molecule serves as an axial ligand in all three tetragonal
coordination polyhedra (at the bottom of the "dish"). The other
axial sites are filled by a water molecule and two carbonyl oxygen
atoms from adjacent trinuclear anions (Cu-O: 2.74, 3.11, 2.71 Å).
Unlike other complexes containing a hydroxycarboxylate with an ionized
hydroxyl group, where the metal-hydroxyl distances are usually shorter
than the metal-carboxyl oxygen distances, the average copper-hydroxyl
oxygen and copper-carboxyl oxygen distances are equal in the tri-μ-
tartronato(3-)tricuprate(II) complex (1.94 Å). This may be due to
the simultaneous binding of two copper ions by each hydroxyl oxygen.
Because of a shortage of reported crystallographic data, geometric
parameters are not reported for this complex in the tables.

Malates. Malic acid, $HOOCCH(OH)CH_2COOH$ (2-hydroxy-1,4-butanedi-
oic acid), in its various ionic forms, is a multidentate ligand capable of binding metal ions in several ways, the most common being
bidentate through the hydroxyl and α-carboxyl groups to give a five-
membered chelate ring and tridentate through the hydroxyl and both
carboxyl groups to give both five- and six-membered chelate rings.
There is little evidence that chelation occurs to give only the six-
membered ring, a not unexpected observation considering the fact that
stability constants for 2- and 3-hydroxypropionates with copper indicate a greater stability for the five-membered ring [209]. Of all of
the metal derivatives of malic acid whose crystal structures have been
reported, only lithium (R)-malate(1-) fails to show chelation of the
metal ion [210]. In the solid state, this compound contains lithium
ions tetrahedrally coordinated by three carboxyl oxygen atoms at distances of 1.917, 1.930, and 1.997 Å and a hydroxyl oxygen atom at
1.973 Å, all belonging to different malate(1-) ions. There has been
a preliminary report of a structure of potassium α-isopropyl malate-
(1-); however, the potassium coordination was not discussed [211].

Proton NMR spectra indicate that of the three conformations
possible for (S)-malate (L-malate) (Fig. 11), the synclinal forms
[124] predominate in aqueous solution (malic acid: 42% -sc, 7% ap,

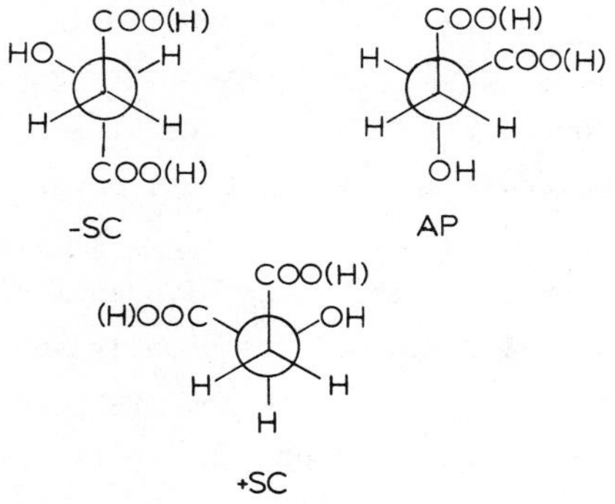

Figure 11   Conformations of (S)-malate.

51% +sc; malate anion: 60% -sc, 3% ap, 37% +sc) [212].   The increase
in the -sc (trans-carboxylate) conformer upon ionization is probably
due to electrostatic destabilization when the ionized carboxylates are
gauche.   A crystal structure of ammonium (S)-malate(1-) (α-carboxyl
group ionized) shows the presence of the -sc conformer [37].   Tridentate chelation in an octahedral complex requires the +sc conformer
for (S)-malate, and it is precisely this conformer whose abundance is
increased upon formation of the zinc complex in aqueous solution (28%
-sc, 2% ap, 70% +sc) [212].   Note, however, that this conformation is
also the one required for hydrogen bonding between the β-carboxyl
group and a coordinated water molecule when the ligand is bidentate,
as found in the crystal structure of (S)-malato(2-)tetraaquomagnesium-
(II) monohydrate [213].   It has been proposed that hydrogen bonding
between the dangling β-carboxyl group and a nitrogen ligator stabilizes the malato(2-)bis(phenanthroline)cobalt(III) cationic complex,
since the corresponding lactate and mandelate complexes cannot be
prepared [194].   Proton NMR studies [132] also show that alkali metal
coordination decreases the proportion of malate with trans carboxylate groups in aqueous solution.   Aqueous solution pressure jump
kinetic results for nickel(II) malate are also best interpreted assuming a tridentate coordination by the ligand [214].

   Carbon-13 NMR spectra of molybdenum(VI) malates shows that in the
presence of a twofold excess of ligand, only one complex is present in

D_2O at pD = 3.9 and that complex has both carboxylates bound [215]. On the other hand, other ^{13}C NMR studies with a fivefold excess of ligand indicate that molybdenum(VI) [and also tungsten(VI)] forms two malate complexes at pH 2, 4, and 6 [126]. At pH 4, aluminum and titanium give only one malate complex, although at higher pH, aluminum malate polymers are formed.

In malato(2-)bis(phenanthroline)cobalt(III) [194,217], the ligand is bidentate through ionized hydroxyl and α-carboxyl groups with the uncoordinated β-carboxyl group protonated. With malate of one optically active form, two diasteromers corresponding to Δ and Λ absolute configurations at the metal can be prepared. When the complex is prepared starting with $[CoCl_2(phen)_2]^+$, little stereoselectivity is observed [217]. On the other hand, stereoselective formation (which may be kinetic in origin) of $\Lambda-[Co((R)-malate)(phen)_2]^+$ results from reaction of $[Co(phen)_2CO_3]^+$ with (R)-malic acid [194].

Diffraction studies show that distorted octahedral coordination polyhedra with bidentate α-hydroxycarboxylate groups are present in (S)-malato(2-)tetraaquomagnesium(II) monohydrate [213] (29), (S)-malato(2-)diaquomanganese(II) monohydrate [218] (30), and bis((S)-malato(1-)copper(II) dihydrate [219] (31). In the latter two

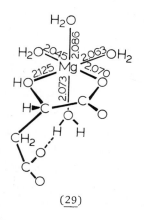

(29)

complexes, the ligand conformations are -sc, with the β-carboxyl groups coordinated to additional metal ions to give polymeric

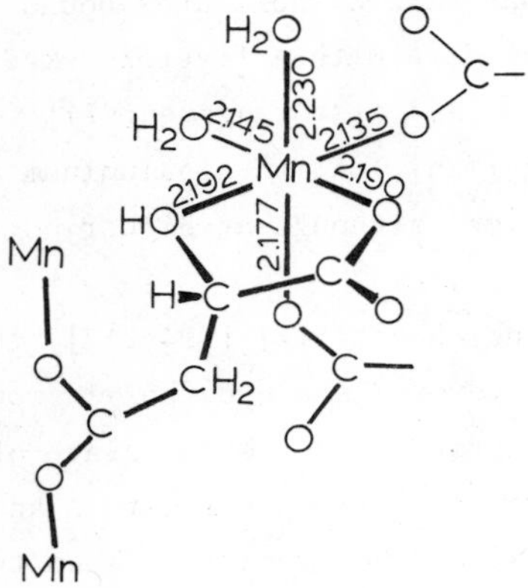

(30)

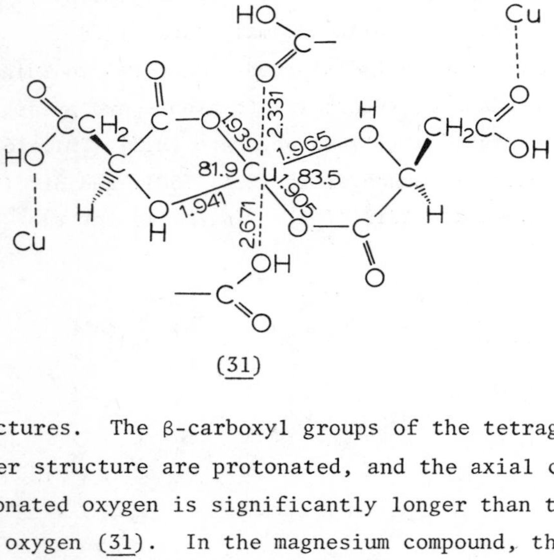

(31)

structures. The β-carboxyl groups of the tetragonally distorted
copper structure are protonated, and the axial contact with the
protonated oxygen is significantly longer than that with the unprotonated oxygen (31). In the magnesium compound, the ligand conformation
is +sc, with the dangling β-carboxyl group (which is not coordinated
to any metal ions) hydrogen bonded to a coordinated water molecule.

In diaquobis((S)-malato(1-)nickel(II) dihydrate [220] and in
tetraaquobis((S)-malato(1-))calcium dihydrate [221], two malate(1-)
ligands are bidentate through the ionized α-hydroxycarboxylate group

to the metal ion. The nickel octahedral bis chelate (32) has a Δ absolute configuration with a twofold rotational symmetry. The calcium ion is eightfold coordinate, with a square antiprismatic geometry (33). Water molecules complete the coordination spheres in

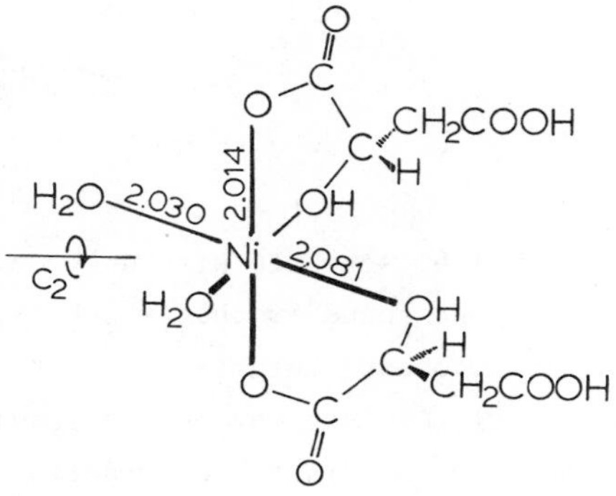

(32)

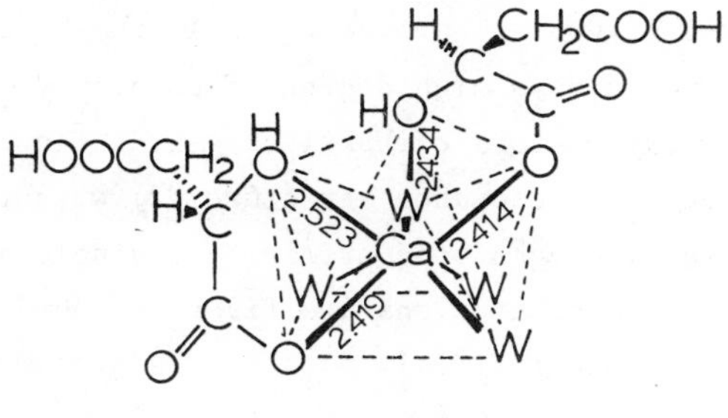

(33)

both compounds. The (S)-malate ion has a -sc (trans carboxyl) conformation.

Bidentate coordination of malate is also present in bis(malato-(2-))borate(III), found as discrete complexes in the potassium salt in the solid state [222]. The hydroxyl group is ionized and, as found in other cases, forms significantly shorter bonds to the central atom than does the coordinated α-carboxylate (34). The β-carboxyl group is protonated and uncoordinated with the ligand conformer being -sc. Surprisingly, of the two enantiomeric coordination

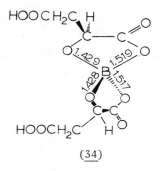

(34)

geometries possible for a tetrahedral bis chelate with unsymmetrical
(AB) bidentates [223,224], the only one found is that which brings
the dangling β-carboxyl groups into closest contact. However, this
stereoselectivity with optically active malate may be the result of
crystal packing forces that select for one diastereomer during crystal growth. The enantiomeric form of the asymmetric envelope chelate
ring conformation (discussed later) is that needed to increase the
distance between β-carboxyl groups. That the central atom-to-oxygen
distances are very short indicates a high degree of covalency in
these bonds. The hydroxyl oxygen-to-α-carboxyl oxygen distances are
correspondingly small [i.e., the resonance structure (2) with the
carbon-to-coordinated carboxyl oxygen bond primarily a single bond
is predominant]. The potassium counterions are fivefold coordinate,
with a trigonal bipyramidal geometry. Only malate oxygen atoms are
involved (K-O distances: 2.61 to 2.70 Å).

Tridentate malate ligands are present in isostructural (S)-
malato(2-)diaquozinc(II) [225,226] and -cobalt(II) [227] monohydrate,
with metal ion bridging by the β-carboxyl groups (35). The conformation about the central carbon-carbon bond is +sc, as required for
tridentate (S)-malate in an octahedral complex. Molecular models
indicate that steric constraints involved in the simultaneous formation of five-membered and six-membered chelate rings in an octahedral
complex by (S)-malate forces the α-hydroxycarboxylate group to be
nonplanar, with a negative dihedral angle as observed in the crystal
structures (-26.8° and -27.3° for the cobalt and zinc compounds).

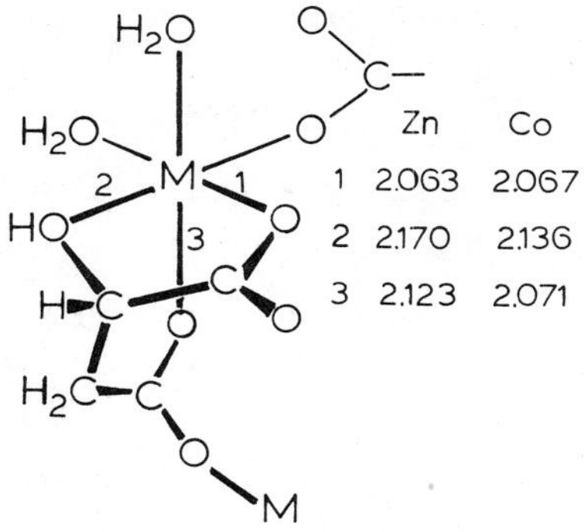
H2O
H2O
M
HO
H
C
H2C
C
O
M
Zn Co
1 2.063 2.067
2 2.170 2.136
3 2.123 2.071

(35)

Note, however, that very nearly planar α-hydroxycarboxylate groups are found in some structurally similar complexes containing tridentate citrate. The structures of the cobalt and zinc malate complexes are strikingly similar to those found for the zinc, cobalt, and nickel aspartates, where the aspartate(2-) ion is tridentate, with a bridging β-carboxyl group [228].

A large negative dihedral angle (-15.8°) is also found for the α-hydroxycarboxylate group of the tridentate ligand in the polymeric structure of calcium (S)-malate(2-) dihydrate [229] (36). The connectivity is similar to that found for the zinc and cobalt malates;

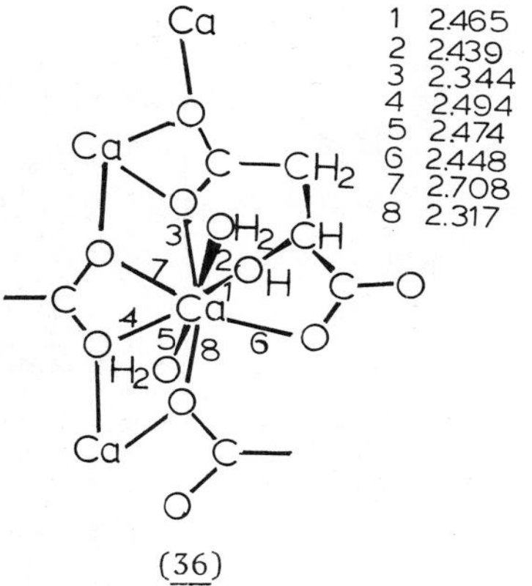
Ca
Ca
CH2
OH2
CH
OH
Ca
H2O
Ca
1 2.465
2 2.439
3 2.344
4 2.494
5 2.474
6 2.448
7 2.708
8 2.317

(36)

however, in the calcium compounds, the β-carboxyl group not only
bridges two metal ions but is bidentate to a third. The calcium ion
and the five coordinating carboxyl oxygen atoms are nearly coplanar,
with one water molecule above the plane and a second water molecule
and a hydroxyl oxygen atom below the plane, giving an irregular
eightfold coordination.

The structures of ammonium bis(malado(3-))undecaoxotetramolyb-
date-(VI) hexahydrate [230] and monohydrate [231] have been determined (the latter to sufficiently low precision that parameters are
not given in the tables). Both hydrates contain nearly identical C_2
symmetry tetranegative, tetranuclear complexes (37). Unlike the

	·GH$_2$O	·H$_2$O
1	2.21	2.50
2	2.00	2.23
3	2.32	2.64
4	2.31	2.59
5	2.38	2.39

(37)

other structures containing tridentate malate ion, in the hexahydrate
the α-hydroxycarboxylate group is nearly planar. The metal-to-ionized
hydroxyl oxygen bond lengths are significantly shorter than the
metal-to-carboxyl oxygen lengths, as found in other structures containing ionized hydroxy groups. The coordination of the malate in
the tetranuclear ion is similar in connectivity to that of the malate
ion in the calcium salt [229]. The ligand is tridentate to one metal
ion, but the β-carboxyl group forms bonds with two additional ions.

These structures indicate that the following geometric features
are exhibited by malate complexes.

1. Normally, malate complexes contain a bidentate α-hydroxycarboxylate group--particularly in malate(1-) complexes.

2. The β-carboxyl group is often found associated with the same metal ion as chelated by the α-hydroxycarboxylate group (giving a tridentate ligand) or with other metal ions (giving polymeric structures). The latter mode of coordination is probably of less importance in solution.

3. When tridentate to one metal ion, (S)-malate exhibits a +sc conformation and a negative α-hydroxycarboxylate group dihedral angle.

4. When the (S)-malate ligand is not tridentate, the overall conformation is usually -sc (carboxyl groups trans).

5. No malate structure has been shown to contain malate groups with ap conformers by X-ray diffraction; however, conformers equivalent to the ap conformer of malate have been found for substituted malates such as (±)-erythro-fluoromalic acid [77] and ms-tartaric acid [61].

6. In the malate(1-) structures that have been solved, the carboxyl group α to the hydroxyl group is ionized.

7. The coordination stereochemistry of malic acid resembles that of the isoelectronic system, aspartic acid [232].

Citrates and Related Compounds. A number of structural studies, primarily by J. P. Glusker and co-workers, have been carried out on citric acid, $HOOCCH_2C(OH)(COOH)CH_2COOH)$, a hydroxytricarboxylic acid, and its derivatives. Must of this work was initiated to help elucidate the stereochemistry of the action of aconitase, an iron-activated enzyme that mediates the interconversion of citrate, cis-aconitate, and (2R:3S)-isocitrate in the Krebs cycle (Fig. 12). Of particular interest is the stereoselective formation of only one isomeric form of isocitrate from citrate in a reaction that formally requires the exchange of a pro-R hydrogen of the pro-R carboxymethyl group and the hydroxyl group [233]. The catalysis by aconitase may involve tridentate chelation by citrate and isocitrate to the iron(II)

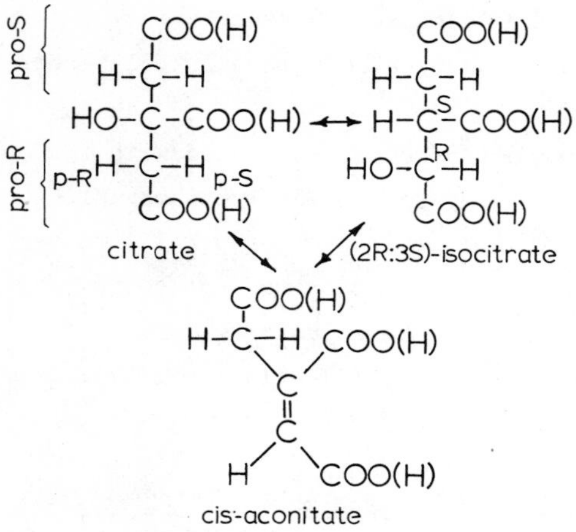

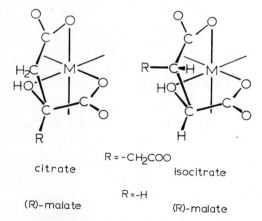

Figure 12 Substrates of aconitase.

required for activation of the enzyme [234], giving complexes whose
proposed geometries [235] resemble those of tridentate R-malate octahedral complexes (Fig. 13).

^{13}C NMR spectra of 1:1 iron(II)-citrate [236] and 1:2 molybdenum-
(VI)-citrate [215] in acid solution show the presence of nonequivalent
β-carboxyl groups, which would be the case if the ligand were tridentate. At higher pH, where the citrate is tetraionized, association

Figure 13 Tridentate chelation by malate, citrate, and isocitrate.

occurs with iron(II) and also with nickel(II) to give polynuclear
species [236]. Binuclear complexes are known to be present in cop-
per(II) citrate solutions from EPR studies [237].

X-ray structure determinations on anhydrous citric acid [73],
citric acid monohydrate [74], ethylenediammonium citrate(2-) [96],
and the alkali metal salts potassium (2R:3S)-isocitrate(1-) [238]
and lithium ammonium citrate(2-) monohydrate [239] show an extended
conformation for the citrate or isocitrate molecules (Fig. 14).
Citrate conformers with a rotated terminal carboxyl group are found
in the alkali metal compounds--isostructural sodium- and lithium
citrate(2-) salts [240] and rubidium citrate(2-) [241]. It is possible that metal ion coordination in the latter compounds stabilizes
the rotated carboxyl group conformers.

All of the alkali metal structures are polymeric. In the isocitrate, the potassium ions are irregular eightfold coordinate with
several types of chelation (each isocitrate is tridentate, bidentate,
and monodentate to different metal ions). A large range of potassium-
oxygen contacts are observed--2.725 to 3.225 Å. Some of these contacts

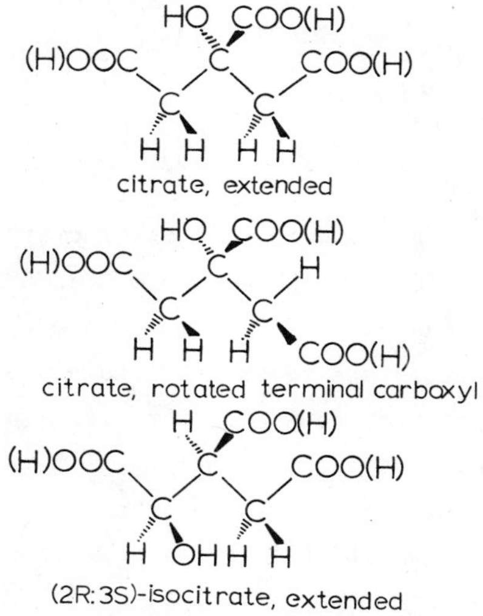

Figure 14 Conformers of citrate and isocitrate found in X-ray structure determinations.

are sufficiently long that the interaction must be considered weak.
In the lithium ammonium compound, the metal ion coordination polyhedra are tetrahedral and the metal ions are not chelated. The
rubidium compound contains ninefold coordinate metal ions, whereas
the isostructural sodium and lithium compounds contain pairs of metal
ions in an octahedral coordination geometry (38). All of the alkali
metal structures exhibit coordination to both protonated and unprotonated carboxyl groups, with the contacts to the latter being generally shorter.

	Na	Li
1	2.305	1.96
2	2.336	2.12
3	2.337	2.17
4	2.348	2.10
5	2.451	2.22
6	2.461	2.49

(38)

Extended citrate conformers are also found in the isostructural
series of compounds $[M(H_2O)_6][M(OOCCH_2C(OH)(COO)CH_2COO)(H_2O)]_2 \cdot 2H_2O$
(39), M = Mg^{2+} [242], Mn^{2+} [243,244], and Fe^{2+} [245]. In these compounds, the citrate(3-) ligands are tridentate to one octahedral metal
ion [giving a structure very similar to that found for zinc [225] and
cobalt [227] malates, where the ligand is tridentate (35)], with the
β-carboxyl group not involved in chelation bridging two other metal
cations [compare with manganese malate (30) [218]] giving polymeric
anions. Discrete hexaquometal cations are also present.

Citrate ligands are present as rotated terminal carboxyl conformers in calcium citrate(2-) trihydrate [246], where discrete binuclear

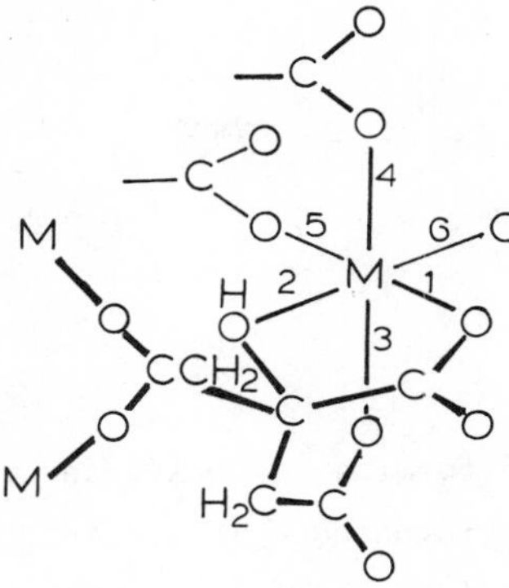

(39)

	Mg	Mn	Fe
1	2.081	2.194	2.130
2	2.118	2.224	2.178
3	2.077	2.181	2.116
4	2.072	2.167	2.116
5	2.019	2.123	2.085
6	2.031	2.140	2.090

complexes containing severely distorted octahedral coordination
polyhedra are present (40), and in citrato(3-)triethylenetetraamine-
cobalt(III) [247a], where the ligand is bidentate through the α-
hydroxycarboxylate group only (41). The latter structure is still
undergoing refinement and, therefore, the structural parameters have
not been included in Table 4.

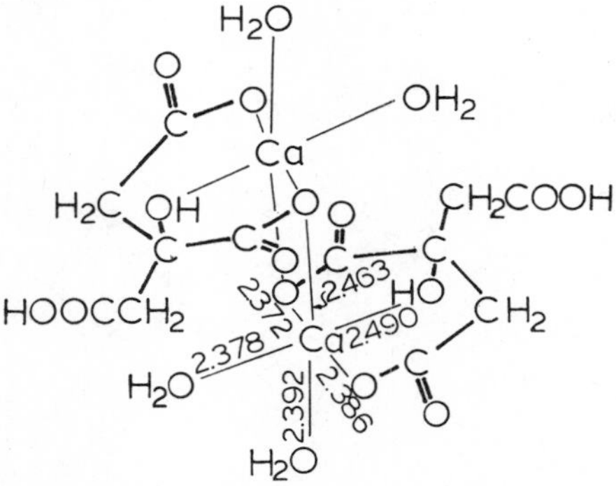

(40)

The hexamminecobalt(III) salt of bis(citrato(3-)antimonate(III)
contains anionic mononuclear complexes, each chelated by two crystallographically nonequivalent citrate ligands through completely ionized
α-hydroxycarboxylate groups [247b]. The β-carboxyl groups of the

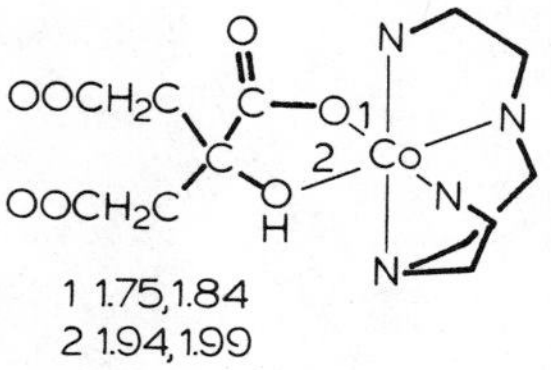
1 1.75,1.84
2 1.94,1.99

(41)

ligands, one of which has the rotated terminal carboxyl and the other, the extended conformation, are not coordinated. The coordination geometry, which can be described as pseudotrigonal bipyramidal with axial carboxyl oxygen atoms and equatorial hydroxyl oxygen atoms [presumably, a lone pair on antimony(III) occupies the third equatorial site], is strikingly similar to that observed in the antimony(III) tartrates (vide infra).

Two citrate(4-) structures have been solved. Copper(II) citrate(4-) dihydrate [248] contains a polymeric network with copper ions in square pyramidal and distorted trigonal bipyramidal sites (42).

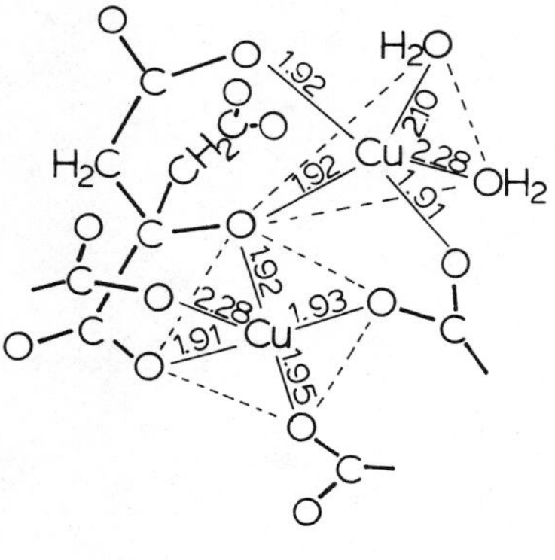

(42)

Tetramethylammonium diaquodihydroxohexakis(citrato(4-))octanickelate(II) 36-hydrate [245] contains discrete centrosymmetric $[Ni_8(OOCCH_2C(O)(COO)CH_2COO)_6(OH)_2(H_2O)_2]^{10-}$ ions containing three crystallographically independent citrate ligands. Two of these ligands exhibit similar coordination, being simultaneously bidentate to three different metal ions. The third independent citrate group is tridentate to one metal ion, bidentate to a second, and monodentate to two others (Fig. 15). In contrast to other, previously discussed α-hydroxycarboxylate complex structures containing ionized hydroxyl groups, the average metal ion-to-ionized hydroxyl oxygen distance is not significantly shorter than the average carboxyl oxygen-to-metal distance in these two citrate(4-) structures. However, in both the copper and nickel compounds, there is multiple metal ion coordination to the ionized hydroxyl oxygens, and this undoubtedly decreases the donor ability of the ligator (in the same way as does protonation). One other anomalous feature of these two citrate(4-) structures is the rather large nonplanarity of the α-hydroxycarboxylate groups (dihedral angles of 18.9° in the copper complex and 16.9°, 30.9°, and 23.0° in the nickel structure). This nonplanarity may be due to the steric requirements imposed by simultaneous coordination to several metal ions by each ligand. The citrate(4-) ligand

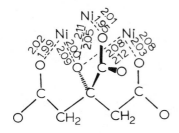

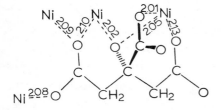

Figure 15   Coordination of citrate ligands in $[Ni_8(OOCCH_2C(O)(COO)-CH_2COO)_6(OH)_2(H_2O)_2]^{10-}$

in the copper compound has a rotated terminal carboxyl conformation,
as does one of the crystallographically independent ligands in the
nickel complex. The other two independent ligands in the latter
structure have extended conformations.

X-ray diffraction studies have been carried out on two of the
four diasteromeric forms of hydroxycitric acid, HOOCCH(OH)C(OH)(COOH)-
CH_2COOH, as the ethylenediamine salts [75]. Although no structural
studies have been reported for a metallo compound of hydroxycitric
acid, a study of the rubidium ammonium salt of the related fluorocit-
ric acid, HOOCCHFC(OH)(COOH)CH_2COOH, has been carried out [249]. A
surprising feature of this structure is the tridentate chelation to
rubidium involving the fluorine (43). The same tridentate coordination to iron has been proposed to explain the inhibition of aconitase by fluorocitrate [250], and it is proposed that hydroxycitrate

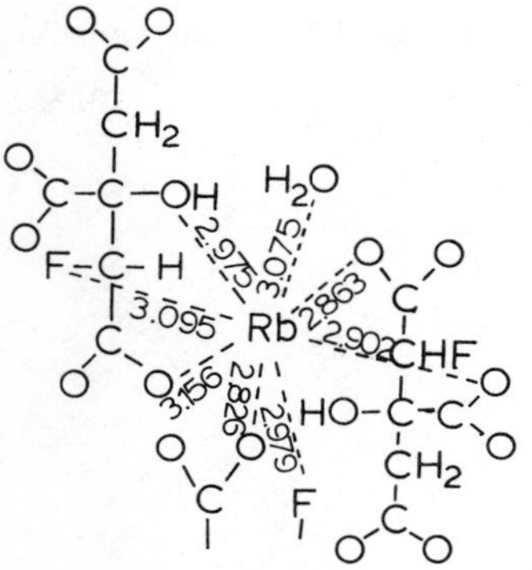

(43)

coordinates to metal ions in a similar fashion [75]. The eightfold
coordination of rubidium is completed by a water molecule and carboxyl oxygen and fluorine atoms of adjacent fluorocitrate anions.
The rubidium ammonium structure exhibits some disorder.

Tartrates and Their Derivatives [4]. The stereochemistry of
the dihydroxydicarboxylic acid, tartaric acid, HOOCCH(OH)CH(OH)COOH,
and its derivatives has been extensively studied. This interest in

tartaric acid stereochemistry is, at least in part, due to the key
role that this compound and its salts played in the development of
the concept of molecular asymmetry by Pasteur and others. Unfortunately, because of the presence of multiple coordinating atoms and
the possibility for several states of ionization, metal tartrate
structural chemistry has been, at least until recently, very confus-
ing. Many of the compounds obtained have been polymeric and of
indefinite composition [5]. In this section, only the more recent,
structurally oriented work is reviewed. A very good review of the,
for the most part, qualitative work prior to 1963 is available [5].
The first crystal structure determination on a metal ion derivative
of tartrate other than alkali metal salts was not reported until
1964 [251].

Because of the presence of the two chiral centers, tartaric
acid exists in three isomeric forms (44-46). Crystal structure

```
      COOH              COOH              COOH
       |                 |                 |
   H►C◄OH            HO►C◄H            HO►C◄H
       |                 |                 |
   HO►C◄H             H►C◄OH            HO►C◄H
       |                 |                 |
      COOH              COOH              COOH

       d                 l               meso

     (44)              (45)              (46)
```

determinations on the free acids [59-61], nonmetallic salts [62-69,
89-93,86-88], and other salts where the tartrate anion is not coordinated to a metal ion [70-72,89-93,252-254] show only -sc and +sc
conformers for d- and l-tartaric acids, respectively, and their
derivatives, and only racemic mixtures of +sc, -sc conformers for
ms-tartrates (Fig. 16). (±)-Erythro-fluoromalic acid, which can be
considered a derivative of ms-tartaric acid, also exhibits the +sc,
-sc conformer mixture [77]. Unlike the malate structures, where
metal ion coordination strongly influences the ligand conformation,
the tartrate conformations found in the presence of coordinating

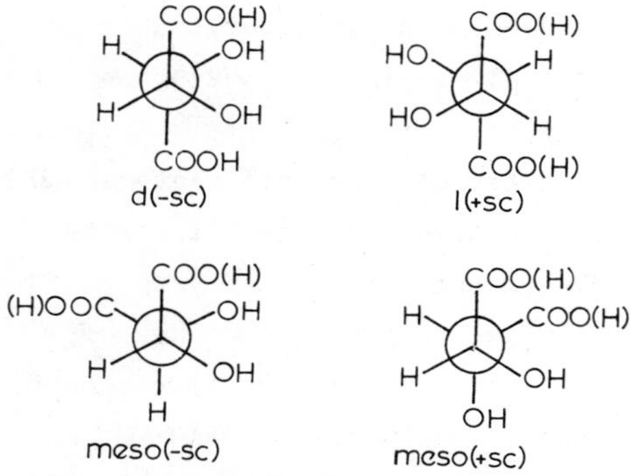

Figure 16   Conformers of tartrate isomers found in X-ray structure determinations.

metal ions in the solid state are unchanged from those determined for free tartrate groups.

A number of structures of "simple" metal tartrate salts have been determined, and these will be briefly discussed here. The lithium ion in lithium ammonium (±)-tartrate(2-) [62] appears to be four-fold coordinated, with bonds to two carboxyl oxygen atoms, a hydroxyl oxygen atom, and a water molecule; however, that some metal-ligand distances are anomalous indicates that one or more atoms (probably the water oxygen is one of these) has been misplaced. The coordination of the lithium in lithium ammonium (+)-tartrate(2-) [255,256] and in the isostructural lithium thallium(I) salt [257] is distorted tetragonal pyramidal, with a chelating α-hydroxycarboxylate group spanning one edge of the quadrilateral base and a water molecule and a carboxyl oxygen occupying the other two basal sites. A carboxyl oxygen at a relatively short distance from the lithium ion occupies the axial position. The two crystallographically nonequivalent thallium(I) ions in the lithium thallium derivative are eightfold and tenfold coordinated, and each lies on a twofold rotation axis. Neither thallium ion is chelated. The average thallium-oxygen bond length for the eight-coordinate thallium ion is less than that for the ten-coordinate ion. A tetrahedral lithium ion containing carboxyl oxygen atoms from two tartrate groups and two water molecules in its coordination sphere is present in lithium tris(ethylenediamine)chromium(III) tartrate(2-) trihydrate [253].

Sixfold coordinate sodium ions with pseudooctahedral geometries are found in sodium (+)-tartrate(2-) dihydrate [258], sodium potassium (+)-tartrate tetrahydrate (both the $P2_12_12_1$ nonferroelectric phase [259,260] and the $P2_1$ ferroelectric phase [261]), and sodium potassium (±)-tartrate tetrahydrate [262]. The potassium ions have coordination numbers of 6 in the (±)-tartrate [262] and both 6 and 8 in the (+)-tartrate [259-261]. The variable coordinations possible for potassium can also be seen from structures of potassium ms-tartrate(2-) dihydrate (two independent potassium ions with six close oxygen contacts for each) [263] and potassium ms-tartrate(1-) (two independent potassium ions, one with seven contacts and the other with eight) [264,265]. The coordination geometries about the potassium ions are highly irregular. All of the hydrated salts have water molecules in the coordination spheres of the potassium or sodium ions.

Rubidium ms-tartrate(2-) dihydrate is isostructural with the potassium salt and has an irregular sixfold coordination about the metal ion [263]. In the (+)-tartrate(1-) compound, on the other hand, the rubidium ion is nine-coordinate [266]. The coordination polyhedron in the isostructural cesium (+)-tartrate(1-) salt has been described as a square-face-capped square antiprism [267]. It is interesting that in the three alkali metal bitartrates whose structures have been solved [264,266,267], only α-hydroxycarboxylate groups containing a protonated carboxyl oxygen atom are involved in chelation.

In the alkali metal tartrates, all of which exhibit polymeric structures, there is a relatively low incidence of chelation. There is also a distinct tendency for the metal-hydroxyl oxygen contacts to be longer than the metal-carboxyl oxygen contacts among these compounds (as in other α-hydroxycarboxylates).

Strontium tartrate(2-) trihydrate and calcium tartrate(2-) tetrahydrate contain metal ions whose coordination polyhedra have been described as distorted dodecahedra [268]. The eightfold coordination includes two chelated α-hydroxycarboxylate groups, water molecules, and contacts with carboxyl oxygen atoms from tartrate ligands chelated to other metal ions. One of the α-hydroxycarboxylate groups in the calcium structure is distinctly nonplanar (dihedral angle, -25.6°).

Two of the strontium-carboxyl oxygen distances reported [268] differ from those calculated from the reported atomic coordinates [Sr-0(5), Sr-(01): reported, 2.26, 2.60 Å; calcd., 2.77, 2.66 Å].

That, taken as a whole, these network structures of the groups IA and IIA tartrates show only a small correlation between the absolute configuration of the carbinol carbon and the α-hydroxycarboxylate group dihedral angle sign may indicate that metal ion coordination strongly influences the α-hydroxycarboxylate conformation. On the other hand, the data for the uncoordinated α-hydroxycarboxylates indicates that the tartrates generally show a less skewed distribution of α-hydroxycarboxylate group dihedral angles (Fig. 6).

Among the metal ion derivatives of tartrates other than the alkaline earth and alkali metal compounds, monodentate, bidentate, and tridentate ligands are found. There is no conclusive proof of a tetradentate chelation by tartrate to a single metal ion--such a coordination being impossibly strained for most metal ion coordination geometries [4]. The relatively uncommon coordination types where tartrate is only monodentate (observed in several of the alkali metal structures just discussed) or is tridentate will be discussed at this time, before proceeding to the very common bidentate coordination.

Two transition metal structures containing tartrate coordinated through a single donor atom have been determined by diffraction techniques. In (+)-tartrato(2-)bis(N-methylethylenediamine)copper(II) dihydrate [94], a tartrate hydroxyl oxygen is coordinated (Cu-O distance, 2.630 Å) at an axial site of an axially elongated octahedral coordination polyhedron. The other axial site is filled by a water molecule. In (2S,3S)-tartrato(2-)bis(N-isopropyl-2-methylpropane-1, 2-diamine)copper(II) n-propanol, a carboxyl oxygen is located in the axial site of a pseudo-square pyramid (Cu-O, 2.23 Å) [95]. In both structures, the copper-tartrate oxygen distances are long and the interaction presumably weak, the diamine ligands providing the primary planar or near-planar coordination. The parameters for the noncoordinating α-hydroxycarboxylate group present in each of these structures are given in Table 3.

Tridentate chelation by tartrate has been observed in only one X-ray study, that of sodium d-tartrato(4-)-ℓ-tartrato(4-)ferrate(III) 14-hydrate [269], where a centrosymmetric octahedral iron(III) coordination polyhedron contains two tartrate groups, each coordinated through a completely ionized α-hydroxycarboxylate group and an additional deprotonated hydroxyl oxygen (47). The other α-hydroxycarboxylate group is bidentate to a sodium ion. A molecular model

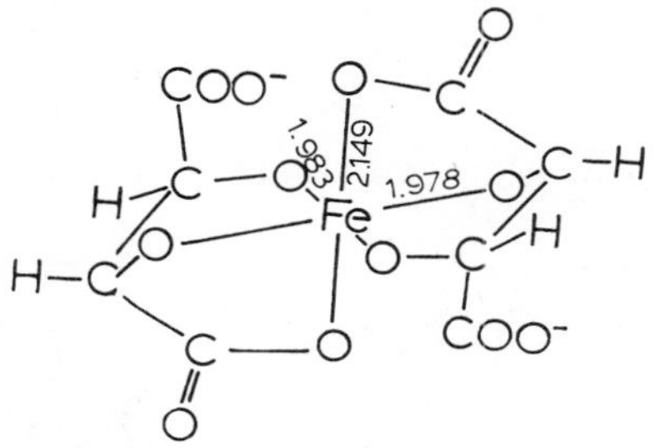

(47)

shows that a large nonplanarity of the α-hydroxycarboxylate group involved in the tridentate chelation (dihedral angle, +24.1° for the d ligand) and the large nonplanarity of the entire chelate ring (Table 5) is forced by the steric constraints of the tridentate coordination. A similar type of chelation is found in some alkaline earth carbohydrate acid complexes (see Sec. III.3).

Tartrate most commonly acts as a bidentate ligand through one or both of its α-hydroxy carboxylate groups (in the latter case, bridging two metal ions). A tris((+)-tartrato(3-))chromate(III) anionic complex has been characterized in aqueous solution by potentiometric and circular dichroic studies [270]. As a result of the sinambic [271] nature of α-hydroxycarboxylate groups, facial and meridional isomers are possible for the tris chelate; there can be two diastereomers for each. Additional isomerism owing to the possibility of mixed-ligand complexes is possible in the racemic chromium(III) tartrate system [272]. Studies have also been reported on mononuclear cobalt and chromium complexes containing a tartrate ligand bidentate

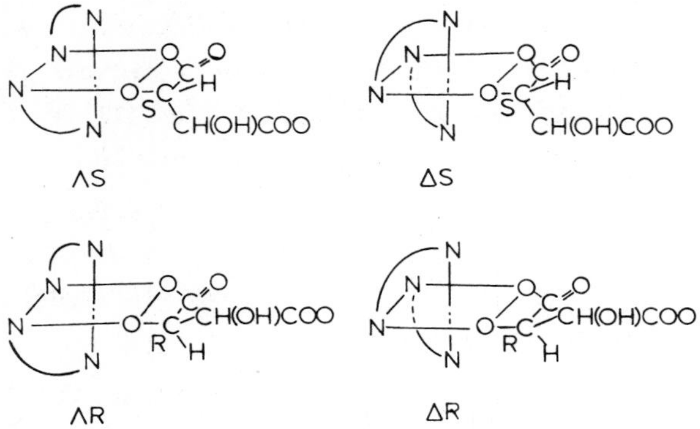

Figure 17 Diastereomers of octahedral bis(diamine)tartrato complexes.

through only one α-hydroxycarboxylate group in addition to two
nitrogen-donor bidentate ligands [139,217,273-276]. Two diastereo-
mers may be separated for $[M(N-N)_2(d\text{-tartrate})]$ or $[M(N-N)_2(\ell\text{-tar-}$
trate)] complexes [139,276], whereas four can be obtained for
$[M(N-N)_2(ms\text{-tartrate})]$ compounds [217] (Fig. 17).

By far the most studied geometry observed for classical metal
ion complexes of tartaric acid is a binuclear structure containing
two bridging tartrate groups, each bidentate toward both metal ions
(48). The stereochemistry of tartrate-bridged complexes has been reviewed [4], and the steric constraints as a function of coordination

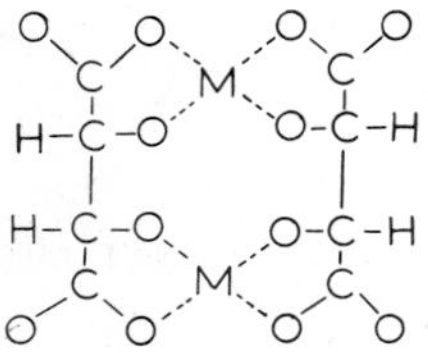

(48)

geometry and isomerism have been analyzed [197]. A binuclear complex
containing a single tartrate bridge has also been reported [138,277].

Excluding isomerism due to chirality in the coordination sphere,
seven isomers are possible for tartrate-bridged binuclear complexes,
depending on the isomeric form of the two bridging ligands.
Nonenantiomeric structures ($\ell\ell$ and ℓ-meso have been excluded) are

illustrated in Fig. 18 (arbitrarily for a planar coordination
geometry). The isomers are not equally strain free, because, in
part, of different degrees of staggering of the tartrate bridges--
differences that depend not only on the isomer but also on the
coordination geometry. For a planar coordination, for example, the
staggering increases in the order meso-meso < d-meso = ℓ-meso < dd =
ℓℓ < dℓ (Fig. 19), and the isomer stability is expected to increase
in the same order.

A number of solution studies have been carried out on tartrate-
bridged binuclear complexes. Both dd and dℓ diastereomers of
vanadyl(IV)tartrate(4-) have been prepared [278], and the presence
of exchange coupling between paramagnetic centers shows that these
and several derivatives have a binuclear structure in solution [279].

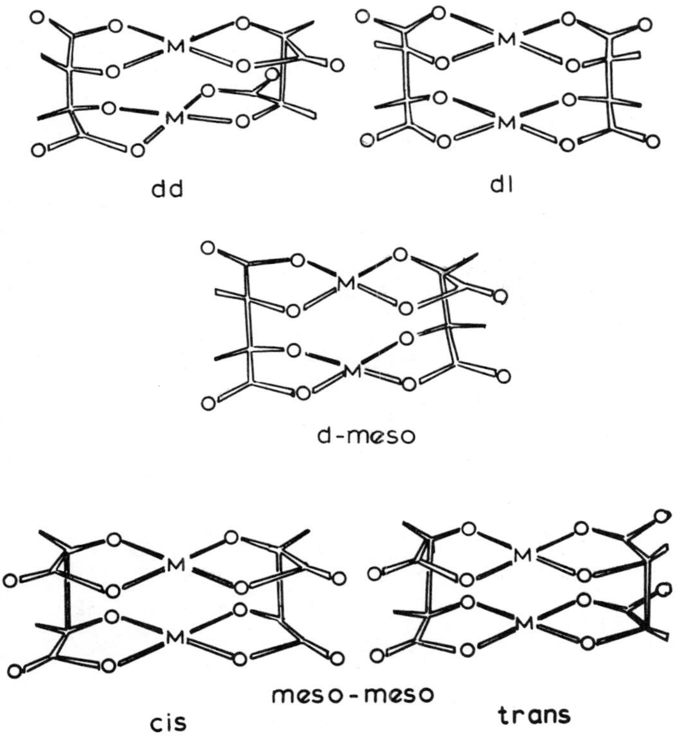

dd   dl

d-meso

meso-meso

cis   trans

Figure 18   Nonenantiomeric isomers of ditartrate-bridged binuclear
complexes.

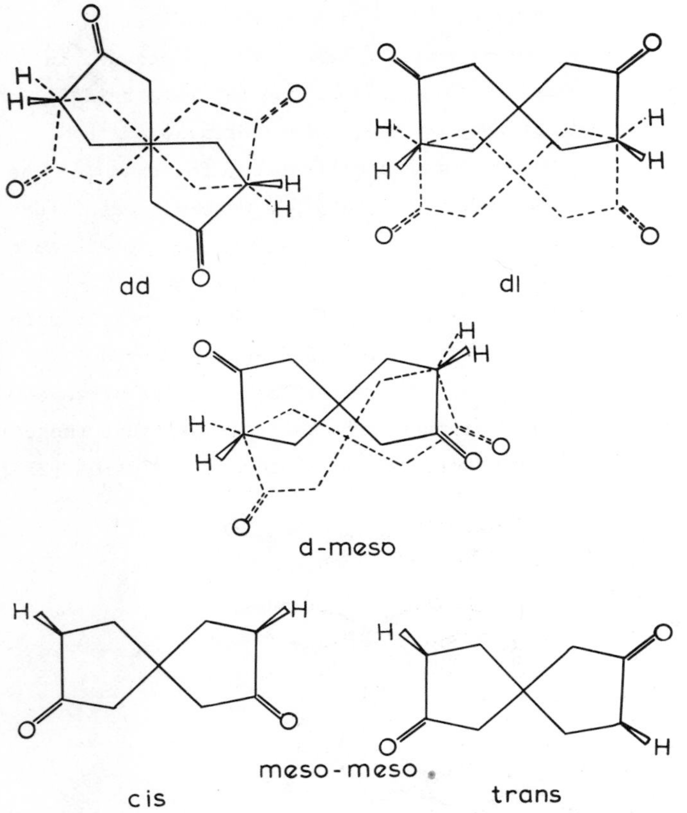

Figure 19 Top views of ditartrate-bridged binuclear complexes with a planar coordination, showing the staggering of the tartrate groups.

Apparently because of better staggering of the bridging tartrate(4-) ligands in the dℓ complex, this diastereomer is more stable than the dd isomer (ΔH = -1.46 kcal mol^{-1} [280]; ΔG = -1.63 kcal mol^{-1}, 25°C [281]). This stereoselectivity has been used to assign the absolute configurations of methyl-substituted tartrates by looking at the formation of mixed-ligand vanadyl(IV) binuclear complexes [282]. The meso-meso diastereomers of the vanadyl(IV) tartrates, with bridges in eclipsed conformations (Fig. 19), are of limited stability [281]. Even in the presence of excess ligand, the dd- and dℓ-vanadyl(IV) tartrate(4-) complexes show only a limited tendency to give mononuclear species [283].

Both meso-meso and dd tartrate-bridged complexes of chromium(III) with 2,2'-bipyridyl or 1,10-phenanthroline completing the coordination sphere have been prepared [141,284]. Like the vanadyl(IV) tartrates [281], binuclear chromium(III) tartrates containing ligands with varying degrees of ionization [in this case, tartrate(2-) and tartrate(3-)] are known [141].

^{13}C NMR studies on antimony(III) and arsenic(III) tartrate(4-), monomethyltartrate(4-), and dimethyltartrate(4-), show the presence of binuclear, bridged complexes in aqueous solution [285]. These studies show not only the formation of meso-meso complexes (for arsenic only) in addition to dd isomers, but the presence of mixed-metal binuclear complexes in mixtures of the arsenic and antimony compounds. Binuclear complexes have also been shown to be principal constituents of copper(II) tartrate solutions by EPR [286,287] and CD [288] spectroscopy.

X-ray structure determinations have been carried out for binuclear tartrate complexes of all of the metal ions discussed above. very recent results are now available on sodium (±)-μ-ms-tartrato(4-)-μ-ms-tartrato(3-)bis(2,2'-bipyridyl)dichromate(III) 6-hydrate [289a] and sodium μ-d-monomethyltartrato(4-)-μ-ℓ-monomethyltartrato(4-)-bis(oxovanadate(IV)) 10-hydrate [289b]. Data for these two recent structures do not appear in the tables.

The tetranegative complexes (49) in the sodium [290] and tetraethylammonium [291] salts of μ-d-tartrato(4-)-μ-ℓ-tartrato(4-)bis-(oxovanadate(IV)) have only crystallographic centers of symmetry but approximate the idealized C_{2h} symmetry expected for dℓ isomers. A centrosymmetric anionic complex is also found in the structure of sodium μ-d-dimethyltartrato(4-)-μ-ℓ-dimethyltartrato(4-)bis(oxovanadate(IV)) 12-hydrate [292]. The methyl-substituted derivative exhibits longer vanadium-ionized hydroxyl oxygen bond lengths (1.96 Å average vs. 1.92 Å average) and a shorter V-V distance (3.4 Å vs. 4.0 Å average) than do the unsubstituted compounds [290,291]. The increase in V-O bond length probably reflects the stabilization of alkoxide ions by alkyl groups [293], whereas the decrease in V-V distance is due to a movement of the vanadyl(IV) ion into the plane of

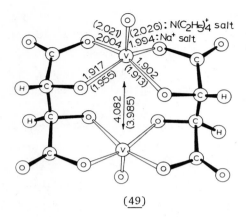

(49)

the four coordinating α-hydroxycarboxylate ligators. Methyl
substitution generally seems to decrease the V-V internuclear distance, according to EPR studies [279]. Like the unsubstituted dℓ
complexes [290,291], the methyl-substituted derivative has one chelated α-hydroxycarboxylate that is distinctly more nonpolar than the
other.

In the only structure that has been solved for a dd isomeric
vanadyl(IV) tartrate binuclear complex [294], the coordination polyhedron is distorted from the pseudo-square pyramidal geometry of the
dℓ isomers toward a trigonal bipyramidal geometry leading to an increased V-V distance (50). The change in coordination geometry probably occurs, at least in part, in order to better stagger the

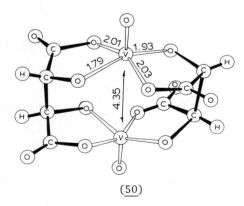

(50)

Figure 20 Coordination about vanadium ions in dℓ (top) and dd, ℓℓ (bottom) tartrate-bridged complexes.

tartrate bridges [4]. The idealized D_3 symmetry anion retains only a crystallographic twofold axis. In the dd complex, the hydroxyl oxygen atoms must be trans as must the carboxyl oxygen atoms (Fig. 20). In the dℓ complex, corresponding groups are cis in the coordination sphere.

Ten structures containing the idealized D_3 symmetry tartrate-(4-)-bridged antimony(III) dinegative anion (51) have been determined

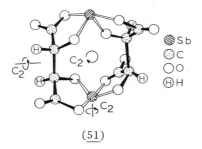

●Sb
◉C
○O
ⒽH

(51)

by diffraction techniques, although unfortunately several are of low precision. The antimony coordination polyhedron approximates a trigonal bipyramid with carboxyl oxygen atoms occupying axial sites and deprotonated hydroxyl oxygen atoms occupying two of the equatorial sites. The third equatorial site is presumably occupied by a stereochemically active lone pair of electrons on antimony. Six of the

reported structures [295-301] show no crystallographic symmetry for
the complex. The remaining four structures [251,302-304] contain
complexes of C_2 symmetry, and at least one structure for each possible twofold axis (51) is known. Only binuclear complexes of the dd
and ℓℓ isomeric forms have been found for antimony(III). Structures
of antimony(III) complexes prepared from racemic tartrate [296,305]
show only mixtures of di-μ-d-tartrato(4-)bis(antimonate(III)) and its
enantiomer. ^{13}C NMR studies also show only dd and ℓℓ isomers in
solutions of racemic tartrate derivatives of antimony(III) [285].
The dd and ℓℓ isomers give tartrate bridges that are better staggered
than those of a dℓ isomer for a trigonal bipyramidal coordination
geometry and are expected to be more stable [4,197].

The geometry of a binuclear tartrate(4-) complex of arsenic(III)
has been determined to be very similar to that of antimony(III) [305].
The As-As intermolecular distance of 4.688 Å is shorter than the Sb-Sb
distance of 5.08 Å (average for all structures) because of shorter
metalloid-donor atom distances. The average values of the metalloid-
to-deprotonated hydroxyl oxygen and metalloid-to-carboxyl oxygen bond
lengths for the tartrates are Sb, 1.99, 2.16 Å; As, 1.80 2.04 Å.
With the exception of the very low precision ammonium antimony(III)
d-tartrate structure [295], the structures of antimony(III) and arse-
nic(III) tartrates contain complexes with highly planar chelate rings.
The chiralities of the small distortions found for the α-hydroxycarboxylate groups show little if any correlation with the chiralities of
the asymmetric carbon atoms--that is, the distribution of α-hydroxycarboxylate dihedral angles for the antimony(III) and arsenic(III)
tartrate(4-)-bridged complexes are not markedly skewed.

Two tartrate-bridged structures of copper(II) have been determined. The sodium salt obtained from racemic copper(II) tartrate
solutions at high pH contains a centrosymmetric tetranegative dℓ
binuclear complex with tartrate(4-) bridges (52) [306,307]. The
copper(II) ions are strictly four-coordinate, and the two coordination polyhedra are slightly concave, giving a short (for a tartrate-
bridged binuclear species) metal-metal distance of 2.9869(7) Å and

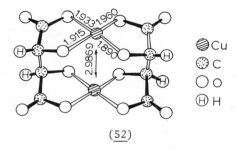

(52)

forcing the α-hydroxycarboxylate groups to be nonplanar (dihedral angles, +18.7°, +19.5° for R,R-tartrate). A distance of 3.3 Å was deduced from EPR studies [287].

Copper ions with axially elongated octahedral coordination poly-hedra bridged by tartrate(2-) groups are found in copper(II) d-tar-trate(2-) trihydrate (53) [308]. Carboxyl oxygen atoms not involved

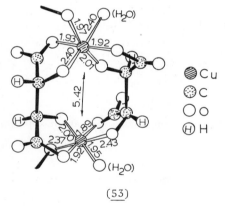

(53)

in chelation link the binuclear clusters together to give a lattice structure. A water molecule is included in each coordination sphere. The binuclear structure is similar to that found for a chromium(III) d-tartrate complex [289a], with the carboxylate oxygen atoms occupy-ing trans sites and the hydroxyl oxygen atoms occupying cis sites in the octahedron.

In copper(II) ms-tartrate(2-) trihydrate [308], each tartrate ligand is bidentate to two different copper ions as in the binuclear species; however, this structure is polymeric (54). There are two

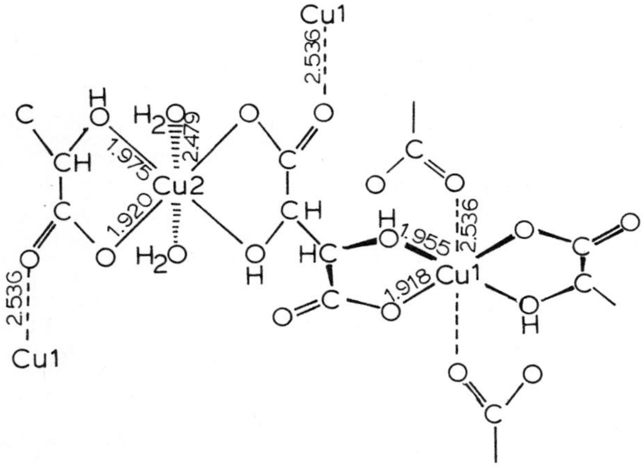

(54)

nonequivalent six-coordinated copper ions, each having an axially
elongated octahedral, centrosymmetric coordination polyhedron. In
addition to two bidentate α-hydroxycarboxylate groups, one coordination sphere includes two water molecules in axial sites, whereas the
other includes two carboxyl oxygen atoms, also in axial sites.

In summary:

1. Tartrate maintains the same conformations about the central
    carbon-carbon bond (Fig. 16) in all structures determined
    to date by diffraction techniques.

2. Tartrate is most commonly bidentate through its α-hydroxy-
    carboxylate groups, often giving polynuclear species, the
    tartrate-bridged binuclear complexes being very common.

3. As in other α-hydroxycarboxylate ligands, metal-to-ionized
    hydroxyl oxygen distances are shorter and metal-to-nonion-
    ized hydroxyl oxygen distances are longer than the metal-to-
    carboxyl oxygen distances in chelated tartrate groups.

4. In general, there is only a marginal correlation between the
    enantiomeric form (the sign of the dihedral angle) of the
    α-hydroxycarboxylate group and the ligand chirality in tar-
    trate complexes.

3. *Complexes of α-Hydroxycarboxylate Ligands Derived from
   or Related to Carbohydrates*

The ligands discussed in this section are primarily carbohydrate acids and related compounds. All contain multiple hydroxyl functions and only one or, at most, two carboxyl groups. Carbohydrate acids where an oxygen α to the carboxylate group is incorporated in a ring are discussed in the section on α-alkoxyacids even when a free α-hydroxyl group is also present. Most of the structural work reported on metal ion complexes of carbohydrate-related acids has been for calcium derivatives, and reviews of calcium carboxylate [9] and calcium carbohydrate [8] structural chemistry are available. Crystal structures of carbohydrates and derivatives (including metal ion complexes of carbohydrate acids) have been surveyed [309-314].

Lanthanide shift ^{1}H NMR studies on the polyhydroxy acids--glyceric, gluconic, and lactobionic acid--show that these ligands are tridentate toward europium(III), with chelation involving the carboxyl and α- and β-hydroxyl groups [315]. The finding that the europium-to-β-hydroxyl oxygen distances are shorter than the europium -to-α-hydroxyl oxygen distances are in conflict with crystal structures (vide infra), which indicate longer metal-to-β-hydroxyl oxygen distances (compared with metal-to-α-hydroxyl oxygen distances in the same structure) whether the β-hydroxyl group is involved in chelation or not. The tridentate chelation observed in the ^{1}H NMR studies is structurally similar to that determined in sodium d-tartrato(4-)-ℓ-tartrato(4-)ferrate(III) 14-hydrate [269]. Spectral studies of polyhydroxycarboxylate complexes of praeseodynium also indicate a γ-hydroxyl participation in chelation [127]. Although the α-hydroxycarboxylate group is expected to be the strongest donor group in polyhydroxycarboxylates, one crystal structure determination of a metal ion derivative, potassium D-gluconate monohydrate (A form), shows no metal ion contact with either the carboxylate group or the α-hydroxyl oxygen [97].

Glycerates and Substituted Glycerates. Glyceric acid, $HOOCCH(OH)CH_2OH$, which exists in two enantiomeric forms, is the simplest of the polyhydroxycarboxylic acids. Formation constants of

the glycerate derivatives, 2,3-dihydroxy-2-methylpropanoate and
2,3-dihydroxy-2-methylbutanoate, with rare earths indicate a tridentate chelation toward the lighter (and larger) lanthanides (La-Sm)
but a bidentate coordination toward the heavier (and smaller) ions
in aqueous solution [316]. ^{1}H NMR in D_2O shows a tridentate coordination of europium(III) by glyceric acid in acid solution with the
β-hydroxyl group gauche to the carboxyl group (dihedral angle, 30°)
[315].

Diffraction studies have been reported for only one metal derivative of glyceric acid--calcium (±)-glycerate(1-) dihydrate [317,318].
The pentagonal bipyramidal coordination polyhedron (55) contains two

(55)

crystallographically nonequivalent glycerate(1-) ions of opposite
enantiomeric form bidentate through α-hydroxycarboxylate groups.
Glycerate ligands of both +sc and -sc (hydroxyl gauche to carboxyl
and hydroxyl trans to carboxyl, Fig. 21) are present. (^{1}H NMR studies [319] indicate that more than one conformation of glycerate
exists in solution, since there are significant proton-proton coupling changes with pH.) The fifth equatorial site of the calcium
ion coordination is occupied by a β-hydroxyl oxygen atom, giving a
polymeric structure. The β-hydroxyl oxygen-to-metal ion distance
is slightly longer than the α-hydroxyl oxygen-to-metal ion distance.
This trend is seen in other structures of polyhydroxycarboxylate
complexes, although the data are very limited.

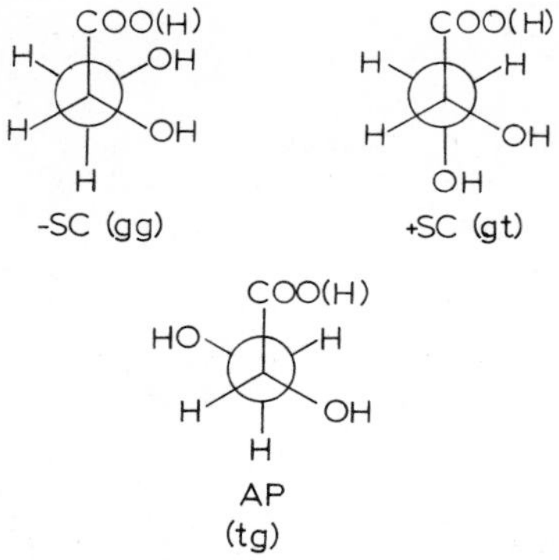

Figure 21 Conformers of glycerate.

A pentagonal bipyramidal coordination polyhedron (56) is also found in a structure determination on cadmium (S)-phosphoglycerate(2-) trihydrate [320]. In this structure, the gauche-gauche (gg, Fig. 21) ligand is coordinated to three metal ions, giving a polymeric structure. The ligand is simultaneously bidentate through its α-hydroxy-carboxylate group and through its two carboxylate oxygen atoms.

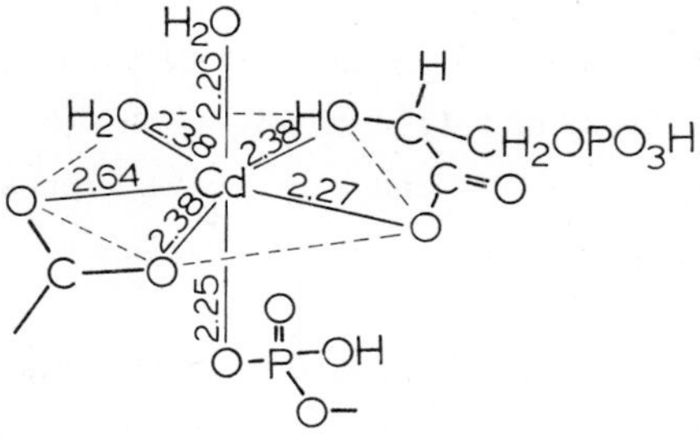

(56)

Unlike the two structures just discussed, the structure of diaquobis((±)-2,3-dihydroxy-2-methylpropanoato(1-))copper(II) [321] shows no coordination by the potential donor group attached to the β-carbon atom. The discrete centrosymmetric complexes contain copper(II) ions in an elongated octahedron of oxygen atoms, with water molecules occupying the more distant axial sites (57). Two

(57)

gauche-gauche ligands of opposite absolute configuration are
bidentate to each metal ion, giving a geometry similar to that found
for copper(II) glycolate and lactate complexes [167].

Gluconates and Derivatives [2]. Complexes of anions of gluconic
acid, $HOOCCH(OH)CH(OH)CH(OH)CH(OH)CH_2OH$ (Fig. 1), have been studied
extensively, although the structures of most are uncertain [2]. Two
idealized conformations of gluconate(1-) have been observed in crystal structures--extended and bent (Fig. 22). In the latter conformer,
which is obtained from the extended conformer by rotations about the
$C(3)-C(4)$ and $C(4)-C(5)$ bonds, the $O(2)-O(4)$ nonbonded steric interaction present in the extended form is relieved. ^{1}H NMR studies show
that in the 1:1 gluconic acid/europium(III) complex, where the ligand
is tridentate through the carboxyl and α- and β-hydroxyl groups, the
gluconate ligand has the extended form (incorrectly called the bent
form in the paper) [315].

The extended conformer of gluconate is found in anhydrous potassium gluconate(1-) [322] and in the A form of potassium D-gluconate
monohydrate [323,324]. In the anhydrous compound (and its isomorphous rubidium salt), the metal ion is surrounded by an irregular
octahedron of oxygen atoms of which four, lying in a roughly square
planar array, are carboxylate oxygens from four different molecules
(K-O distances, 2.71, 2.73, 2.78, 2.94 Å) and two are oxygens of a
β-hydroxyl group (from one of the same molecules whose carboxyl oxygen atom was coordinated, K-O distance, 3.03 Å) and a δ-hydroxyl
group (from a fifth molecule, K-O distance, 2.76 Å). Somewhat unexpected is the lack of participation of the α-hydroxyl oxygen in

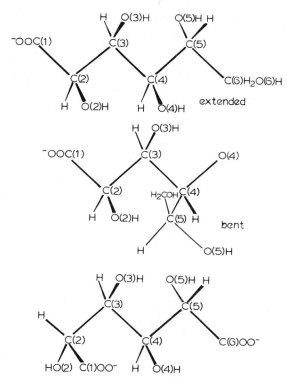

Figure 22 Conformers of gluconate (top and middle) and glucarate (bottom) found in X-ray studies.

coordination. In the monohydrated compound (A form), neither the α-hydroxyl oxygen atom nor the carboxyl oxygen atoms lie within coordinating distance of the metal ion (parameters for their structure are, therefore, in Table 3). The potassium ions are surrounded by two water molecules (K-O, 2.738, 3.076 Å) and six gluconate hydroxyl oxygen atoms (K-O, 2.612 to 3.239 Å) in a distorted dodecahedral geometry. The cell dimensions and density for lead gluconate(1-) [325] indicate that the gluconate ions in this structure also have an extended conformation. The lead ions lie on twofold axes.

The bent form of gluconate is found in the B dimorph of potassium D-gluconate (1-) monohydrate. In this structure, eight oxygen

atoms surround the potassium ions at distances of 2.685 to 3.003 Å.
A ninth oxygen is located at 3.338 Å. The α-hydroxycarboxylate group
is chelated.

Diffraction studies show that in the solid state, manganese(II)
D-gluconate(1-) dihydrate contains metal ions with distorted octahedral coordinations bridged by gluconate ligands attached by
carboxylate oxygens to one metal ion and by terminal hydroxyl oxygen
atoms to another [326]. Each coordination is completed by two water
molecules and a chelated α-hydroxycarboxylate group of a gluconate
ligand crystallographically distinct from the bridging ligands (58).

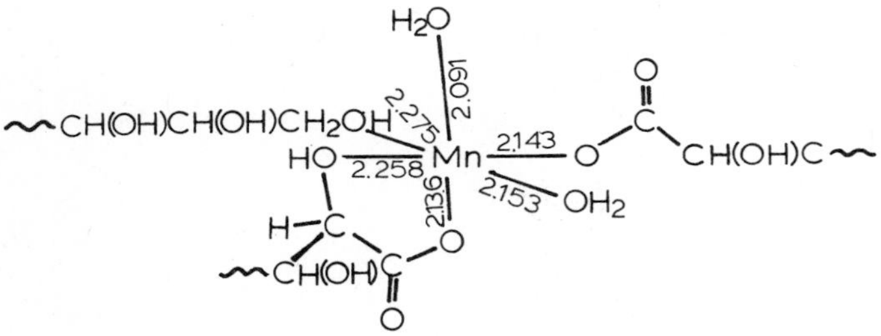

(58)

Although the difference in conformation between the two crystallographically unique gluconate ions has been stressed in the literature
[326], in fact, both conformers resemble the bent form of Fig. 22.
Polarographic and magnetic susceptibility studies indicate a dimeric
form for manganese(II) gluconate in aqueous solution; however, the
structure proposed for this polyhydroxycarboxylate complex of manganese (and those proposed for other similar manganese(II) complexes
[327]) is based on little or no structure-specific data [328].

The crystal structure of a phospho derivative of gluconate,
sodium 6-phospho-D-gluconate(3-) dihydrate, has been determined [329].
The coordination about the sodium ions is unusual in that each of the
three crystallographically nonequivalent metal ions has a different
coordination number. Ions with coordination numbers of 4 (distorted
tetrahedral), 5 (distorted tetragonal pyramidal), and 6 (distorted
octahedral) are present. The five-coordinate sodium ion is chelated

by the α-hydroxycarboxylate group of the phosphogluconate ion. The
organic molecule has the bent conformation (Fig. 22).

In the calcium bromide salt of lactobionic acid [330], the bent
conformation gluconate moiety links eight-coordinate calcium ions by
chelating to one ion through the carboxyl, α-hydroxyl, and β-hydroxyl
groups and chelating to a second through the terminal and adjacent
hydroxyl oxygens (59). The galactose substituent is not involved in
calcium coordination. The square antiprismatic coordination geometry
is similar to that found for all other calcium carbohydrate structures discussed in this and the following section.

(59)

Two structures of salts of D-glucaric acid, $HOOCCH(OH)CH(OH)CH-$
$(OH)CH(OH)COOH$, have been determined. In the monopotassium salt,
where neither α-hydroxyl oxygen is coordinated, square antiprismatic
metal ion coordinations by six carboxyl and two hydroxyl oxygen atoms
are found [331]. The dicarboxylate anion has a configuration analogous to that of the bent conformer of gluconate. Square antiprismatic
coordination is also found in the solid-state structure of calcium D-
glucarate(2-) tetrahydrate [332]. Here the glucarate anion has the
unusual conformation shown at the bottom of Fig. 22, where rotation
about the C2-C3 bond of an extended form has occurred. Moreover, one
of the α-hydroxycarboxylate groups is distinctly nonplanar, as a result of rotation of the carboxyl group (dihedral angle, -51°). This
α-hydroxycarboxylate is not chelated. Interestingly, the same

carboxyl group of the glucarate ion in the potassium salt [331]
exhibits a larger-than-normal rotation in the same direction (-18.0°).

   Other Carbohydrate Acids. Structures of four additional carbohydrate acid compounds, all calcium or strontium salts, have been
determined. All the metal ion coordination geometries are eight-
coordinate distorted square antiprismatic.

   Tridentate chelation of the metal ions is present in the isostructural calcium and strontium salts of the five-carbon polyhydroxy
acid, arabonic acid, $HOOCCH(OH)CH(OH)CH(OH)CH_2OH$ [333]. The metal
ions lie on a twofold crystallographic axis. The coordination geometry is very similar to that determined for a europium(III) bis-
(glucarate) complex in aqueous solution by ^{1}H NMR [331].

   In crystals of isostructural calcium [334] and strontium [335]
salts of 3-D-glucoisosacharic acid, $HOOCC(OH)(CH_2OH)CH_2CH(OH)CH_2OH$,
eightfold-coordinated metal ions chelated by two α-hydroxycarboxylate
groups lie on twofold symmetry axes. The square antiprismatic coordination is completed by four hydroxyl oxygen atoms from ligands chelated to other metal ions. Extended Hückel molecular orbital
calculations on the 3-D-glucoisosacharate(1-) ion [334] predict not
only the planarity of the α-hydroxycarboxylate group but also the
observed conformation with respect to rotation about the C(3)-C(4)
bond.

IV. THE α-HYDROXYCARBOXYLATE CHELATE RING

The geometric parameters as determined by diffraction studies for
chelate rings formed by α-hydroxycarboxylate groups are given in
Table 5. Once again, all structures, whether or not for transition
metal complexes, are reported. Some of the data given in Table 4
(the absolute configuration and the dihedral angle) are also presented in Table 5 for ready reference and also as a means of cross-
referencing structures. (A comparison of the dihedral angle values
and the reference number allows a correlation between data reported
in Tables 4 and 5 for the same α-hydroxycarboxylate group.)

The distinctly nonmetallic acceptors B(III) and C(!) are also included in Table 5 for completeness. The carbon "chelate ring" is for the α-hydroxycarboxylate grouping found in the cyclic ester-acetal of glycolic acid and 2-methly-4,5,6-trichlorocylohex-2-en-1-one [336] (60) and might be considered a chelate ring formed for the

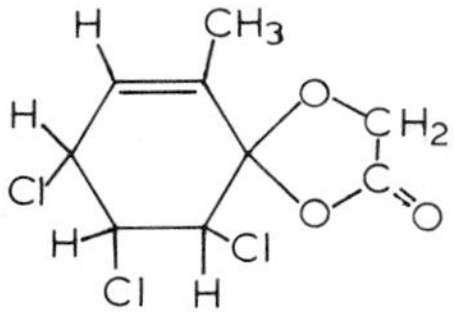

(60)

largest covalency in the donor-acceptor bonds. The alkali metal ion chelates, of course, represent the other extreme, whereas the transition metal ions are intermediate in covalency.

Table 5 also contains average values of the geometric parameters for chelate rings for each type of acceptor ion. The averages are divided into two groups according to whether or not the hydroxyl oxygen bears a proton. Admittedly, the averaging of values for different ligands is questionable; however, the results are useful in some correlations (vide infra).

The data in Table 5 illustrate some interesting trends. The distribution of dihedral angles for dissymmetric chelated α-hydroxy-carboxylates is much less skewed than the distribution observed for the free ligands (Fig. 5). In fact, for the tartrates, there seems to be little or no correlation between the sign of the α-hydroxycarbox-ylate dihedral angle and the chirality of the carbinol carbon. Free tartrate groups also showed much less skewing of the dihedral angle distribution than other uncomplexed α-hydroxycarboxylates (Fig. 6). For other chelated α-hydroxycarboxylates, there appears to be a small preference for positive dihedral angles (Fig. 3) to be adopted by α-hydroxycarboxylate groups having a hydrogen-bearing carbinol carbon of R absolute configuration (and conversely). Although the data are limited, the results in Table 5 indicate that about 65% of the time,

the dihedral angle sign is correlated with the carbinol chirality
for nontartrate groups in the direction indicated. As pointed out
earlier, this slightly preferred direction of α-hydroxycarboxylate
conformational dissymmetry will place the nonhydrogen carbinol
substituent in a more axial position (contrary to what is usually
assumed [194]) as long as the chelate ring has a puckered conformation (which, as will be seen, is often not the case).

Chelation of an α-hydroxycarboxylate group can occur without proton loss by the hydroxyl oxygen (and does in most cases examined in
Table 5); however, in all cases where a crystal structure determination has been made, ionization of the hydroxyl group of a coordinated
α-hydroxycarboxylate is always accompanied by chelation. In nearly
all of the structures whose parameters are reported in Table 5, the
metal-to-hydroxyl oxygen bond length is longer than the metal-to-carboxyl oxygen bond length in chelate rings where the α-hydroxyl group
is not ionized. The reverse is generally true when the chelate ring
contains an ionized hydroxyl group. The decrease in metal-hydroxyl
oxygen bond length upon ionization of the hydroxyl group is striking
and gives rise to strong crystal field perturbations in α-hydroxycarboxylate complexes [4].

Whether ionized or not, the hydroxyl oxygen-metal bond length
increases with alkyl substitution at the carbinol carbon. Compare
bis(glycolato(1-))copper(II) (1.93 Å), aquobis(DL-lactato(1-))cop-
per(II) (1.97 Å), and diaquobis(2-hydroxy-2-methylpropionato(1-))-
copper(II) (2.01 Å) [167] or compare the vanadyl(IV) dℓ-tartrate
binuclear complexes (1.92 Å average) [290,291] with the methyl-
substituted derivative (1.97 Å average) [292]. Gas-phase acidities
of alcohols are known to increase with alkyl substitution [293]--a
fact that not only shows a corresponding decrease in alkoxide ion
basicity but also indicates a decrease in the basicity of the unionized alcohols.

The data in Table 5 show a strong dependence of the O-M-O angle
and the M-O bond lengths. This is shown most clearly in the plot of
the average values of the O-M-O angle for the various acceptors as a

function of the sum of the average acceptor-hydroxyl oxygen and acceptor-carboxyl oxygen bond lengths given in Fig. 23. That the correlation of larger angle with decreasing bond length holds equally well for the most ionic to the most covalent M-O bonds and for chelate rings with ionized as well as un-ionized hydroxyl groups indicates that the O-M-O angle is determined primarily by steric considerations. A similar dependence is found for amino acid complexes [337]. Interestingly, most of the other geometric parameters for α-hydroxycarboxylate chelate rings show only a marginal interdependence. There is, however a small and irregular increase in the C-O-M angles with increasing M-O distance.

The values for the distances of the two chelate ring carbon atoms to the O-M-O plane given in Table 5 are highly variable, and the average values for the various acceptor species probably have little meaning because of scatter; however, the individual values indicate two important facts. First, in almost every case (the exceptions being denoted by those instances where the distances have

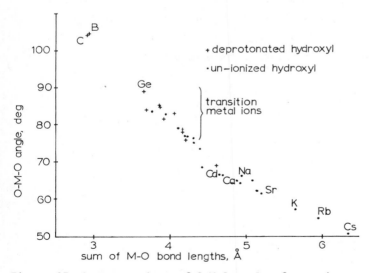

Figure 23 Average values of O-M-O angles for various acceptors M plotted as a function of the average M-O bond lengths for α-hydroxycarboxylate chelate rings (data taken from Table 5).

opposite signs in Table 5), the two carbon atoms lie on the same side
of the O-M-O plane. Second, in the majority of α-hydroxycarboxylate
chelate rings, the carboxyl carbon lies farther from the O-M-O plane
than the carbinol carbon. The results show clearly that the chelate
ring conformation exhibited by most α-hydroxycarboxylates is an asymmetric envelope (Fig. 24). The other possible form, the puckered
conformer sometimes observed for amino acid chelates [338], occurs
only rarely in α-hydroxycarboxylates. One measure of the nonplanarity of an asymmetric envelope chelate ring is the angle between the
O-M-O plane and the O-C-C-O plane. This angle is given in Table 5
for all α-hydroxycarboxylate chelate rings. The large scatter in
values of this ring folding angle, even among complexes of the same
ligand or the same acceptor, indicates that the degree of folding of
the α-hydroxycarboxylate chelate ring into an asymmetric envelope
conformation is highly dependent on steric restrictions imposed by
polydentate coordination and/or molecular packing. Two trends are,
however, observed. First, the magnitude of the folding angle appears
to increase as the coordination bond lengths increase, although there
is a large amount of scatter in the data. Second, there is a tendency for a chelate ring containing only one hydrogen atom on a carbinol
carbon to fold such that the nonhydrogen substituent is more
equatorial than would otherwise be the case (61). This is

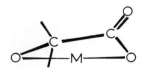

asymmetric envelope

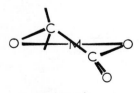

puckered

Figure 24 Asymmetric envelope and puckered conformers of five-
membered α-hydroxycarboxylate chelate rings.

particularly true when the metal ion contains axial substituents.
This type of folding can, of course, offset the effect of a dissymmetric α-hydroxycarboxylate group conformation, which often tends
to force the nonhydrogen carbinol substituent toward an axial position, as was pointed out earlier.

(61)

V. COMPLEXES OF α-ALKOXYCARBOXYLATES

In this section is discussed the structural chemistry of complexes
containing a ligand in which an oxygen atom adjacent to a carboxyl
group is bonded to a second carbon atom. This includes alkoxy- and
aryloxycarboxylates as well as a number of α-carboxyl cyclic ethers
found in some lactones and carbohydrate derivatives. The term
alkoxycarboxylate will often be used to designate all of these compounds.

A. α-Alkoxycarboxylic Acids and Esters

Bonding parameters for free α-alkoxycarboxylic acids and their esters as determined from diffraction studies are given in Table 6.
Surprisingly, there have been no structures reported for uncoordinated α-oxocarboxylates with simple aliphatic groups on the α-oxygen;
however, a number of structures have been reported for relatively
simple α-aryloxycarboxylic acids. The data in Table 6 indicate a
much lower tendency for the α-alkoxycarboxylate grouping to be planar
than is the case for α-hydroxycarboxylates. However, the total number of structures of free α-alkoxycarboxylates whose parameters are
given in Table 6 is relatively small, and many have the α-oxygen atom
incorporated in a ring, which may cause steric problems with
formation of a planar grouping. In those compounds where the carbon
alpha to the carboxyl group contains a single hydrogen atom, the α-

alkoxycarboxylate group dihedral angle distribution shows the same
skewing as found for α-hydroxycarboxylates (Fig. 5). Groups with α-
carbon atoms of R chirality tend to have positive dihedral angles
(Fig. 3), and conversely.

An alkoxy group increases the acidity of an α-carboxyl group to
an even greater extent than does a hydroxyl group. This can be seen
in the relative pK_a values of methoxyacetic, glycolic, and acetic
acids (Table 1). The same trend is also evident in the pK_1 and pK_2
values for oxydiacetic (2.81 and 3.94 [357]), tartaric, and succinic
acids (Table 1). As pointed out earlier, this decrease in pK_a is ascribed to a solvation effect [29]. Although IR spectra indicate that
in a nonpolar solvent there is extensive hydrogen bonding between
carboxylic acid groups and adjacent alkoxy or aryloxy oxygens [358-
360], it is unlikely that this has any effect on aqueous solution pK_a
values (which would be increased by such bonding).

B. Complexation by Alkoxycarboxylates

α-Alkoxycarboxylates, like α-hydroxycarboxylates, can chelate in
aqueous solution (as shown by differential proton relaxation studies
on copper(II) systems [150,151]) and in the solid state. There are,
however, two major differences. First, α-alkoxy groups obviously do
not ionize (under normal conditions) upon chelation. Second, α-
alkoxy groups appear to be less basic toward metal ions than α-
hydroxyl groups, as indicated by solid-state structures (vide infra)
and stability constant compilations [150]. A decrease in basicity
is not unexpected, since this follows the trend discussed in the previous section for alkyl substituents on carbinol carbons of α-
hydroxycarboxylates.

In the sections that follow, the known structures of α-alkoxycarboxylate complexes as determined primarily by diffraction studies are
discussed by ligand type. The geometries of α-alkoxycarboxylate
groups in metallic compounds are tabulated in Table 6 along with those
for the free ligands. Geometric parameters for α-alkoxycarboxylate
chelate rings are listed in Table 7.

C. α-Alkoxycarboxylate Structures

   1. *Complexes of Simple α-Alkoxycarboxylates*

   Methoxyacetates and Ethoxyacetates. Methoxyacetic acid, CH_3OCH_2COOH, is perhaps the simplest of the α-alkoxycarboxylic acids, and several structural studies of its complexes, primarily with copper(II), have been carried out by C. K. Prout and his co-workers. These studies show that methoxyacetate can coordinate either as a mondentate ligand through a carboxylate oxygen or as a bidentate ligand through the α-alkoxycarboxylate group.

   The centrosymmetric complexes (62) found in structural studies on diaquobis(methoxyacetato)copper(II) [167] and diaquobis(methoxyace-tato)nickel(II) [361] are isostructural, although there are important

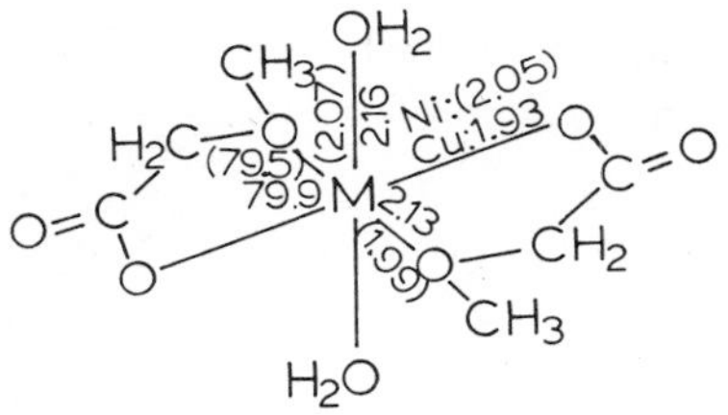

(62)

differences in the exact coordination geometries. The coordination spheres in both cases are compressed octahedra; however, the compression in the copper complex is along the copper-carboxylate oxygen bonds and is distinctly larger than the compression in the nickel complex, which is along the nickel-methoxy oxygen bonds. In the nickel complex, the metal-methoxy oxygen bonds (1.99 Å) are not only the shortest bonds in the coordination sphere but are shorter than nickel-hydroxyl oxygen bonds observed in chelated nickel(II) α-hydroxycarboxylates, where the α-hydroxyl group is not deprotonated (2.080 Å [190], 2.081 Å [220]). On the other hand, in the copper(II) complex, the metal-methoxy oxygen bond (2.13 Å), although not the longest bond in the coordination sphere, is longer than the metal-hydroxyl oxygen bonds observed for chelated copper(II) α-hydroxycarboxylates

where the hydroxyl oxygen is not axial in an elongated octahedron
(average, 1.97 Å, Table 5).

The results on these two structures are ambiguous in predicting
the relative donor strength of α-alkoxy and α-hydroxyl groups toward
metal ions. Admittedly, the copper(II) ion is susceptible to Jahn-
Teller distortion, which makes structural data less easily interpretable for copper(II) systems. On the other hand, most of the
structural results for α-alkoxycarboxylates [unfortunately, primarily
for copper(II) complexes] indicate a decreased donor ability for an
α-alkoxy group compared with an α-hydroxyl group.

Unlike the compressed octahedral structure observed for the
diaquo derivative, the centrosymmetric bis(methoxyacetato)bis(pyridine)copper(II) complex [362] has an elongated octahedral coordination (63). The copper-methoxy oxygen bond lengths (2.36 Å) are

(63)

somewhat longer than the copper-hydroxyl oxygen bond lengths observed
for other elongated octahedral CuN_2O_4 systems containing chelated α-
hydroxycarboxylate groups with axial hydroxyl oxygens (2.310 Å [189],
2.28, 2.26 Å [191]).

In the centrosymmetric, elongated octahedral bis(methoxyacetato)-
tetrakis(imidazole)copper(II) complex (64), the methoxy acetate ligands are monodentate through carboxylate oxygen atoms [363]. The
axial copper-carboxylate oxygen bond lengths are among the longest

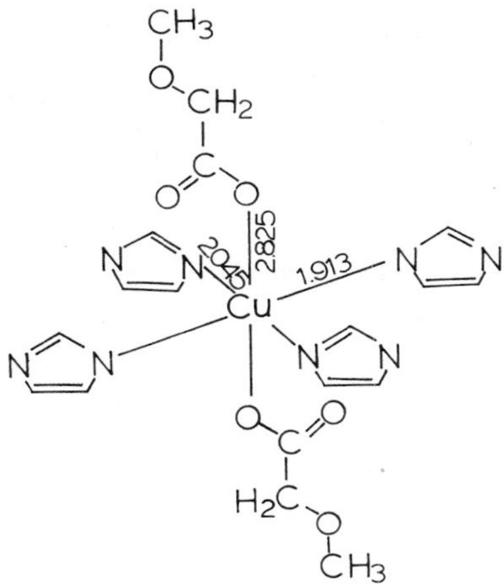

(64)

observed for any copper(II) complex and imply a very weak interaction. Bonding parameters indicate that the two imidazole molecules at a coordinating distance of 1.913 Å are deprotonated. The locations of the protons needed for charge balance in this structure are uncertain.

In the solid state, the centrosymmetric geometry determined [364] for the diaquobis(ethoxyacetato)copper(II) complex (65) is similar to that found for the corresponding methoxyacetato compound (62) except

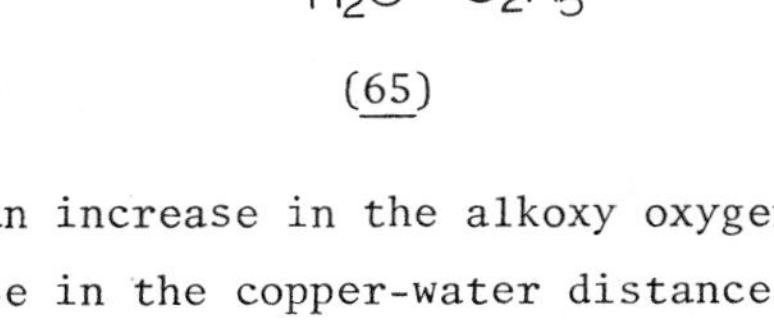

(65)

for an increase in the alkoxy oxygen-metal ion bond length and a decrease in the copper-water distance. The longer metal ion-alkoxy oxygen bond length may indicate a decreased basicity for the alkoxy oxygen. The ethoxide ion is known to be intrinsically less basic than the methoxide ion [293].

A comparison of the structures of the three copper(II) methoxyacetate complexes [167,362,363] indicate that in the presence of strongly donating ligands such as pyridine and imidazole, the copper-methoxy oxygen interaction becomes weaker and the alkoxyacetate group may even become nonchelating.

All of the alkoxyacetate chelate rings found in these structure determinations have asymmetric envelope conformations with α-alkoxycarboxylate groups that are slightly nonplanar to almost perfectly planar (dihedral angles from 15.7 to 0.1°). Surprisingly, in every case, the alkyl substituent on the alkoxy oxygen lies on the same side of the α-alkoxyacetate plane as the metal ion in these asymmetric envelope conformers (Fig. 25). However, since in most cases the geometry about the alkoxy oxygen is not far from trigonal planar and the chelate rings do not deviate greatly from planarity, the distinction between the alkyl group lying on the same or opposite side is not great. In the phenoxyacetate chelates (vide infra), the larger phenoxy group usually lies opposite the metal ion in the chelate rings whose structures have been determined.

Aryloxyacetates. A number of structures of derivatives of free phenoxyacetic acid, $C_6H_5OCH_2COOH$, have been reported (Table 6). The structure of the sodium salt has also been determined [365]. The sodium ion in this structure is surrounded by six oxygen atoms (four from carboxyl groups at 2.301-2.492 Å and one each from a phenoxy group at 2.658 Å and a water molecule at 2.607 Å) in a very irregular polyhedron that includes one chelated aryloxycarboxylate group. The sodium-phenoxy oxygen contact is slightly longer than the average sodium hydroxyl oxygen contact observed for chelated α-hydroxycarboxylates (2.56 Å, Table 5).

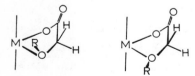

Figure 25 Proximal (left) and distal (right) configurations of alkyl groups (relative to the metal ion) in asymmetric envelope α-alkoxycarboxylate chelate rings.

X-ray structure determinations have been carried out on seven copper(II) phenoxyacetates or substituted phenoxyactates. The two crystallographically nonequivalent diaquobis(phenoxyacetato)copper-(II) complexes found in one solid-state structure [167] have very similar geometries. One molecule (66) has a crystallographic center of symmetry, whereas the other (67) lies on a general position in the

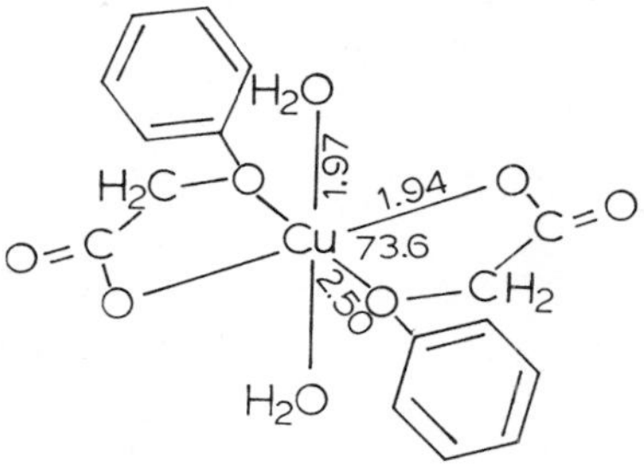

(66) 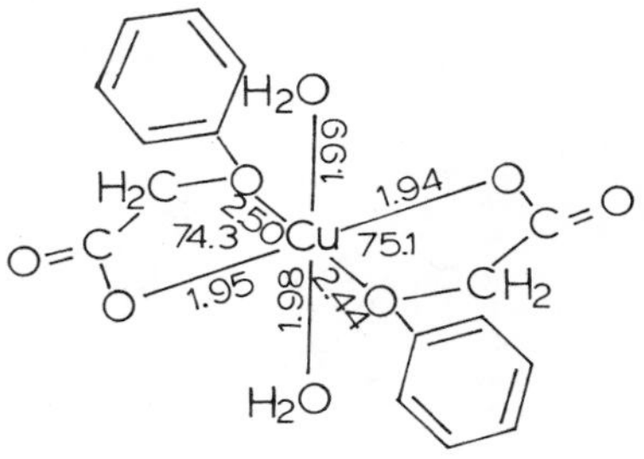   (67)

crystal. A similar geometry is found for diaquobis(p-methoxyphenoxy-acetato)copper(II) [366]. That the copper-phenoxy oxygen distances in these complexes are longer than the copper-alkoxy oxygen distances found for the alkoxyacetates discussed in the preceding section indicates that in order of decreasing basicity of the alkoxy oxygen in $ROCH_2COO^-$ ligands, $R = CH_3 > R = C_2H_5 > R = C_6H_5$. The phenyl rings lie near the plane of the aryloxycarboxylate group, so that each bidentate ligand is nearly planar. In the free substituted phenoxyacetic acids whose crystal structures have been determined, angles of 4.2 to 85.2° are found between the phenyl and carboxylic acid planes [342], showing that the phenyl ring and aryloxycarboxylate groups can range from nearly coplanar to nearly mutually perpendicular. In the phenoxyacetate complexes these two groups are often approximately coplanar.

Among the remaining crystal structures of copper aryloxyacetates, only anhydrous copper(II) phenoxyacetate [367] has aryloxycarboxylate groups chelated through the aryloxy oxygen and a carboxylate oxygen. This structure contains discrete hexanuclear clusters of S_6 symmetry with two types of phenoxyacetate ligands. One type of ligand bridges metal ions through the carboxylate grcup only, whereas the other forms a five-membered chelate ring to one metal ion while coordinating to two others through carboxylate oxygen atoms.   The coordination polyhedron (68) is a rather distorted elongated octahedron.

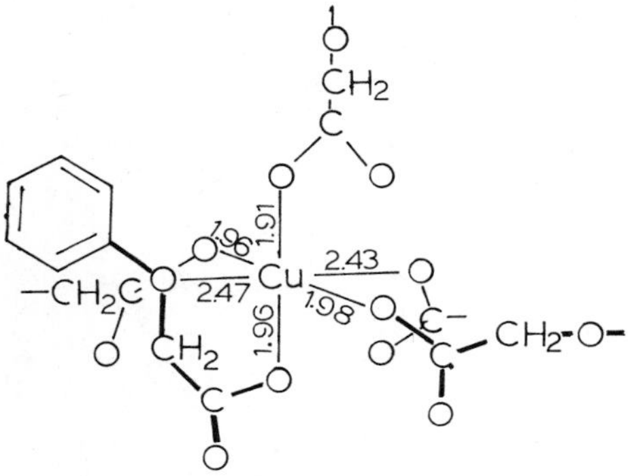

(68)

It may be of some interest that in every chelate ring but two (for the structures sodium phenoxyacetate hemihydrate [365] and diaquobis(p-methoxyphenoxyacetato)copper(II) [366]) in the crystal structures for simple alkoxycarboxylates in Table 7, a positive dihedral angle is obtained for a bidentate alkoxyacetate group having an R absolute configuration at the alkoxy oxygen atom and conversely (priority [17]: metal ion > methylene carbon > alkyl or aryl substituent > lone pair).   These results must be viewed with caution, however, as the alkoxy or aryloxy oxygen is nearly trigonal planar in many cases, so that relatively minor differences in positions of the atoms distinguish between R and S chiralities.

Bridging of two metal ions by carboxylate groups is also observed for the binuclear complex tetra-μ-(2,4-dichlorophenoxyacetato)-bis(dioxanecopper(II)) [368].   Here, discrete binuclear clusters (69) contain copper(II) ions in a square pyramidal coordination with four carboxylate oxygen atoms and a dioxane oxygen.   The centrosymmetric molecules contain two crystallographically nonequivalent phenoxyacetate groups.

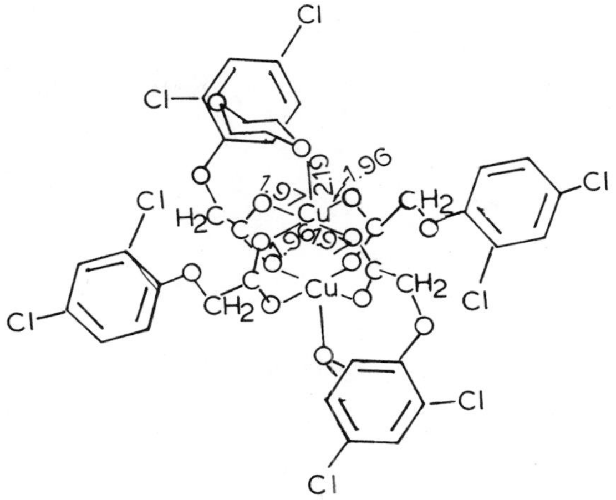

(69)

Diaquobis(p-methoxyphenoxyacetato)copper(II) dihydrate [366] (70), triaquobis(phenoxyacetato)copper(II) [369] [average dimensions for two crystallographically nonequivalent molecules are given in (71)], and aquobis(phenoxyacetato)bis(pyridine)copper(II) [362] (72)

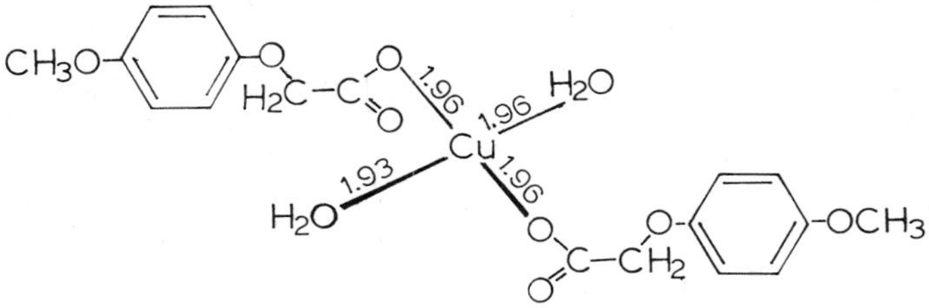

(70)

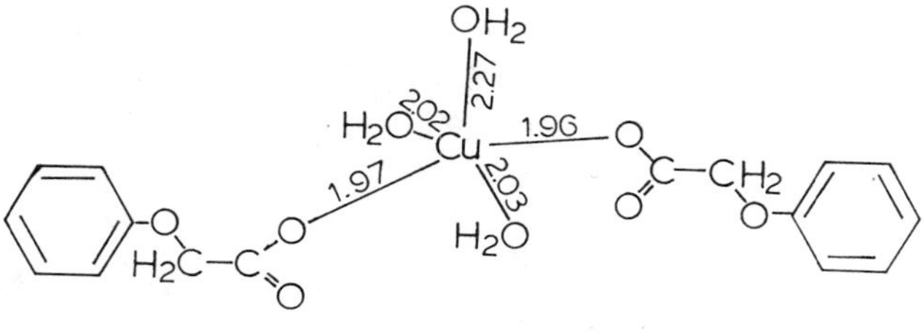

(71)

(72)

all contain aryloxyacetate ligands monodentate through carboxylate
oxygens in trans arrangements. The first structure has a square
planar geometry with a crystallographic mirror plane relating the two
aryloxycarboxylates. The latter two complexes have quadrilateral pyramidal structures with no crystallographic symmetry. The conformations
of the aryloxycarboxylate groups in (70) are such that the methylene
carbon-ether oxygen bond eclipses the carboxyl carbon-oxygen bond involving the oxygen bound to copper. In (72), all aryloxycarboxylate
groups have conformations differing by 180° from those in (70), with
the methylene carbon-ether oxygen bonds eclipsing the carbonyl groups.
In (71), each coordination sphere contains both types of conformers.
Differential proton relaxation studies [151] indicate that both monodentate and bidentate phenoxyacetate groups are present in solutions
of copper(II) phenoxyacetate.

2. *Complexes of Oxydiacetate*

A number of structure determinations have been carried out on
oxydiacetic acid, $HOOCCH_2OCH_2COOH$, and its metal derivatives. The
structural chemistry of this alkoxydicarboxylate resembles that of
iminodiacetic acid with which it is isoelectronic [370]. As the free
acid, the oxydiacetate molecule is twisted about each methylene carbon-ether oxygen bond (while maintaining C_2 symmetry), so that the
two α-alkoxycarboxylate moieties, whose nonhydrogen atoms are separ-
ately planar, are not mutually planar [347,371]. In most metal
derivatives, on the other hand, the nonhydrogen atoms of the entire
oxydiacetate group are found to be coplanar or nearly so. This

coplanarity is often accompanied by tridentate chelation.   Thus in
sodium and potassium oxydiacetate(1-), planar tridentate ligands are
present in pentagonal bipyramidal coordination polyhedra (73, 74)

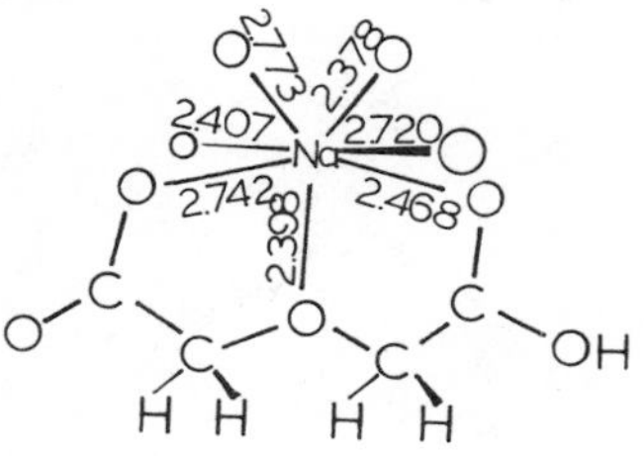

(73) 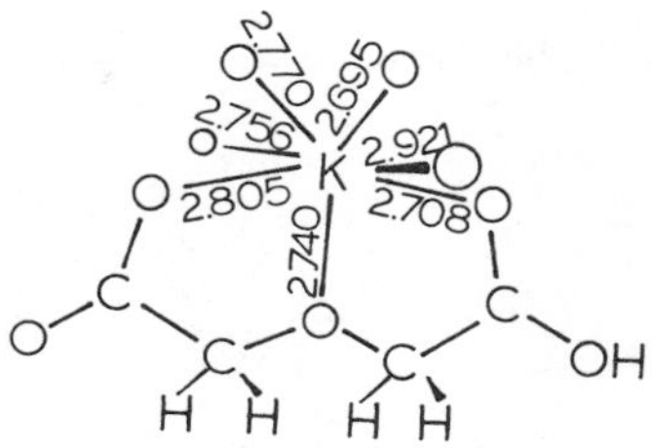   (74)

which are completed by oxygen atoms from tridentate ligands in other
polyhedra [372].   Every oxygen atom of the oxydiacetate(1-) groups in
these structures, including the protonated carboxyl oxygen atoms, is
in contact with a metal ion.

Lithium oxydiacetate(1-), on the other hand, contains oxydiacetate ions that are bidentate and nonplanar [373].   The lithium ions
have a trigonal bipyramidal coordination geometry involving all of
the oxygen atoms of the oxydiacetate(1-) ions except the protonated
carboxyl oxygen (75).   It is believed that tridentate chelation does
not occur in this compound because of the small size of the lithium
ion [373].

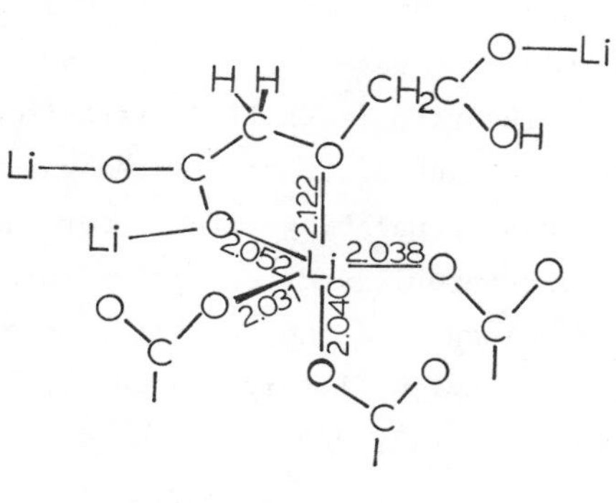

(75)

The crystal structure of rubidium oxydiacetate(1-) shows the
presence of nonchelating, yet planar, oxydiacetate ions [374,375].
The rubidium eightfold coordination has C_2 symmetry (76). The geometry is octahedral with additional coordination sites in the center
of two of the triangular faces.

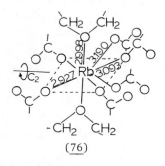

(76)

Only one other structure in which oxydiacetate ligands are not
tridentate to a metal ion has been reported. In orthorhombic cad-
mium(II) oxydiacetate(2-) trihydrate, layers having the composition
$Cd(OOCCH_2OCH_2COO) \cdot 2H_2O$ are held together by noncoordinating water
molecules [376]. The eightfold coordination about each cadmium atom
contains bidentate α-alkoxycarboxylate and carboxylate groups. The
geometry is distorted pentagonal bipyramidal, with coordination distances ranging from 2.269 to 2.637 Å. An eighth oxygen atom is situ-
ated at a somewhat longer distance, 2.841 Å.

This structure is one of three that Boman has reported as part
of a study on cadmium dicarboxylate hydrates [377]. The structure
of the monoclinic form of the trihydrate [actually, bis(triaquooxydi-
acetato(2-)cadmium(II))] [378] contains discrete, centrosymmetric,
binuclear complexes made up of pentagonal bipyramidal coordination
polyhedra with tridentate oxydiacetate ligands occupying three of
the equatorial sites and water molecules occupying axial sites (77).
A similar structure is found for cadmium(II) oxydiacetate(2-) 3-1/2
hydrate [379]. Here crystallographically nonequivalent cadmium
atoms are held together in pairs, which are themselves linked to
give chains. The coordination polyhedra are pentagonal bipyramidal

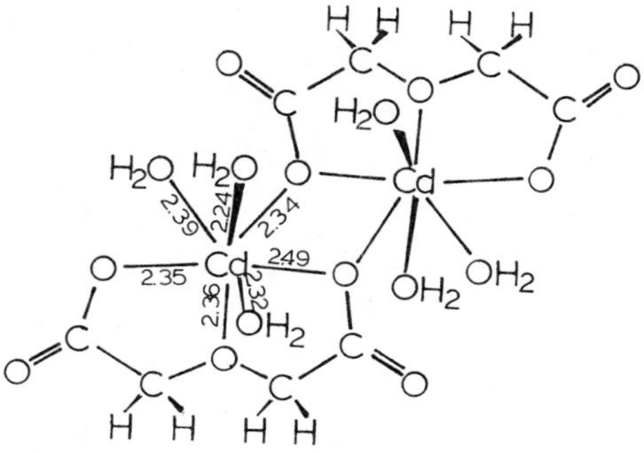

(77)

with equatorial, planar, tridentate oxydiacetate ligands and axial water molecules.

Planar, tridentate oxydiacetate ligands are present in pentaaquooxydiacetato(2-)calcium(II) monohydrate, where discrete eight-coordinate complexes are found in the X-ray structure determination [380]. Raman studies show that, in solution, calcium oxydiacetate contains ligands with the same conformation as found for the solid [380].

The crystal structure of $Na_2[UO_2(oxydiacetate(2-))_2] \cdot 2H_2O$ [381] shows the presence of two crystallographically nonequivalent, centrosymmetric, hexagonal bipyramidal coordinations (78). Unfortunately, the atomic coordinates have not been published, so some bonding parameters are missing in Tables 6 and 7. A pentagonal bipyramidal geometry is observed for uranium(VI) in the other uranyl(VI) oxydiacetate structure that has been reported [38]. In uranyl(VI) oxydiacetate(2-), nearly planar tridentate oxydiacetate ligands bond to three of the equatorial sites in the C_2 symmetry complex (79). The other two equatorial sites are filled by nonchelating carboxyl oxygen atoms from oxydiacetate groups tridentate to other metal ions in a polymeric network.

The polymeric structure of copper(II) oxydiacetate(2-) hemihydrate [383] is unusual in that it contains tridentate oxydiacetate

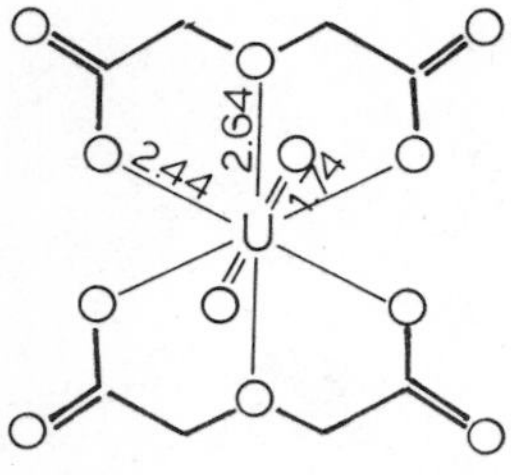

(78) 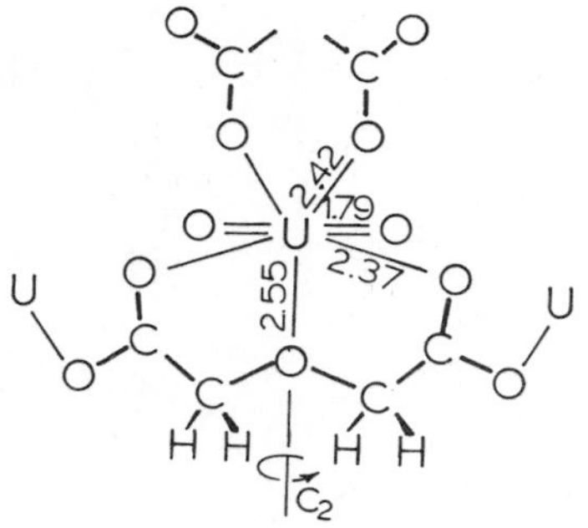   (79)

ligands that are nonplanar--a conformation required by the coordination of the ligand to three facial sites in the axially elongated
octahedral coordination polyhedron (80). As is usually the case with
copper(II) alkoxycarboxylates, the alkoxy oxygen occupies an axial
position in this structure. The water molecule is disordered.

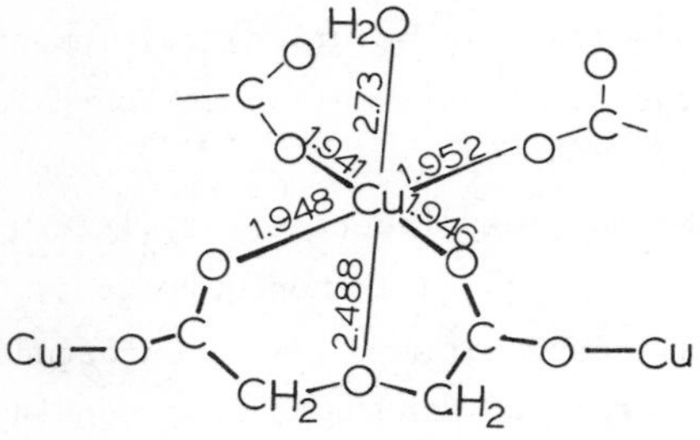

(80)

The structures of $Na_3[M(OOCCH_2OCH_2COO)_3] \cdot 6H_2O$ (M = Nd(III) [384,
385], Gd(III) [384], Yb(III) [384,385], and Ce(III) [386]) show the
presence of nine-coordinate, distorted, face-centered-trigonal prismatic tris chelates with crystallographic D_3 symmetry (81). The two
faces of the discrete trigonal prisms are rotated by 13-20° with respect to each other. A low-precision structure reported for
"$Na_3[Nd(OOCCH_2OCH_2COO)_3] \cdot 6H_2O$" [387] may actually be for the sodium
perchlorate double salt discussed above. The chelate ring conformations of these complexes differ from those observed for most other

(81)

α-hydroxycarboxylates. Although nearly planar, the deviations from planarity of the chelate rings are most commonly toward that of a puckered conformer rather than an asymmetric envelope (Fig. 24, Table 7).

The C_2 symmetry tris chelate in erbium(III) tris(oxydiacetato-(2-))erbate(III) hexahydrate [388] and the C_1 symmetry tris chelate in sodium tris(oxydiacetato(4-))cerate(III) 9-hydrate [389,390] are similar in geometry to the D_3 tris chelates found in the sodium tris(oxydiacetato(2-))lanthanide/sodium perchlorate mixed salts. In addition to the nine-coordinate tris chelate, the erbium(III) structure [388] contains an eight-coordinate metal ion surrounded by four water molecules and four nonchelating carboxyl oxygen atoms from nearest-neighbor tris chelates, producing a lattice structure.

3. α-Carboxyl Cyclic Ethers

α-Carboxyl cyclic ethers (82) are usually produced in the lactonization of polyhydroxypolycarboxylic acids or upon formation of hemiacetals from uronic acids (Fig. 1). They are carbohydrates or closely related to carbohydrates. Since no structures of transition metal compounds of α-hydroxyl cyclic ethers have been reported, those structures that have been determined (alkali metal and alkaline earth compounds) will be only briefly surveyed. α-Carboxyl cyclic ether

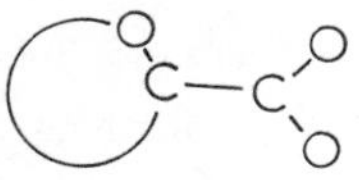

(82)

groups often exhibit large deviations from planarity. Chelated
groups are usually more planar.

Structures have been determined for three glucuronates (as the
hemiacetals): potassium D-glucuronate(1-) dihydrate [391,392],
rubidium D-glucuronate(1-) dihydrate [392], and calcium D-glucuro-
nate(1-) bromide trihydrate [393]. The metal ions are coordinated
to all oxygens of the glucuronate ion except for the ether oxygen
and one hydroxyl oxygen in the alkali metal structures. In the calcium compound, the eight-coordinated tetragonal antiprismatic coordination polyhedron includes a chelation by an α-alkoxycarboxylate
grouping.

Ninefold-coordinate alkaline earth complexes with trigonal symmetry are found in calcium and strontium sodium galacturonate(1-)
hexahydrate [394,395]. Each group II metal ion is coordinated to
three water molecules and three galacturonate ions chelated through
the carboxyl group and the ring oxygen. Chelation to the metal ion is
found for a strontium disaccharide containing a galacturonate residue
[396]. In this structure, two independent molecules are present,
each containing two α-alkoxycarboxylate groups, one of which has an
sp^2 α-carbon. Chelation of europium(III) involving the ring oxygen
and carboxylate group of galacturonate in aqueous solution has been
indicated by NMR studies [397,398].

The calcium salts of (-)-hydroxycitric acid lactone [399] and
(+)-allohydroxycitric acid lactone [400] contain molecules with both
α-hydroxycarboxylate and α-alkoxycarboxylate groups. In the latter,
the ring oxygen is actually a lactone oxygen atom. Only the α-
hydroxycarboxylate group is chelated; the lactone oxygen is not
coordinated. In the potassium salt of isocitrate(1-), on the other
hand, the ring oxygen (also a lactone oxygen) and the α-carboxyl
group are bidentate to potassium [401].

The calcium salt of 4,10-dioxo-5-methoxy-2,8-dicarboxy-4H,10H-
benzodipyran as the 9-hydrate [402] and the sodium salt of 8-allyl-
5-(3-methylbutoxy)-4-oxo-4H-1-benzopyran-2-carboxylic acid as the
monohydrate [403] contain, respectively, two and one α-carboxyl

cyclic ether groups, none of which is chelated. All three groups have an sp^2 carbon alpha to the carboxyl group.

In the calcium salt of the lactol of 5-keto-D-gluconic acid [404], eightfold-coordinate metal ions lie on twofold axes. The ligand is tridentate through the α-alkoxycarboxylate group and hydroxyl group.

VI. COMPLEXES OF α-THIOCARBOXYLATES

This section briefly covers the structural chemistry of complexes that contain ligands in which there are sulfur atoms attached to carbons alpha to carboxyl groups. This includes both α-mercapto and α-thioether groups.

A. Complexation by α-Thiocarboxylates

α-Thiocarboxylic acids differ from α-hydroxycarboxylic acids in three important characteristics: The mercapto group is significantly more acidic than the hydroxyl group, so that in most complexes this group is ionized [48]; the sulfur atom is much larger and therefore gives a larger coordination distance than oxygen; and the α-thiocarboxylate ligands generally bond more readily with "soft" Lewis acids. Bonding parameters for α-thiocarboxylates, whether containing α-mercapto or α-thioether groups, are given in Table 8. Parameters for chelate rings are given in Table 9.

B. α-Thiocarboxylate Structures

1. α-Mercaptocarboxylates

No structures determined by diffraction methods have been reported for uncomplexed α-mercaptocarboxylates.

The simplest α-mercaptocarboxylic acid is mercaptoacetic (thioglycolic) acid, HOOCCH$_2$SH. The structures of the mercaptoacetato(2-)bis-(ethylenediamine)cobalt(III) cation and the corresponding chromium(III) complex have been reported [405]. Both complexes are octahedral, with the α-mercaptocarboxylate chelating to give a five-membered ring (83,

84). In the cobalt complex, there appears to be a significant lengthening (by ~.04 Å) of the Co-N bond opposite the sulfur atom. No significant "trans effect" is observed in the chromium(III) compound.

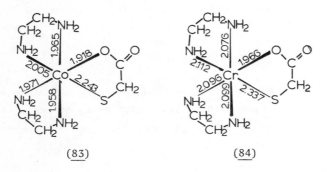

(83)   (84)

Sheets of highly distorted octahedral coordination polyhedra are present in iron(II) mercaptoacetate(2-) monohydrate [406]. The coordination polyhedra (85) contain five-membered chelate rings formed by the mercaptoacetate ligand.

(85)

Hydrogen bis(mercaptoacetato(2-))antimonate(III) is the only additional mercaptoacetate complex for which a structure determination has been carried out [407]. The pseudo-trigonal bipyramidal coordination geometry with a stereochemically active lone pair of electrons in an equatorial site is very similar to the coordination geometry found for the binuclear antimony(III) tartrate complexes [4].

The proton is believed to reside on or be shared by carboxyl oxygen atoms.

A chelated thio derivative of mercaptoacetate is present in the anion complex found in tetrabutylammonium bis(thiomercaptoacetato-(2-))oxotechnetate(V) [408]. The rectangular pyramidal coordination polyhedron (86) has crystallographic mirror-plane symmetry. The ligands and the chelate rings are distinctly nonplanar.

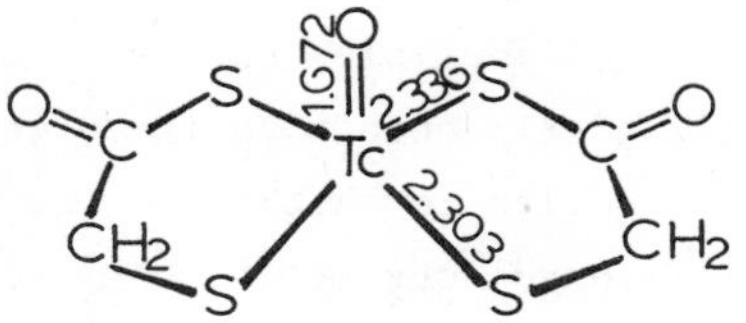

(86)

A centrosymmetric cluster consisting of an octahedron of six copper(II) ions around an icosahedron of 12 sulfur atoms (from the mercaptocarboxylates) surrounding a cube of eight copper(I) ions surrounding a chloride ion is found in a low-precision structure determination for $Tl_5[Cu_6^{II}Cu_8^I(SC(CH_3)_2COO)_{12}Cl]\cdot{\sim}12H_2O$ [409]. The planar copper(II) coordination contains two chelating α-mercaptocarboxylate groups (87). Similar $Cu_8^IS_{12}$ cores have been found in several other copper compounds [409].

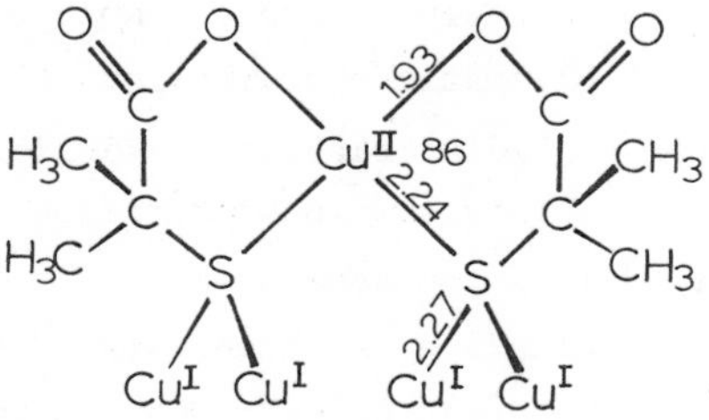

(87)

2. α-Thioether Carboxylates

The structures of two uncomplexed α-thioether carboxylates have been determined: those of thiodiacetic (thiodiglycolic) acid,

HOOCCH$_2$SCH$_2$COOH [410], and racemic thiodilactic acid, HOOCCH(CH$_3$)-
SCH(CH$_3$)COOH [411]. In the latter structure, the carboxyl groups
are rotated so that the α-thiocarboxylate groups in this structure
are distinctly nonplanar. The thiodiacetic acid structure contains
nearly planar molecules with C$_2$ symmetry. The closely related oxydiacetic acid is found to be nonplanar but to have C$_2$ symmetry in the
solid state [347,371]. Three structure determinations on thiodiacetate complexes have been reported.

The triaquothiodiacetato(2-)zinc(II) complex has an octahedral
geometry with the thiodiacetate ligand tridentate to three facial
sites [412] (88). The geometry is similar to that found for the
coordination polyhedra in copper(II) oxydiacetate(2-) hemihydrate

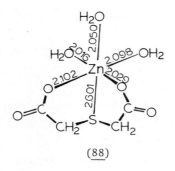

(88)

[383] (80) and in cadmium(II) thiodiacetate(2-) monohydrate [413],
although these two structures do not contain discrete complexes.

A distorted tricapped trigonal prismatic coordination is found
for the metal ion in neodynium(III) thiodiacetate(2-) chloride tetrahydrate [414]. A crystallographic twofold axis passes through the
sulfur atom of the tridentate ligand and the neodynium ion.

A low-precision structure determination has been carried out on
a nickel(II) complex containing the interesting tetradentate ligand
ethylenedithiodiacetate(2-), ⁻OOCCH$_2$SCH$_2$CH$_2$SCH$_2$COO⁻ [415]. The distorted octahedral coordination polyhedron has twofold rotational
symmetry (89).

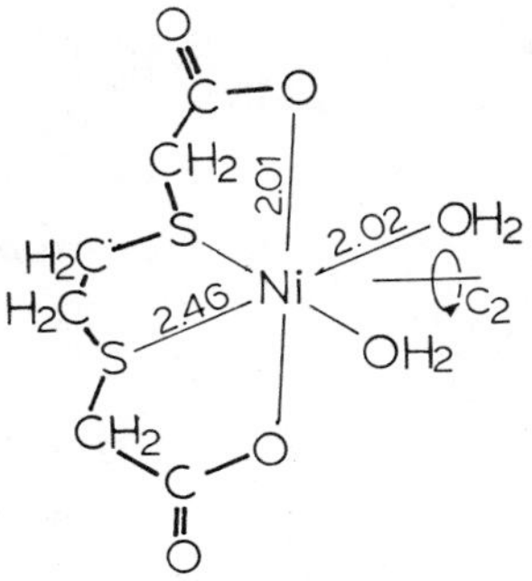

(89)

VII. RECENT WORK

A brief survey of some work published since the initial writing of this chapter is given here.

A ^{1}H NMR study of tungsten malic acid complexes [416] shows the presence of four complexes (one apparently binuclear) in aqueous solution in the pH range 3-6. Additional ^{1}H NMR studies [417] of a number of diamagnetic lactates indicates complexation through the carboxylate group only, through the hydroxyl oxygen only, or through bidentate coordination of the α-hydroxycarboxylate grouping depending on solution pH.

A completed X-ray diffraction study of uranyl(VI) glycolate(1-), about which preliminary X-ray results had been published [184], shows the presence of two types of glycolate ligands [418]. One unique glycolate bridges two uranium atoms through the carboxylate group, whereas the other chelates one uranium atom through the α-hydroxycarboxylate grouping while binding a second uranium atom through the carbonyl oxygen not involved in chelation.

The syntheses and separations of isomeric macrocyclic polyethers containing two carboxylate functions have been reported, and crystal structures of calcium and strontium complexes have been determined [419]. Each complex contains chelated α-alkoxycarboxylate residues as well as additional coordinated ether oxygen atoms.

Two bidentate ligands chelated to an axially elongated
octahedral copper atom are found in the structure of diaquobis(p-
chlorophenoxyacetato)copper(II) [420]. Water molecule oxygen atoms
occupy the other two sites. The phenoxy oxygen atoms are coordinated in the axial positions. This same study shows that diaquobis-
(phenoxyacetato)zinc(II) and diaquobis(p-chlorophenoxyacetato)zinc(II)
are isostructural, with only carboxyl and water molecule oxygen atoms
coordinated to the zinc atoms. All phenoxycarboxylate ligands in
these structures are planar.

The structure of diaquobis(glycolato(1-)manganese(II), isomorphous with the zinc compound [168,169], has been reported [421].

Alkali metal ions chelated by α-alkoxycarboxylate groups are
found in the structures of the acid salts of rubidium and sodium
with α-methoxy-α-phenylacetate [422].

The following structures on nonmetal α-hydroxy- and α-alkoxycarboxylates have been reported: bulgarsenine hydrogen (+)-tartrate
[423], the tartrate salt of thyrotropin-releasing hormone (structural
details not given) [424], (-)-methylbenzylammonium salt of (+)-
erythrofluorocitric acid as the diethyl ester (extended conformer
similar to that found in the rubidium ammonium salt of the nonester-
ified racemate [249]) [425], p-nitrophenoxy acetic acid [426], (±)-2-
(4-chlorophenoxy)propionic acid [427], (±)-2-(2-chlorophenoxy)propionic
acid [428], and a β-D-glucopyranuronic acid derivative [429]. The
last structure has two crystallographically independent molecules.
In one the α-alkoxycarboxylate group is distinctly nonplanar (dihedral angle 54.7°).

ACKNOWLEDGMENTS

I appreciate communications of unpublished or to-be-published data
from V. S. Yadava, A. J. J. Sprenkels, T. Matsui, Robert Job, Jenny P.
Gluster, Sine Larsen, Hayami Yoneda, A. T. H. Lenstra, Birthe Jensen,
Jan A. Kanters, Raymond E. Davis, T. C. van Soest, and K. Matsumoto.
Helpful correspondence from Hans C. Freeman, Carl-Erik Boman, and
L. Johansson is gratefully acknowledged.

Table 3 Geometries of Free α-Hydroxycarboxylate Groups[a]

H-O3
O1
C1—C2
O2(H,R)

Protonated Carboxyl Group

Absolute config.[b]	Dihed. angle	Distances, Å					Angles, deg				Ref.
		C1-C2	C2-O1	C2-O2	C1-O3	O1-O3	C2-C1-O3	C1-C2-O1	O1-C2-O2	C1-C2-O2	
CH₂(OH)COOH, glycolic acid											49
--	5.6	1.503(1)	1.205(1)	1.314(1)	1.403(1)	2.737(1)	112.08(8)	124.73(9)	122.83(10)	112.44(9)	
--	3.0	1.505(1)	1.201(2)	1.310(1)	1.406(2)	2.740(2)	112.33(9)	124.60(10)	123.48(11)	111.91(9)	
(CH₃)₂C(OH)COOH, α-hydroxy-isobutyric acid											50
--	13.3	1.518(3)	1.197(3)	1.307(3)	1.422(3)	2.667(3)	108.3(2)	123.8(2)	123.9(2)	112.2(2)	
C₈H₁₇CH(OH)COOH, (±)-2-hydroxydecanoic acid											51
S	-7.2	1.514(7)	1.224(6)	1.312(5)	1.421(5)	2.681(3)	109.9(3)	122.8(3)	123.4(4)	113.9(4)	
C₆H₅CH₂CH(OH)COOH, S(-)-phenyl-3-lactic acid											52
S	-2.5	1.502(5)	1.206(4)	1.317(5)	1.415(5)	2.686(4)	110.4(3)	123.7(4)	124.2(4)	112.1(3)	
C₆H₅CH(OH)COOH, (±)-mandelic acid											53
R	+23.9	1.487(7)	1.211(5)	1.314(5)	1.427(6)	2.727(6)	110.7(4)	123.8(4)	122.8(4)	113.4(4)	
C₆H₅CH(OH)COOH, L(+)-mandelic acid											54,55
(S)	+3.7	1.55(3)	1.20(1)	1.30(3)	1.43(3)	2.66(3)	109(1)	122(2)	125(2)	113(1)	
(S)	-5.5	1.54(3)	1.20(2)	1.31(3)	1.41(3)	2.65(3)	109(1)	122(2)	125(3)	112(1)	

Table 3 (Continued)

Absolute config.[b]	Dihed. angle	Distances, Å					Angles, deg				Ref.
		C1-C2	C2-O1	C2-O2	C1-O3	O1-O3	C2-C1-O3	C1-C2-O1	O1-C2-O2	C1-C2-O2	
BrC₆H₄CH(OH)COOH, L-o-bromomandelic acid											54,55
(S)	-1.4	1.55	1.15	1.31	1.46	2.61	103	126	130	104	
(S)	-21.6	1.52	1.15	1.31	1.41	2.74	111	126	125	109	
FC₆H₄CH(OH)COOH, L-m-fluoromandelic acid											54,55
R	+3.3	1.50(3)	1.25(1)	1.27(3)	1.41(2)	2.63(3)	111(1)	120(2)	123(2)	117(1)	
FC₆H₄CH(OH)COOH, L-p-fluoromandelic acid											54,55
R	-2.0	1.52(3)	1.21(2)	1.32(3)	1.42(3)	2.65(3)	110(1)	121(2)	126(2)	113(2)	
C₆H₅CH(OH)CH(OH)COOH, (+)(2S,3S)-phenylglyceric acid											56
S	-11.8	1.51(1)	1.21(1)	1.33(1)	1.42(1)	2.68(1)	108.6(6)	124.0(7)	123.4(7)	112.6(6)	
c S	-4.0	1.49(1)	1.31(1)	1.23(1)	1.45(1)	2.65(1)	114.2(6)	115.2(7)	123.1(7)	121.6(7)	
C₆H₅CH(OH)CH(OH)COOH, (-)(2S,3R)-phenylglyceric acid											57
S	-3.8	1.523(2)	1.206(2)	1.309(2)	1.405(2)	2.691(2)	110.7(1)	122.8(1)	126.3(2)	110.9(1)	
HOOCCH(OH)COOH, tartronic acid[d]											58
S	-17.5	1.531(3)	1.222(3)	1.316(3)	1.396(3)	2.753(3)	112.1(2)	122.7(2)	125.4(2)	111.9(2)	
R	+14.0	1.527(3)	1.221(3)	1.308(3)	1.396(3)	2.676(3)	111.4(2)	120.2(2)	125.9(2)	114.0(2)	
HOOCCH(OH)CH(OH)COOH·H₂O, (±)-tartaric acid monohydrate											59
S	+0.2	1.44	1.22	1.28	1.49	2.68	112	123	124	113	
S	+25.2	1.52	1.20	1.33	1.36	2.69	110	120	121	116	

HOOCCH(OH)CH(OH)COOH, (+)-tartaric acid

Config.	[α]										Ref.
R	+4.8	1.526(2)	1.215(2)	1.307(2)	1.404(2)	2.660(1)	108.2(1)	123.7(1)	125.4(1)	110.8(1)	60
R	+4.8	1.526(2)	1.199(2)	1.315(2)	1.413(2)	2.752(1)	111.8(1)	124.6(1)	125.8(1)	109.6(1)	

HOOCCH(OH)CH(OH)COOH·H₂O, ms-tartaric acid, triclinic monohydrate

Config.	[α]										Ref.
R	-5.7	1.517(12)	1.203(12)	1.321(12)	1.409(12)	2.640(12)	107.4(8)	123.9(8)	124.6(8)	111.4(8)	61
S	-4.2	1.517(12)	1.206(12)	1.323(12)	1.418(12)	2.633(12)	106.9(8)	124.2(8)	124.5(8)	111.3(8)	

HOOCCH(OH)CH(OH)COOH·H₂O, ms-tartaric acid, monoclinic monohydrate

Config.	[α]										Ref.
C	-9.9	1.521(12)	1.302(12)	1.213(12)	1.415(12)	2.575(12)	111.0(8)	114.2(8)	120.9(8)	120.9(8)	61
R	+7.0	1.511(12)	1.223(12)	1.304(12)	1.414(12)	2.652(12)	107.9(8)	123.8(8)	124.7(8)	111.4(0)	

HOOCCH(OH)CH(OH)COOH, ms-tartaric acid, anhydrous

Config.	[α]										Ref.
R	-8.8	1.520(20)	1.222(20)	1.307(20)	1.413(20)	2.598(20)	106.9(14)	121.8(14)	125.3(14)	113.0(14)	61
S	+11.8	1.520(20)	1.212(20)	1.341(20)	1.414(20)	2.676(20)	108.2(14)	124.3(14)	122.5(14)	113.0(14)	

NH₄[HOOCCH(OH)CH(OH)COO], ammonium hydrogen (±)-tartrate

Config.	[α]										Ref.
R	+10.7	1.56	1.24	1.30	1.48	2.79	108	126	125	108	62

NH₄[HOOCCH(OH)CH(OH)COO], ammonium hydrogen (+)-tartrate

Config.	[α]										Ref.
R	+8.7	1.524(4)	1.220(4)	1.311(4)	1.416(4)	2.708	110.5	123.0	124.5	112.5	63

C₉H₁₄NO₃·HC₄H₄O₆, (-)-adrenoline hydrogen (+)-tartrate

Config.	[α]										Ref.
R	+0.5	1.517(4)	1.213(4)	1.301(4)	1.416(3)	2.682(3)	109.7(2)	123.4(3)	124.9(3)	111.6(2)	64

C₁₃H₁₈NO·HC₄H₄O₆·H₂O, (+)-[(-)-1-methyl-3-benzoylpiperdine hydrogen (+)-tartrate] monohydrate

Config.	[α]										Ref.
R	+1.2	1.521(5)	1.199(5)	1.317(5)	1.424(5)	2.646(3)	108.0(3)	123.4(3)	124.9(3)	111.7(3)	65

C₁₅H₂₂NO·HC₄H₄O₆, (-)-[(-)-1-methyl-2-ethyl-3-benzoylpiperdine hydrogen (+)-tartrate] 66

Table 3 (Continued)

Absolute config.b	Dihed. angle	Distances, Å					Angles, deg				Ref.
		C1-C2	C2-O1	C2-O2	C1-O3	O1-O3	C2-C1-O3	C1-C2-O1	O1-C2-O2	C1-C2-O2	
C25H33N2O2·HC4H4O6, dextromoramide hydrogen (+)-tartrate											
R	-4.0	1.530(3)	1.217(3)	1.297(3)	1.412(3)	2.716(3)	111.4(2)	121.7(2)	124.5(2)	113.7(2)	67
c R	-12.3	1.542(7)	1.298(6)	1.211(6)	1.409(5)	2.661(5)	113.4(4)	114.7(5)	125.1(6)	120.1(5)	
C7H16NO2·HC4H4O6, acetylcholine hydrogen (-)-tartrate											
R	+3.9	1.536(4)	1.218(3)	1.279(3)	1.402(3)	2.631(4)	110.5(2)	119.0(2)	126.6(3)	114.3(2)	68
C7H16NO2·HC4H4O6, acetylcholine hydrogen (±)-tartrate											
R	-1.0	1.525(4)	1.210(3)	1.284(3)	1.403(3)	2.638(4)	110.1(2)	120.4(2)	126.2(3)	113.4(2)	68
C2H9N2·HC4H4O6·2H2O, ethylenediamine hydrogen (+)-tartrate dihydrate											
R	+2.8	1.509(4)	1.209(4)	1.322(4)	1.421(4)	2.700(4)	109.7(3)	124.8(3)	123.3(3)	111.9(3)	69
[Co(C2O4)(C2H8N2)2]·HC4H4O6·H2O, (+)-oxalatobis(ethylenediamine)cobalt(III) hydrogen (+)-tartrate monohydrate											
R	+14.4	1.52(2)	1.21(2)	1.29(2)	1.43(2)	2.76(2)	113(1)	122(1)	126(2)	112(1)	70
[Co(C2O4)(C2H8N2)2]·HC4H4O6·2H2O, (-)-oxalatobis(ethylenediamine)cobalt(III) hydrogen (+)-tartrate dihydrate											
R	+3.0	1.509(8)	1.211(7)	1.294(8)	1.419(7)	2.703(6)	110.7(5)	123.8(5)	124.9(6)	111.3(5)	71
[Co(C2H5NO2)2(C2H8N2)]·HC4H4O6·3H2O, (+)-trans-0-ethylenediaminebis(glycinato(1-))cobalt(III) hydrogen (+)-tartrate trihydrate											
R	+4.8	1.508(8)	1.226(10)	1.328(8)	1.429(7)	2.721(7)	110.8(5)	124.0(6)	124.2(6)	111.8(5)	72
[Co(C2H5NO2)2(C2H8N2)]·HC4H4O6·H2O, (-)-trans-0-ethylenediaminebis(glycinato(1-))cobalt(III) hydrogen (+)-tartrate monohydrate											
R	+0.5	1.499(13)	1.211(16)	1.330(12)	1.428(11)	2.730(13)	110.5(7)	125.7(10)	121.4(11)	112.9(9)	72

Compound	Config										

HOOCCH₂C(OH)(COOH)CH₂COOH, citric acid (anhydrous)

	--	12.3	1.535(3)	1.202(3)	1.316(3)	1.407(3)	2.700(3)	109.8(2)	123.0(2)	124.4(2)	112.5(2)	73

HOOCCH₂C(OH)(COOH)CH₂COOH·H₂O, citric acid monohydrate

| | -- | 3.3 | 1.533(4) | 1.200(4) | 1.330(3) | 1.431(3) | 2.650(3) | 107.8(2) | 123.1(3) | 125.5(3) | 111.4(2) | 74 |

C₂H₁₀N₂(OOCCH(OH)C(OH)(COOH)CH₂COO), ethylenediamine(2+) hydrogen (2S,3S)-2-hydroxycitrate

| | S | -13.8 | 1.533(4) | 1.234(3) | 1.290(4) | 1.416(3) | 2.687(4) | 110.5(2) | 120.7(2) | 125.4(3) | 113.9(2) | 75 |

HOOCCH(OH)C₄H₅O₄·H₂O, D-glucaro-1,4-lactone monohydrate

| | S | +5.5 | 1.511(6) | 1.212(4) | 1.312(5) | 1.426(4) | 2.678(5) | 109.1(3) | 123.9(3) | 124.4(4) | 111.7(3) | 76 |

HOOCCH(OH)CHFCOOH, (±)-erythro-fluoromalic acid

| | S | +5.9 | 1.518(4) | 1.196(4) | 1.303(4) | 1.409(4) | 2.660(4) | 110.1(2) | 122.2(3) | 126.3(3) | 111.5(3) | 77 |

HOOCCH(OH)CH₂CH(CH₃)(NH₃)Cl, (2RS,4RS)-2-hydroxy-4-aminovaleric acid hydrochloride

| | S | -18.2 | 1.518(3) | 1.206(6) | 1.321(6) | 1.421(6) | 2.728(3) | 109.9(3) | 124.4(4) | 123.9(3) | 111.7(4) | 78 |

HOOCCH(OH)CH₂CH₂NH₃Cl·H₂O, (±)-2-hydroxy-4-aminobutyric acid hydrochloride monohydrate

| | R | -3.1 | 1.521(2) | 1.208(2) | 1.317(2) | 1.409(2) | 2.654(1) | 107.8(1) | 124.2(1) | 124.3(1) | 111.5(1) | 79 |

Average[e]

| | | | 1.519(12) | 1.209(11) | 1.310(11) | 1.418(13) | 2.685(34) | 109.7(14) | 123.1(13) | 124.6(11) | 112.1(13) | |

Cyclic Structures

```
      ┌──────O──────┐
CH₂CH(OH)CH(OH)CH(OH)C(OH)COOH·H₂O, 2-keto-L-gulonic acid monohydrate
```

| | (R) | -7.8 | 1.539(3) | 1.197(2) | 1.308(3) | 1.381(2) | 2.712(3) | 111.6(2) | 122.7(2) | 125.0(3) | 112.3(2) | 80 |

357

Table 3 (Continued)

Absolute config.b	Dihed. angle	Distances, Å					Angles, deg				Ref.
		C1-C2	C2-O1	C2-O2	C1-O3	O1-O3	C2-C1-O3	C1-C2-O1	O1-C2-O2	C1-C2-O2	

HOCH2CH(OH)CH(OH)CHCH(NHCOCH3)CH(OH)CH2C(OH)COOH·2H2O, β-D-N-acetylneuramic acid dihydrate

| | | | | | | | | | | | 81 |
| S | +12.4 | 1.531(6) | 1.198(7) | 1.290(7) | 1.400(6) | 2.621(5) | 106.9(3) | 122.9(5) | 122.7(5) | 114.4(4) | |

C(NO2)CH2CHC(NO2)CH2CHCH2C(OH)COOH, 2,3:4,5-dimethano-2,4-dinitro-1-hydroxycyclohexane-1-carboxylic acid

| | | | | | | | | | | | 82 |
| c S | +32.3 | 1.540(4) | 1.308(4) | 1.218(4) | 1.416(4) | 2.670(4) | 110.9(3) | 114.2(3) | 123.0(3) | 122.8(3) | |

α,β-Unsaturated

HOOC(OH)C=C(OH)COOH·2H2O, dihydroxyfumaric acid dihydrate

| | | | | | | | | | | | 83 |
| -- | 4.0 | 1.49(2) | 1.28(1) | 1.24(1) | 1.39(2) | 2.58(1) | 113.6(7) | 114.3(3) | 126.6(8) | 119.0(8) | |

Ionized Carboxyl Group

C6H14N·C6H5CH(OH)COO, trans-2R,5R-dimethylpyrrolidinium (S)-mandelate

| | | | | | | | | | | | 84 |
| S | -23.1 | 1.522(10) | 1.260(8) | 1.249(9) | 1.415(9) | 2.604(8) | 109.9(6) | 115.7(6) | 124.7(6) | 119.6(6) | |

C8H11N·C6H5CH(OH)COO, 1-phenylethylammonium mandelate

| | | | | | | | | | | | 85 |
| S | -11.5 | 1.532(6) | 1.231(6) | 1.274(6) | 1.412(6) | 2.664(4) | 110.5(4) | 120.0(4) | 125.0(4) | 115.0(7) | |

(NH4)2[OOCCH(OH)CH(OH)COO], ammonium tartrate

											86
S	-6.3	1.540(13)	1.227(13)	1.274(13)	1.411(13)	2.638(13)	110.8(9)	118.3(9)	124.8(9)	116.8(9)	
S	+10.8	1.544(13)	1.243(13)	1.262(13)	1.414(13)	2.671(13)	111.2(9)	118.6(9)	125.1(9)	115.9(9)	

C9H22N2O2·C4H4O6, γ-aminobutyric acid choline ester (±)-tartrate

											87
R	-9.3	1.59(3)	1.25(2)	1.26(2)	1.44(2)	2.59(2)	108(1)	116(2)	128(2)	117(2)	
R	+1.8	1.55(3)	1.28(2)	1.21(2)	1.44(2)	2.60(3)	112(2)	113(2)	129(2)	117(2)	

C2H10N2·C4H4O6, ethylenediamine (+)-tartrate

											88
(S)	-9.9	1.539(3)	1.246(4)	1.255(4)	1.406(4)	2.721(4)	113.4(2)	119.4(2)	123.8(2)	116.7(2)	
(S)	-19.5	1.539(3)	1.254(4)	1.244(4)	1.407(4)	2.644(4)	111.2(2)	116.9(2)	123.9(3)	119.1(3)	

[Co(C2H8N2)3]·C4H4O6·Cl·5H2O, tris(ethylenediamine)cobalt(III) chloride (+)-tartrate pentahydrate

											89
R	+19.5	1.536(2)	1.249(2)	1.251(2)	1.421(1)	2.709(1)	112.4(1)	118.4(1)	126.1(1)	115.5(1)	
R	+15.6	1.531(2)	1.246(2)	1.254(2)	1.419(1)	2.712(1)	112.1(1)	119.6(1)	124.3(1)	115.9(1)	

[Co(C5H15N3)2]Cl·C4H4O6·5.4H2O, (+)-bis[1,1,1-tris(aminomethyl)ethane]cobalt(III) chloride (+)-tartrate hydrate

											90
R	+1.3	1.531(4)	1.244(6)	1.250(6)	1.411(5)	2.628(4)	111.7(4)	117.2(4)	126.7(3)	116.2(4)	
R	-9.0	1.531(4)	1.249(4)	1.254(4)	1.423(4)	2.630(4)	110.8(3)	117.0(0)	125.6(4)	116.8(3)	

[Cu(C7H18N2)2(HOCH(CH3)COO)]CH3CH(OH)COO·H2O, L-lactato(1-)bis(N-isopropyl-2-methylpropane-1,2-diamine)copper(II) L-lactate 1-hydrate

											91
S	-9.1	1.534(21)	1.258(17)	1.250(18)	1.433(19)	2.677(13)	111.3(12)	118.8(12)	125.5(13)	115.4(12)	

[Ni(C10H8N2)3]2Cl2·C4H4O6·nH2O, (+)-tris(2,2'-bipyridyl)nickel(II) chloride (+)-tartrate hydrate

											92
R	+12.7	1.55(2)	1.22(2)	1.23(2)	1.44(2)	2.68(1)	110(1)	121(1)	126(2)	114(1)	
R	+10.6	1.59(2)	1.26(2)	1.24(2)	1.42(2)	2.62(1)	107(1)	118(2)	127(2)	115(2)	

[Pt(C2H8N2)2]2·C4H4O6, bis(ethylenediamine)platinum(II) (+)-tartrate

											93
R	+14.6	1.549(19)	1.237(14)	1.245(13)	1.422(12)	2.632(9)	110.7(9)	116.5(10)	126.3(13)	117.2(10)	
R	+4.2	1.540(15)	1.239(12)	1.258(13)	1.413(17)	2.657(11)	112.1(11)	117.6(10)	126.3(9)	116.0(9)	

[Cu(C3H10N2)2(C4H4O6)]·2H2O, (+)-tartrato(2-)bis(N-methylethylenediamine)copper(II) dihydrate

	94

Table 3 (Continued)

Absolute config.[b]	Dihed. angle	Distances, Å					Angles, deg				Ref.
		C1-C2	C2-O1	C2-O2	C1-O3	O1-O3	C2-C1-O3	C1-C2-O1	O1-C2-O2	C1-C2-O2	
R	-2.2	1.531(13)	1.244(13)	1.252(12)	1.411(10)	2.714(11)	114.6(7)	118.3(9)	125.3(10)	116.2(8)	95

[Cu($C_7H_{18}N_2$)$_2$($C_4H_4O_6$)]·C_3H_7OH,(2S,3S)-tartrato(2-)bis(N-isopropyl-2-methylpropane-1,2-diamine)copper(II) n-propanol

Absolute config.[b]	Dihed. angle	C1-C2	C2-O1	C2-O2	C1-O3	O1-O3	C2-C1-O3	C1-C2-O1	O1-C2-O2	C1-C2-O2	Ref.
S	-0.1	1.54(3)	1.24(4)	1.25(2)	1.44(3)	2.63(1)	111.4(8)	116.6(9)	126.2(9)	117.1(9)	62

NH_4[HOOCCH(OH)CH(OH)COO], ammonium hydrogen (±)-tartrate

R	-2.5	1.55	1.21	1.26	1.50	2.73	113	118	126	115	63

NH_4[HOOCCH(OH)CH(OH)COO], ammonium hydrogen (+)-tartrate

R	-1.9	1.531(4)	1.263(4)	1.261(4)	1.415(4)	2.688	113.0	118.2	124.8	117.0	64

$C_9H_{14}NO_3$·$HC_4H_4O_6$, (-)-adrenoline hydrogen (+)-tartrate

R	-0.2	1.520(4)	1.272(3)	1.234(3)	1.421(3)	2.640(3)	110.6(2)	118.7(2)	124.6(3)	116.6(2)	65

$C_{13}H_{18}NO$·$HC_4H_4O_6$·H_2O, (+)-[(-)-1-methyl-3-benzoylpiperidine hydrogen (+)-tartrate] monohydrate

R	-5.6	1.551(5)	1.245(5)	1.238(5)	1.404(5)	2.704(3)	114.1(3)	117.3(3)	125.8(3)	116.9(3)	66

$C_{15}H_{22}NO$·$HC_4H_4O_6$, (-)-[(-)-]-methyl-2-ethyl-3-benzoylpiperdine hydrogen (+)-tartrate]

R	-3.0	1.527(3)	1.242(3)	1.271(3)	1.414(3)	2.693(2)	113.0(2)	119.1(2)	125.0(2)	115.9(2)	68

$C_7H_{16}NO_2$·$HC_4H_4O_6$, acetylcholine hydrogen (+)-tartrate

R	+1.9	1.535(3)	1.223(3)	1.275(3)	1.402(4)	2.613(4)	110.1(2)	118.5(2)	126.2(3)	115.2(2)	68

$C_7H_{16}NO_2$·$HC_4H_4O_6$, acetylcholine hydrogen (±)-tartrate

R	+3.1	1.533(3)	1.239(3)	1.269(3)	1.405(3)	2.584(4)	109.9(2)	117.0(2)	126.4(3)	116.6(2)	67

$C_{25}H_{33}N_2O_2$·$HC_4H_4O_6$, dextromoramide hydrogen (+)-tartrate

											Ref
R	+5.5	1.536(7)	1.246(6)	1.259(6)	1.409(6)	_2.578(6)_	109.8(4)	116.4(5)	127.8(5)	115.8(5)	69

$C_2H_9N_2 \cdot HC_4H_4O_6 \cdot 2H_2O$, ethylenediamine hydrogen (+)-tartrate dihydrate

| R | -2.7 | 1.526(4) | 1.249(4) | 1.242(4) | 1.417(4) | 2.682(4) | 111.4(3) | 117.9(3) | 124.5(3) | 117.6(3) | 70 |

$[Co(C_2O_4)(C_2H_8N_2)_2] \cdot HC_4H_4O_6 \cdot H_2O$, (+)-oxalatobis(ethylenediamine)cobalt(III) hydrogen (+)-tartrate monohydrate

| R | -2.1 | 1.52(2) | 1.25(2) | 1.24(2) | 1.42(2) | _2.61(2)_ | 114(1) | 115(1) | 127(1) | 118(1) | 71 |

$[Co(C_2O_4)(C_2H_8N_2)_2] \cdot HC_4H_4O_6 \cdot 2H_2O$, (-)-oxalatobis(ethylenediamine)cobalt(III) hydrogen (+)-tartrate dihydrate

| R | +4.9 | 1.548(8) | 1.234(8) | 1.253(7) | 1.422(9) | 2.606(6) | 110.1(5) | 116.8(5) | 126.8(6) | 116.4(5) | 72 |

$[Co(C_2H_5NO_2)_2(C_2H_8N_2)] \cdot HC_4H_4O_6 \cdot 3H_2O$, (+)-trans-0-ethylenediaminebis(glycinato(1-))cobalt(III) hydrogen (+)-tartrate trihydrate

| R | +3.5 | 1.532(8) | 1.272(7) | 1.217(7) | 1.425(7) | _2.678(6)_ | 113.2(4) | 117.1(5) | 126.2(6) | 116.7(5) | 72 |

$[Co(C_2H_5NO_2)_2(C_2H_8N_2)] \cdot HC_4H_4O_6 \cdot H_2O$, (-)-trans-0-ethylenediaminebis(glycinato(+)cobalt(III) hydrogen (+)-tartrate monohydrate

| R̄ | -3.0 | 1.530(12) | 1.262(11) | 1.269(11) | 1.428(11) | _2.697(7)_ | 111.2(7) | 120.4(8) | 124.0(8) | 115.4(7) | 96 |

$C_2H_{10}N_2 \cdot HOOCCH_2C(OH)(COO)CH_2COO$, ethylenediamine (2+) hydrogen citrate

| -- | 9.6 | 1.543(2) | 1.261(2) | 1.248(2) | 1.429(2) | _2.591(1)_ | 107.7(1) | 118.1(1) | 124.6(1) | 117.3(1) | 75 |

$C_2H_{10}N_2[OOCCH(OH)C(OH)(COO)CH_2COOH] \cdot H_2O$, ethylenediamine (2S,3R)-2-hydroxycitrate monohydrate

| S | _-13.5_ | 1.522(5) | 1.232(3) | 1.280(4) | 1.412(3) | _2.665(3)_ | 109.2(2) | 121.8(3) | 124.8(3) | 113.4(2) | |
| R | _+2.4_ | 1.542(3) | 1.236(3) | 1.276(2) | 1.422(5) | _2.554(2)_ | 106.1(2) | 119.2(2) | 123.2(2) | 117.7(2) | 37 |

$NH_4[OOCCH(OH)CH_2COOH]$, ammonium hydrogen (S)-malate

| S | _-7.3_ | 1.537(3) | 1.236(2) | 1.258(2) | 1.413(2) | _2.669(2)_ | 111.8(2) | 118.6(2) | 125.3(2) | 116.1(2) | 75 |

$C_2H_{10}N_2[OOCCH(OH)C(OH)(COOH)CH_2COO]$, ethylenediamine(2+) hydrogen (2S,3S)-2-hydroxycitrate

| S | _-15.4_ | 1.555(3) | 1.255(3) | 1.257(3) | 1.403(3) | _2.706(3)_ | 112.0(2) | 118.5(2) | 124.7(2) | 116.8(2) | 97 |

$K[OOCCH(OH)CH(OH)CH(OH)CH(OH)CH_2OH] \cdot H_2O$, potassium D-gluconate monohydrate, "A" form

| R | _+6.2_ | 1.534(3) | 1.246(4) | 1.259(3) | 1.420(3) | _2.579(4)_ | 109.5(2) | 116.7(2) | 125.4(2) | 117.9(2) | |

Table 3 (Continued)

Absolute config.[b]	Dihed. angle	Distances, Å					Angles, deg				Ref.
		C1-C2	C2-O1	C2-O2	C1-O3	O1-O3	C2-C1-O3	C1-C2-O1	O1-C2-O2	C1-C2-O2	
Average[e]		1.539(10)	1.246(11)	1.254(12)	1.420(10)	2.651(41)	111.1(14)	118.0(13)	125.6(10)	116.4(10)	98

Esterified Carboxyl Group

$CH_3OOCCH(OH)CH(OH)COOCH_3$, ms-tartaric acid dimethyl ester

Absolute config.[b]	Dihed. angle	C1-C2	C2-O1	C2-O2	C1-O3	O1-O3	C2-C1-O3	C1-C2-O1	O1-C2-O2	C1-C2-O2	Ref.
S	-4.5	1.511(8)	1.190(8)	1.341(8)	1.411(7)	2.717(8)	109.8(5)	126.2(6)	124.2(6)	109.5(5)	
R	-0.2	1.530(8)	1.198(7)	1.329(7)	1.418(7)	2.713(5)	109.9(4)	124.4(5)	125.7(5)	109.9(4)	99

$CH_3OOCCH(OH)CH_2CHCH(OH)CH(OH)CO$, 2,3,4,6-tetrahydroxy-1,7-heptanedioic acid γ-monolactone methyl ether

| S | +23.8 | 1.534(9) | 1.218(8) | 1.308(8) | 1.403(9) | 2.726(7) | 111.2(7) | 121.0(8) | 124.8(8) | 114.1(7) | 100 |

$CH_3OOCC(OH)CH_2CH(OH)CH(HNC(O)CH_3)CH(C_3H_7O_3)\cdot H_2O$, N-acetylneuramic acid methyl ester monohydrate

| S | -25.6 | 1.541(3) | 1.192(3) | 1.317(3) | 1.403(3) | 2.712(3) | 109.3(2) | 122.5(2) | 124.2(2) | 113.3(2) | 101 |

$(C_6H_5)_2C(OH)COOC_7H_{12}N\cdot HBr$, quinuclidinyl benzilate hydrobromide

| | 7.21 | 1.524(14) | 1.186(12) | 1.359(11) | 1.409(11) | 2.726(10) | 109.9(7) | 125.8(9) | 123.5(10) | 110.7(8) | 102 |

$(C_4H_3S)_2C(OH)COOC_7H_{12}N$, quinuclidinyl di-α,α'-thienylglycolate

| | 15.2 | 1.556(6) | 1.190(5) | 1.329(5) | 1.395(5) | 2.719(5) | 109.5(3) | 123.7(4) | 125.4(4) | 111.2(3) | 103 |

$(C_6H_5)(C_6H_{11})C(OH)COOC_4H_7N(CH_3)_2\cdot Br$, 3(2-cyclohexyl-2-hydroxy-2-phenylacetoxy)-1,1-dimethylpyrrolidinium bromide

Config[b]	[α]										Ref
S	−21.8	1.52(1)	1.20(1)	1.36(1)	1.41(1)	2.71(1)	107.9(6)	125.2(8)	122.0(8)	112.9(6)	104
S	−23.7	1.556(15)	1.197(13)	1.352(13)	1.416(13)	2.72(1)	107.1(8)	124.7(8)	122.9(8)	112.3(8)	105
c S	+42.2	1.540(10)	1.198(10)	1.324(10)	1.423(9)	2.670(8)	109.9(6)	111.2(6)	123.1(6)	125.6(6)	106
S	+20.8	1.54(1)	1.20(1)	1.36(1)	1.41(1)	2.703(7)	108.9(5)	124.0(6)	123.7(6)	112.2(6)	107
c --	46.5	1.532(8)	1.209(8)	1.340(10)	1.427(7)	2.656(5)	108.5(5)	110.2(5)	122.5(5)	127.3(6)	
Average[e,f]		1.534(13)	1.197(7)	1.342(16)	1.409(5)	2.717(6)	109.3(10)	124.0(10)	124.4(11)	111.6(13)	

Compounds:

104 — (C5H9)(C6H5)C(OH)COOC4H7N(CH3)2·Br, glycopyrronium bromide

105 — (C5H9)(C4H7S)C(OH)COOCH2CH2N(C2H5)2(CH3)·Br, diethyl-(2-hydroxyethyl)-methylammonium bromide α-cyclopentyl-2-thienylglycolate

106 — (C6H5)(C4H7S)C(OH)COOC4H7N(CH3)2·Br, 3-hydroxy-1,1-dimethylpyrrolidinium bromide α-phenyl-2-thiopheneglycolate

107 — (C6H5)2C(OH)COOCH2CH2NH(C2H5)2·Cl, 2-diethylaminoethylbenzilate hydrochloride

[a] Underlined distances, angles, and esd's were calculated from reported coordinates.

[b] Absolute configuration of carbinol carbon. Parentheses denote that the absolute configuration as calculated from the reported coordinates is opposite to that which the compound should have.

[c] The protonated or esterified carboxyl oxygen lies on the same side as the α-hydroxyl oxygen.

[d] The "absolute configuration" was established by setting the priority of the carboxyl carbon not in the α-hydroxycarboxylate under consideration equal to that of a methyl group.

[e] Unweighted average with average absolute deviation for all structures except those where a protonated or esterified carboxyl oxygen lies on the same side as the α-hydroxyl oxygen.

[f] The cyclic structure N-acetylneuraminic acid methyl ester monohydrate [100] was omitted from the average.

Table 4 Geometries of α-Hydroxycarboxylate Groups in Complexes and Metal Salts[a]

$$\underset{O2}{\overset{O1}{(H)O3-C1-C2}}$$

Absolute config.[b]	Dihed. angle	Distances, Å					Angles, deg				Ref.
		C1-C2	C2-O1	C2-O2	C1-O3	O1-O3	C2-C1-O3	C1-C2-O1	O1-C2-O2	C1-C2-O2	
Protonated Hydroxyl Group											
Li(HOCH2COO), lithium glycolate(1-)											
c	7.8	1.525(2)	1.255(2)	1.254(2)	1.415(2)	2.677(2)	112.8(10)	118.1(10)	125.0(10)	116.9(10)	163
Li(HOCH2COO)·H2O, lithium glycolate(1-) monohydrate[d]											
c	0	1.54(1)	1.22(1)	1.26(1)	1.39(1)	2.65(1)	107.9(6)	120.4(6)	124.3(6)	115.3(6)	164
K(HOCH2COOH)(HOCH2COO), potassium hydrogen bisglycolate[e]											
c,f	6.2	1.528(12)	1.214(12)	1.284(12)	1.398(12)	2.773(9)	113.3(8)	123.0(8)	129.4(8)	111.6(8)	165
	3.2	1.511(13)	1.245(11)	1.242(11)	1.418(13)	2.733(8)	112.3(7)	121.4(8)	121.5(8)	117.0(7)	
Rb(HOCH2COOH)(HOCH2HCOO), rubidium hydrogen bisglycolate											
c,f	7.9	1.471(16)	1.212(14)	1.308(14)	1.435(14)	2.763(12)	113.8(10)	124.8(10)	123.0(9)	112.3(10)	166
	3.6	1.519(16)	1.241(14)	1.264(14)	1.426(14)	2.687(12)	111.8(10)	120.3(10)	123.5(9)	116.1(10)	
Cu(HOCH2COO)2, bis(glycolato(1-))copper(II)											
c	2.5	1.53(2)	1.27(2)	1.24(2)	1.43(2)	2.57(2)	106.7(11)	118.3(12)	124.6(13)	117.0(12)	167
[Zn(HOCH2COO)2(H2O)2], diaquobis(glycolato(1-))zinc(II)											
c	0	1.521(10)	1.271(9)	1.248(8)	1.415(9)	2.604(7)	109.6(6)	118.1(6)	124.9(6)	117.0(6)	168,169
c	+6.2[g]	1.546(10)	1.266(8)	1.221(8)	1.431(9)	2.646(7)	109.9(6)	118.3(6)	125.0(6)	116.7(6)	

Er(HOCH₂COO)₃·2H₂O, erbium(III) triglycolate(1-) dihydrate

c	--	4.7	1.53(2)	1.24(2)	1.32(2)	1.36(2)	2.57(1)	108.5(11)	118.8(14)	123.0(13)	118.2(13)	173
c	--	0.2	1.54(2)	1.27(2)	1.24(2)	1.50(2)	2.54(1)	105.0(11)	117.6(11)	127.1(13)	115.3(13)	
c	--	7.4	1.48(2)	1.29(2)	1.25(2)	1.47(2)	2.54(1)	108.8(12)	116.4(11)	123.8(11)	119.4(13)	

Sc(HOCH₂COO)₃·2H₂O, scandium(III) triglycolate(1-) dihydrate

c	--	3.7	1.50(5)	1.20(7)	1.27(8)	1.42(9)	2.50(4)	101(4)	124(6)	122(4)	112(5)	174
c	--	2.8	1.42(7)	1.26(9)	1.29(6)	1.47(7)	2.43(4)	102(5)	121(4)	121(4)	116(6)	
c	--	7.1	1.42(6)	1.32(9)	1.24(6)	1.45(7)	2.49(4)	110(6)	115(5)	119(4)	125.6(6)	

[Er(HOCH₂COO)(OCH₂COO)(H₂O)]·H₂O, aquoglycolato(1-)glycolato(2-)erbium(III) monohydrate

| c | -- | 10.1 | 1.491(24) | 1.195(18) | 1.237(15) | 1.444(21) | 2.532(17) | 111.9(15) | 113.7(13) | 128.5(14) | 117.8(15) | 172 |

Gd(HOCH₂COO)₃, gadolinium(III) triglycolate(1-)

c	--	+16.7ʰ	1.473(20)	1.293(20)	1.264(20)	1.440(15)	2.657(15)	113.2(11)	117.0(13)	118.9(15)	122.6(14)	175,177
c	--	+29.0ʰ	1.539(12)	1.240(9)	1.269(5)	1.441(11)	2.618(7)	105.1(7)	119.5(7)	126.2(7)	110.5(8)	
c	--	+30.8ʰ	1.630(23)	1.249(21)	1.230(23)	1.464(15)	2.584(12)	102.7(10)	114.9(15)	133.0(18)	111.2(14)	

La(HOCH₂COO)₃, lanthanum(III) triglycolate(1-)

c	--	-17.7ʰ	1.441(13)	1.322(12)	1.315(13)	1.437(10)	2.696(10)	112.5(7)	120.6(8)	115.3(9)	122.3(9)	177
c	--	-13.6ʰ	1.489(8)	1.263(5)	1.259(6)	1.372(8)	2.632(6)	112.0(5)	118.8(5)	124.5(5)	115.0(5)	
c	--	-5.1ʰ	1.556(15)	1.252(12)	1.222(12)	1.443(12)	2.591(9)	110.3(8)	114.7(9)	129.0(10)	112.6(9)	

Eu(HOCH₂COO)₃, europium(III) triglycolate(1-)

c	--	0.1	1.58(5)	1.36(3)	1.27(3)	1.41(4)	2.74(3)	108(3)	122(2)	122(2)	116(2)	176
c	--	7.4	1.48(4)	1.32(3)	1.31(3)	1.46(4)	2.72(3)	111(2)	123(2)	124(2)	113(2)	
c	--	3.8	1.51(4)	1.31(3)	1.29(3)	1.44(4)	2.69(3)	112(2)	120(2)	120(2)	120(2)	

Table 4 (Continued)

Absolute config.b	Dihed. angle	Distances, Å					Angles, deg				Ref.
		C1-C2	C2-O1	C2-O2	C1-O3	O1-O3	C2-C1-O3	C1-C2-O1	O1-C2-O2	C1-C2-O2	
[Cu(HOCH(CH₃)COO)₂(C₆H₁₆N₂)], bis(L-lactato(1-))(N,N,N',N'-tetramethylethylenediamine)copper(II)											189
c S	+0.7	1.544(12)	1.254(12)	1.246(11)	1.416(11)	2.646(9)	108.8(7)	120.4(8)	124.7(8)	114.9(8)	
[Ni(HOCH(CH₃)COO)₂(C₆H₁₆N₂)], bis(L-lactato(1-))(N,N,N',N'-tetramethylethylenediamine)nickel(II)											190
c S	-0.5	1.528(9)	1.253(8)	1.267(8)	1.430(8)	2.624(6)	109.2(5)	119.3(6)	124.7(6)	115.9(6)	
[Cu(HOCH(CH₃)COO)₂(C₁₀H₂₄N₂)]·1/2H₂O, bis(L-lactato(1-))(N,N,N',N'-tetraethylethylenediamine)copper(II) hemihydrate											191
c S	-0.6	1.56(3)	1.23(3)	1.26(3)	1.40(3)	2.65(2)	109(2)	120(2)	122(2)	118(2)	
S	+3.3	1.55(4)	1.23(3)	1.25(3)	1.47(5)	2.65(4)	108(2)	121(2)	125(2)	114(2)	
c S	-6.6	1.58(3)	1.26(3)	1.19(3)	1.43(3)	2.62(3)	110(2)	115(2)	126(2)	119(2)	
S	-18.0	1.55(4)	1.23(4)	1.26(3)	1.43(4)	2.66(3)	108(2)	122(3)	124(3)	114(3)	
[Zn(HOCH(CH₃)COO)₂(H₂O)₂]·H₂O, (D-lactato(1-))(L-lactato(1-))diaquozinc(II) monohydrate											192
c S	-11.9ʲ	1.585(18)	1.229(18)	1.241(17)	1.399(16)	2.672(13)	111.8(10)	116.5(12)	127.1(13)	116.4(12)	
c R	+9.8ʲ	1.586(18)	1.226(18)	1.248(16)	1.410(18)	2.605(14)	106.8(11)	119.7(11)	123.8(12)	117.3(12)	
[Cu(HOCH(CH₃)COO)₂(H₂O)], aquobis(DL-lactato(1-))copper(II)											167
c S	+3.6	1.59(4)	1.25(3)	1.23(3)	1.40(4)	2.56(3)	106.3(22)	117.3(21)	124.6(22)	117.9(22)	
c R	+3.5	1.53(4)	1.24(3)	1.26(3)	1.40(4)	2.54(3)	106.6(22)	118.7(22)	122.4(22)	118.9(22)	
[Cu(C₇H₁₈N₂)₂(HOCH(CH₃)COO)]CH₃CH(OH)COO·H₂O, L-lactato(1-)bis(N-isopropyl-2-methylpropane-1,2-diamine)copper(II) L-lactate(1-) 1-hydrate											91
S	-1.4	1.528(18)	1.244(16)	1.257(15)	1.443(18)	2.575(16)	108.3(11)	117.8(11)	125.7(12)	116.5(11)	

[Cu(HOC(CH₃)₂COO)₂(H₂O)₂], diaquobis(2-hydroxy-2-methylpropionato(1-))copper(II)

c	--	0.0	1.61(4)	1.32(4)	1.24(4)	1.45(4)	2.58(3)	103(2)	118(2)	124(3)	118(2)	167

Ca(OH)(HOCH(C₆H₅)COO)·3CH₃COOH, calcium mandelate(1-) hydroxide 3-acetic acid

c	R	+19.1	1.525(16)	1.296(11)	1.225(11)	1.428(8)	2.633(10)	111.0(10)	115.5(12)	125.1(11)	119.4(10)	201

K(HO(C₆H₅)₂COO), potassium benzilate

c	--	49.0	1.54(1)	1.27(1)	1.24(1)	1.40(1)	2.76(1)	111.5(7)	113.4(7)	124.6(9)	121.9(9)	204

KH(OOCCH(OH)COO)(α), potassium hydrogen tartronate (α form)

	Rʲ	-6.8	1.531(4)	1.223(3)	1.277(3)	1.417(3)	2.634(3)	110.1(2)	119.3(2)	126.2(2)	114.6(2)	206,207
c	Sʲ	-23.0	1.516(4)	1.281(3)	1.226(3)	1.417(3)	2.656(3)	112.7(2)	115.0(3)	124.5(2)	120.5(2)	

KH(OOCCH(OH)COO)(β), potassium hydrogen tartronate (β form)

	Rʲ	+5.4	1.519(4)	1.218(4)	1.302(4)	1.406(3)	2.736(4)	111.7(2)	123.8(3)	124.1(3)	112.1(2)	206,207
c	Sʲ	-21.4	1.532(4)	1.233(4)	1.277(4)	1.406(3)	2.688(4)	111.5(2)	118.7(3)	125.2(3)	116.1(2)	

Li(OOCCH(OH)CH₂COOH), lithium (R)-malate(1-)

R	-5.9	1.526(2)	1.254(2)	1.255(2)	1.427(2)	2.731(2)	114.1(2)	119.2(2)	124.9(2)	115.8(2)	210

[Mg(OOCCH(OH)CH₂COO)(H₂O)₄]·H₂O, (S)-malato(2-)tetraaquomagnesium(II) monohydrate

c	S	+2.7	1.538(8)	1.263(7)	1.246(7)	1.449(7)	2.572(6)	108.0(4)	117.0(5)	125.1(6)	117.9(5)	213

[Mn(OOCCH(OH)CH₂COO)(H₂O)₂]·H₂O, (S)-malato(2-)diaquomanganese(II) monohydrate

c	S	-1.2	1.551(6)	1.266(6)	1.244(6)	1.453(6)	2.644(4)	108.5(4)	119.2(4)	124.9(4)	115.9(4)	218

[Cu(OOCCH(OH)CH₂COOH)₂]·2H₂O, bis((S)-malato(1-))copper(II) dihydrate

c	S	+11.9	1.512(5)	1.275(5)	1.261(5)	1.443(5)	2.577(5)	106.4(3)	119.5(3)	121.6(3)	118.8(3)	219
c	S	+25.2	1.541(5)	1.281(5)	1.237(5)	1.436(5)	2.544(5)	104.3(3)	116.7(3)	123.7(3)	119.5(3)	

Table 4 (Continued)

Absolute config.[b]	Dihed. angle	Distances, Å					Angles, deg				Ref.
		C1-C2	C2-O1	C2-O2	C1-O3	O1-O3	C2-C1-O3	C1-C2-O1	O1-C2-O2	C1-C2-O2	
[Ni(OOCCH(OH)CH₂COOH)₂(H₂O)₂]·2H₂O, diaquobis((S)-malato(1-))nickel(II) dihydrate											220
C S	-4.7	1.532(6)	1.273(5)	1.221(5)	1.433(5)	2.599(5)	109.4(3)	117.4(4)	123.5(4)	119.4(4)	
[Ca(OOCCH(OH)CH₂COOH)₂(H₂O)₄]·2H₂O, tetraaquobis((S)-malato(1-))calcium(II) dihydrate											221
C S	-8.4	1.533(3)	1.254(3)	1.250(3)	1.420(3)	2.632(2)	116.3(3)	118.1(3)	124.9(3)	117.0(3)	
C S	+14.7	1.534(3)	1.265(3)	1.223(3)	1.430(3)	2.534(2)	107.6(3)	115.9(3)	125.1(3)	116.9(3)	
[Zn(OOCCH(OH)CH₂COO)(H₂O)₂]·H₂O, (S)-malato(2-)diaquozinc(II) monohydrate											225
C S	-26.8	1.531(5)	1.266(5)	1.241(5)	1.436(5)	2.610(3)	107.2(3)	117.2(3)	123.9(4)	118.9(3)	
[Co(OOCCH(OH)CH₂COO)(H₂O)]·H₂O, (S)-malato(2-)diaquocobalt(II) monohydrate											227
C S	-27.3	1.534(6)	1.262(5)	1.237(5)	1.435(5)	2.610(3)	107.8(3)	116.4(3)	124.4(4)	119.2(3)	
Ca(OOCCH(OH)CH₂COO)·2H₂O, calcium (S)-malate(2-) dihydrate											229
C S	-15.8	1.529(13)	1.243(11)	1.256(10)	1.434(10)	2.667(9)	111.1(8)	118.3(7)	124.8(9)	116.9(8)	
K(HOOC(OH)CH(COO)CH₂COOH), potassium (2R:3S)-isocitrate(1-)											238
c,f R	-5.5	1.527(7)	1.218(7)	1.301(7)	1.416(7)	2.614(7)	106.6(5)	122.8(5)	124.5(5)	112.7(5)	
Li(NH₄)(HOOCCH₂C(OH)(COO)CH₂COO)·H₂O, lithium ammonium citrate(2-) monohydrate											239
--	4.3	1.541(2)	1.248(2)	1.244(2)	1.428(2)	2.584(2)	107.4(1)	118.8(1)	124.8(1)	116.4(1)	
Na(HOOCCH₂C(OH)(COO)CH₂COOH), sodium citrate(1-)											240
C --	16.9	1.556(5)	1.252(5)	1.253(5)	1.433(5)	2.638(3)	108.2(4)	118.5(4)	126.3(4)	115.3(4)	
Li(HOOCCH₂C(OH)(COO)CH₂COOH), lithium citrate(1-)											240

												Ref
c	--	19.4	1.54(2)	1.25(2)	1.27(2)	1.45(2)	2.59(1)	106.4(8)	118.0(10)	126.5(9)	115.4(9)	
RbH$_2$(OOCCH$_2$C(OH)(COO)CH$_2$COO), rubidium citrate(2-)												241
c	--	15.1	1.58(10)	1.28(9)	1.28(9)	1.39(9)	2.70(10)	112(5)	117(6)	131(7)	112(6)	
[Mg(H$_2$O)$_6$][Mg(OOCCH$_2$C(OH)(COO)CH$_2$COO)(H$_2$O)]$_2$·2H$_2$O, hexaquomagnesium(II) aquocitrato(3-)magnesiumate(II) dihydrate												242
c	--	3.7	1.555(2)	1.260(2)	1.250(2)	1.443(2)	2.595(1)	108.4(1)	117.0(1)	124.1(1)	118.9(1)	
[Mn(H$_2$O)$_6$][Mn(OOCCH$_2$C(OH)(COO)CH$_2$COO)(H$_2$O)]$_2$·2H$_2$O, hexaquomanganese(II) aquocitrato(3-)manganate(II) dihydrate												243,244
c	--	0.6	1.558(2)	1.258(2)	1.256(2)	1.434(2)	2.636(2)	109.4(1)	118.0(1)	123.9(1)	118.1(2)	
[Fe(H$_2$O)$_6$][Fe(OOCCH$_2$C(OH)(COO)CH$_2$COO)(H$_2$O)]$_2$·2H$_2$O, hexaquoiron(II) aquocitrato(3-)ferrate(II) dihydrate												245
c	--	1.6	1.544(4)	1.266(3)	1.248(3)	1.442(3)	2.624(3)	109.6(2)	117.6(2)	123.9(3)	118.5(2)	
Ca(OOCCH$_2$C(OH)(COO)CH$_2$COOH)·3H$_2$O, calcium citrate(2-) trihydrate												246
c	--	16.3	1.531(9)	1.265(8)	1.242(9)	1.451(8)	2.573(4)	107.0(4)	117.4(5)	124.4(5)	118.1(5)	
Rb(NH$_4$)H(COOCHFC(OH)(COO)CH$_2$COO)·2H$_2$O, rubidium ammonium fluorocitrate(2-) dihydrate												249
c	--	1.3	1.548(9)	1.252(6)	1.251(7)	1.427(6)	2.634(4)	110.4(4)	117.3(5)	124.5(6)	118.2(4)	
LiNH$_4$(OOCCH(OH)CH(OH)COO)·H$_2$O, lithium ammonium (±)-tartrate(2-) monohydrate[k]												62
R		-6.0	1.54(2)	1.25(3)	1.24(2)	1.42(3)	2.65(2)	112(2)	117(2)	122(2)	120(2)	
R		-12.0	1.52(4)	1.28(8)	1.26(5)	1.46(2)	2.59(1)	107(3)	118(3)	122(3)	119(6)	
LiNH$_4$(OOCCH(OH)CH(OH)COO)·H$_2$O, lithium ammonium (+)-tartrate(2-) monohydrate												256
R		+7.1	1.522(2)	1.251(2)	1.253(2)	1.415(2)	2.582(2)	110.1(1)	116.4(1)	125.9(2)	117.7(2)	
R		-1.6	1.525(2)	1.253(2)	1.257(2)	1.419(2)	2.589(2)	109.1(1)	118.1(2)	125.4(2)	116.5(1)	
LiTℓ(OOCCH(OH)CH(OH)COO)·H$_2$O, lithium thallium(I) tartrate(2-) monohydrate												257
R		+9.8	1.528(6)	1.234(6)	1.260(6)	1.409(6)	2.565(4)	109.3(4)	116.6(3)	126.4(3)	117.0(4)	
c		-2.9	1.534(5)	1.242(5)	1.253(7)	1.418(7)	2.609(4)	109.1(3)	118.7(3)	125.7(3)	115.5(2)	

Table 4 (Continued)

Absolute config.[b]	Dihed. angle	Distances, Å					Angles, deg				Ref.
		C1-C2	C2-O1	C2-O2	C1-O3	O1-O3	C2-C1-O3	C1-C2-O1	O1-C2-O2	C1-C2-O2	
$Na_2(COOCH(OH)(OH)COO) \cdot 2H_2O$, sodium (+)-tartrate(2-) dihydrate											258
R	+7.5	1.540(4)	1.240(4)	1.249(4)	1.414(4)	2.601(2)	110.0(3)	116.9(3)	127.5(3)	115.5(3)	
C R	+18.0	1.526(4)	1.251(4)	1.248(4)	1.420(4)	2.708(2)	113.0(3)	118.3(3)	125.7(3)	115.8(3)	
$NaK(OOCCH(OH)(OH)COO) \cdot 4H_2O$, sodium potassium (+)-tartrate(2-) tetrahydrate, nonferroelectric phase											259,260
R	+11.7	1.50	1.21	1.31	1.40	2.60	112	117	130	112	
R	+24.1	1.58	1.06	1.41	1.48	2.63	103	126	131	102	
$NaK(OOCCH(OH)(OH)COO) \cdot 4H_2O$, sodium potassium (+)-tartrate(2-) tetrahydrate, ferroelectric phase											261
R	+7.6	1.49	1.28	1.28	1.48	2.64	111	118	125	117	
R	+9.7	1.55	1.22	1.28	1.43	2.60	110	116	126	118	
R	+1.6	1.55	1.30	1.22	1.39	2.54	109	114	130	117	
R	+19.5	1.54	1.28	1.23	1.42	2.75	115	117	127	116	
$NaK(OOCCH(OH)(OH)COO) \cdot 4H_2O$, sodium potassium (±)-tartrate(2-) tetrahydrate											262
R	+32.3	1.58	1.29	1.19	1.40	2.78	114	114	117	118	
R	+12.1	1.58	1.28	1.28	1.40	2.80	111	123	125	111	
$K_2(OOCCH(OH)(OH)COO) \cdot 2H_2O$, potassium ms-tartrate(2-) dihydrate											263
S	-14.0	1.50(2)	1.30(2)	1.30(2)	1.40(2)	2.75(2)	112(2)	123(2)	120(2)	117(2)	
R	+6.9	1.51(2)	1.27(2)	1.28(2)	1.42(2)	2.68(3)	114(2)	119(2)	124(2)	117(2)	

K(OOCCH(OH)CH(OH)COOH), potassium ms-tartrate(1-)

	R	-2.8	1.526(2)	1.219(2)	1.324(1)	1.419(1)	2.687(1)	110.3(1)	122.3(1)	124.8(1)	112.8(1)	264
f,ℓ	S	+6.3	1.535(2)	1.297(1)	1.231(2)	1.414(2)	2.785(1)	113.7(1)	121.2(1)	121.5(1)	117.3(1)	
	R	-2.6	1.530(2)	1.271(1)	1.255(1)	1.413(1)	2.690(1)	113.7(1)	117.5(1)	125.1(1)	117.4(1)	
c,f	S	-8.1	1.529(2)	1.229(1)	1.297(1)	1.423(2)	2.730(1)	111.5(1)	122.4(1)	124.6(1)	113.0(1)	

Rb₂(OOCCH(OH)CH(OH)COO), rubidium ms-tartrate(2-) dihydrate

	S	-4.3	1.51	1.24	1.25	1.38	2.63	112	118	127	115	263
	R	+1.0	1.52	1.25	1.28	1.40	2.68	113	118	123	119	

Rb(OOCCH(OH)CH(OH)COOH), rubidium (+)-tartrate(1-)

c,f	R	+8.0	1.52	1.20	1.22	1.44	2.73	112	123	130	108	266
	R	-6.7	1.57	1.20	1.24	1.45	2.66	113	115	133	113	

Cs(OOCCH(OH)CH(OH)COOH), cesium (+)-tartrate(1-)

c,f	R	+10.1	1.523(3)	1.210(3)	1.312(3)	1.422(3)	2.721(2)	110.5(2)	123.7(2)	124.1(2)	112.2(2)	267
	R	-4.2	1.537(3)	1.264(3)	1.247(3)	1.399(3)	2.697(2)	113.9(2)	117.7(2)	125.2(2)	117.1(2)	

Sr(OOCCH(OH)CH(OH)COO)·3H₂O, strontium (+)-tartrate trihydrate

c	R	+8.8	1.56(3)	1.27(3)	1.27(3)	1.42(3)	2.70(3)	110(2)	120(2)	124(2)	116(2)	268
c	R	+12.7	1.54(3)	1.23(3)	1.25(3)	1.45(3)	2.65(3)	108(2)	121(2)	125(2)	113(2)	

Ca(OOCCH(OH)CH(OH)COO)·4H₂O, calcium (+)-tartrate tetrahydrate

c	R	-25.6	1.50(1)	1.27(1)	1.26(1)	1.44(1)	2.61(1)	110(1)	116(1)	123(1)	122(1)	268
c	R	-1.4	1.50(1)	1.31(1)	1.24(1)	1.43(1)	2.65(1)	112(1)	118(1)	124(1)	118(1)	

[Cu(C₃H₁₀N₂)₂(C₄H₄O₆)]·2H₂O, (+)-tartrato(2-)bis(N-methylethylenediamine)copper(II) dihydrate

94

Table 4 (Continued)

Absolute config.[b]	Dihed. angle	Distances, Å					Angles, deg				Ref.
		C1-C2	C2-O1	C2-O2	C1-O3	O1-O3	C2-C1-O3	C1-C2-O1	O1-C2-O2	C1-C2-O2	
[Cu(C₇H₁₈N₂)₂(C₄H₄O₆)]·C₃H₇OH, (2S,3S)-tartrato(2-)bis(N-isopropyl-2-methyl]propane-1,2-diamine)copper(II) n-propanol											
R	-1.7	1.557(14)	1.249(12)	1.238(13)	1.429(12)	2.634(9)	111.4(7)	116.0(9)	126.9(9)	117.1(8)	95
S	+3.8	1.55(3)	1.25(3)	1.21(3)	1.43(4)	2.63(1)	110.4(9)	116.8(9)	127.1(8)	116.1(9)	
Cu(d-C₄H₄O₆)·3H₂O, copper(II) d-tartrate(2-) trihydrate											308
c R	-4.3	1.56(3)	1.27(2)	1.22(3)	1.41(3)	2.73(2)	110(2)	122(2)	124(2)	114(2)	
c R	+9.7	1.54(3)	1.25(3)	1.28(3)	1.43(2)	2.58(2)	108(2)	117(2)	125(2)	118(2)	
c R	+4.2	1.51(3)	1.26(3)	1.25(2)	1.45(2)	2.55(2)	107(1)	118(2)	125(2)	117(2)	
c R	-2.0	1.52(3)	1.29(3)	1.23(3)	1.42(2)	2.70(2)	113(2)	119(2)	123(2)	118(2)	
Cu(ms-C₄H₄O₆)·3H₂O, copper(II) ms-tartrate(2-) trihydrate											308
c R	-6.9	1.528(5)	1.282(4)	1.225(4)	1.436(4)	2.600(3)	109.1(3)	117.3(3)	124.1(3)	118.6(3)	
c S	+9.1	1.535(5)	1.292(5)	1.220(5)	1.437(4)	2.592(3)	109.5(3)	115.9(3)	125.1(4)	118.9(3)	
Ca(OOCCH(OH)CH₂OH)₂·2H₂O, calcium (±)-glycerate(1-) dihydrate											317
c S	-1.2	1.531(4)	1.252(4)	1.248(4)	1.421(3)	2.575(2)	108.6(2)	117.6(2)	123.5(2)	118.8(3)	
c S	+1.0	1.524(4)	1.239(4)	1.267(3)	1.424(3)	2.565(3)	107.6(2)	118.7(2)	124.0(3)	117.3(3)	
Cd(OOCCH(OH)CH₂OPO₃H)·3H₂O, cadmium (S)-phosphoglycerate(2-) trihydrate											320
c S	+3.2	1.56(4)	1.34(3)	1.19(4)	1.41(3)	2.55(2)	110(2)	112(3)	128(3)	120(2)	
[Cu(OOCC(CH₃)(OH)CH₂OH)₂(H₂O)₂], diaquobis((±)-2,3-dihydroxy-2-methylpropanoato(1-))copper(II)											321
c R	-5.3	1.52(2)	1.30(1)	1.22(1)	1.46(1)	2.57(1)	106.6(9)	118.2(10)	122.1(10)	119.7(10)	

322 K(OOCCH(OH)CH(OH)CH(OH)CH(OH)CH₂OH), potassium gluconate(1-)

S	-8.2	1.55(4)	1.28(4)	1.23(4)	1.44(4)	2.68(4)	108(2)	121(2)	122(2)	116(2)

324 K(OOCCH(OH)CH(OH)CH(OH)CH(OH)CH₂OH)·H₂O, potassium D-gluconate(1-) monohydrate "B" form

c										
R	+5.3	1.535(3)	1.232(4)	1.263(4)	1.409(4)	2.671(6)	112.2(2)	118.8(2)	125.2(3)	116.0(2)

326 Mn(OOCCH(OH)CH(OH)CH(OH)CH(OH)CH₂OH)₂·2H₂O, manganese(II) D-gluconate(1-) dihydrate

c										
R	+3.9	1.535(8)	1.247(11)	1.251(11)	1.441(11)	2.587(4)	107.2(6)	119.2(7)	125.4(7)	115.4(7)
R	-7.4	1.506(16)	1.266(10)	1.252(10)	1.445(8)	2.596(6)	110.6(7)	116.6(8)	127.3(9)	116.1(8)

329 Na₃(OOCCH(OH)CH(OH)CH(OH)CH(OH)CH₂OPO₃)·2H₂O, sodium 6-phospho-D-gluconate(3-) dihydrate

c										
R	+15.3	1.537(4)	1.256(4)	1.246(4)	1.421(4)	2.723(3)	113.0(2)	118.7(3)	125.9(3)	115.3(2)

330 Ca(C₁₂H₂₁O₁₂)Br·4H₂O, calcium lactobionate(1-) bromide tetrahydrate

c										
R	+12.0	1.569(9)	1.254(9)	1.245(10)	1.406(8)	2.667(7)	111.0(5)	117.4(6)	126.5(6)	116.2(6)

331 K(OOCCH(OH)CH(OH)CH(OH)CH(OH)COOH), potassium D-glucarate(1-)ᵐ

c										
R	+12.4	1.54(3)	1.28(2)	1.30(2)	1.43(2)	2.77(2)	113(1)	120(2)	126(2)	114(2)
S	-18.0	1.53(3)	1.30(2)	1.30(2)	1.45(2)	2.75(2)	112(1)	120(2)	124(2)	116(2)

332 Ca(OOCCH(OH)CH(OH)CH(OH)CH(OH)COO)·4H₂O, calcium D-glucarate(2-) tetrahydrate

c										
R	+0.7	1.534(16)	1.277(17)	1.236(17)	1.469(18)	2.613(12)	110.0(10)	116.4(12)	126.0(10)	117.6(11)
S	-51.1	1.584(16)	1.237(18)	1.279(19)	1.449(16)	2.870(11)	111.4(9)	116.2(12)	128.1(11)	115.6(12)

333 Ca(OOCCH(OH)CH(OH)CH(OH)CH(OH)CH₂OH)₂·5H₂O, calcium arabonate(1-) pentahydrate

c										
(R)	+24.6	1.57(5)	1.22(5)	1.30(5)	1.39(5)	2.65	108(3)	118(3)	112(3)	117(3)

333 Sr(OOCCH(OH)CH(OH)CH(OH)CH(OH)CH₂OH)₂·5H₂O, strontium arabonate(1-) pentahydrate

c										
(R)	+16.1	1.58	1.19	1.34	1.35	2.61	106	124	124	113

334 Ca(OOCC(OH)(CH₂OH)CH₂CH(OH)CH₂OH)₂, calcium 3-D-glucoisosacharate(1-)

Table 4 (Continued)

Absolute config.b	Dihed. angle	Distances, Å					Angles, deg				Ref.
		C1-C2	C2-O1	C2-O2	C1-O3	O1-O3	C2-C1-O3	C1-C2-O1	O1-C2-O2	C1-C2-O2	
c S	+22.2	1.55(3)	1.21(2)	1.28(3)	1.48(2)	2.60(2)	104.3(12)	119.9(16)	125.0(18)	112.9(14)	335

Sr(OOCC(OH)(CH2OH)CH2CH(OH)CH2OH)2,strontium 3-D-glucoisosacharate(1-)

Absolute config.b	Dihed. angle	C1-C2	C2-O1	C2-O2	C1-O3	O1-O3	C2-C1-O3	C1-C2-O1	O1-C2-O2	C1-C2-O2	Ref.
c S	+13.8	1.54(2)	1.25(2)	1.27(2)	1.44(2)	2.60(1)	107.7(8)	118.4(9)	125.2(10)	116.4(9)	

Ionized Hydroxyl Group

[Er(HOCH2COO)(OCH2COO)(H2O)]·H2O, aquoglycolato(1-)glycolato(2-)erbium(III) monohydrate

Absolute config.b	Dihed. angle	C1-C2	C2-O1	C2-O2	C1-O3	O1-O3	C2-C1-O3	C1-C2-O1	O1-C2-O2	C1-C2-O2	Ref.
c --	3.2	1.536(19)	1.255(19)	1.258(18)	1.396(19)	2.618(14)	112.9(13)	115.4(13)	125.8(12)	118.8(13)	172

K[CrO(OC(CH3)(C2H5)COO)2]·H2O, potassium bis(2-hydroxy-z-methylbutyrato(2-))oxochromate(V) monohydrate

Absolute config.b	Dihed. angle	C1-C2	C2-O1	C2-O2	C1-O3	O1-O3	C2-C1-O3	C1-C2-O1	O1-C2-O2	C1-C2-O2	Ref.
c S	+8.9	1.49(2)	1.29(2)	1.22(2)	1.46(2)	2.47(2)	107.1(15)	114.4(16)	121.3(16)	124.1(16)	195
c S	+4.6	1.50(2)	1.28(2)	1.26(2)	1.44(2)	2.47(2)	105.7(14)	116.3(14)	123.2(15)	120.6(15)	

Na(C2H5)4N[VO(OC(C6H5)2COO)2]·2C3H7OH, sodium tetraethylammonium bis(benzilato(2-))oxovanadate(IV) diisopropanol

Absolute config.b	Dihed. angle	C1-C2	C2-O1	C2-O2	C1-O3	O1-O3	C2-C1-O3	C1-C2-O1	O1-C2-O2	C1-C2-O2	Ref.
c --	+12.4	1.57(2)	1.25(2)	1.26(2)	1.44(2)	2.518(9)	103(1)	118(1)	124(1)	118(1)	196
c --	-11.3	1.53(2)	1.31(2)	1.23(2)	1.41(2)	2.550(10)	109(1)	115(1)	122(1)	123(1)	

[Ge(OCH(C6H5)COO)2(H2O)2]·2H2O, diaquobis((±)-mandelato(2-))germanium(IV) dihydrate

Absolute config.b	Dihed. angle	C1-C2	C2-O1	C2-O2	C1-O3	O1-O3	C2-C1-O3	C1-C2-O1	O1-C2-O2	C1-C2-O2	Ref.
c R	-1.8	1.54(1)	1.31(1)	1.22(1)	1.44(1)	2.58	108.9(7)	115(1)	123.6(8)	121(1)	202

K[B(OOCCH(O)CH2COOH)2]·H2O, potassium bis(malato(2-))borate(III) monohydrate

222

c												
c	S	+10.1	1.523(6)	1.311(5)	1.208(4)	1.430(4)	2.317(4)	104.6(3)	108.6(3)	125.0(4)	126.4(4)	
c	S	+9.3	1.484(11)	1.308(7)	1.224(8)	1.443(7)	2.337(6)	105.0(5)	110.9(5)	121.9(7)	127.2(6)	230

$(NH_4)_4[(MoO_2)_4O_3(OOCCH(O)CH_2COO)_2] \cdot 6H_2O$, ammonium bis(malato(3-))undecaoxotetramolybdate(VI) hexahydrate

c	R	+1.6	1.57(6)	1.25(6)	1.17(6)	1.50(6)	2.58	114.9	106.8	129.2	121.7	247b

$[Co(NH_3)_6][Sb(OOCCH_2C(O)(COO)CH_2COOH)_2] \cdot 5H_2O$, hexamminecobalt(III) bis(citrato(3-))antimonate(III) 5-hydrate

c	--	-3.4	1.53	1.28	1.24	1.42	2.61	110.0	117.3	123.1	119.6	
c	--	1.4	1.52	1.28	1.24	1.43	2.63	110.5	117	123.6	119.9	248

$Cu_2(OOCCH_2C(O)(COO)CH_2COO) \cdot 2H_2O$, copper(II) citrate(4-) dihydrate

c	--	18.9	1.50(2)	1.30(2)	1.23(2)	1.47(2)	2.58(1)	108(1)	117(1)	121(1)	122(1)	

$[N(CH_3)_3]_{10}[Ni_8(OOCCH_2C(O)(COO)CH_2COO)_6(OH)_2(H_2O)_2] \cdot 36H_2O$, tetramethylammonium diaquodihydroxohexakis(citrato(4-))octanickelate(II) 36-hydrate

c	--	16.9	1.53(2)	1.25(2)	1.27(2)	1.44(2)	2.64(1)	108(1)	120(1)	123(1)	117(1)	117(1)
c	--	30.9	1.54(2)	1.25(2)	1.29(2)	1.44(2)	2.69(1)	108(1)	119(1)	122(1)	118(1)	
c	--	23.0	1.58(2)	1.27(2)	1.26(2)	1.40(2)	2.74(1)	112(1)	117(1)	124(1)	118(1)	245

$Na_5[Fe(d-C_4H_2O_6)(\ell-C_4H_2O_6)] \cdot 14H_2O$, sodium d-tartrato(4-)-$\ell$-tartrato(4-)ferrate(III) 14-hydrate

c	R	+24.1	1.546(5)	1.276(4)	1.237(5)	1.420(5)	2.584(3)	108.5(2)	114.4(2)	125.4(3)	119.8(3)	
c	R	-2.0	1.541(6)	1.248(5)	1.270(4)	1.404(4)	2.735(3)	114.2(3)	119.3(3)	124.2(3)	116.2(3)	269

$Na_4[(VO)_2(d-C_4H_2O_6)(\ell-C_4H_2O_6)] \cdot i2H_2O$, sodium μ-d-tartrato(4-)-μ-ℓ-tartrato(4-)bis(oxovanadate(IV)) 12-hydrate

c	R	+15.0	1.523(11)	1.312(9)	1.229(9)	1.407(10)	2.585(8)	110.6(7)	114.5(5)	124.1(6)	121.2(6)	
c	S	-1.0	1.534(11)	1.277(10)	1.227(10)	1.41u(9)	2.536(8)	109.9(7)	113.9(5)	125.3(4)	120.8(6)	290

$[N(C_2H_5)_4]_4[(VO)_2(d-C_4H_2O_6)(\ell-C_4H_2O_6)] \cdot 8H_2O$, tetraethylammonium μ-d-tartrato(4-)-μ-ℓ-tartrato(4-)bis(oxovanadate(IV)) 8-hydrate

c	S	-20.2	1.524(5)	1.292(5)	1.235(5)	1.394(4)	2.580(4)	110.5(3)	114.3(3)	124.3(3)	121.4(4)	
c	R	+5.2	1.528(4)	1.298(4)	1.244(5)	1.389(4)	2.570(4)	110.9(3)	115.2(3)	124.1(3)	120.7(3)	291

Table 4 (Continued)

Absolute config.[b]	Dihed. angle	Distances, Å					Angles, deg				Ref.
		C1–C2	C2–O1	C2–O2	C1–O3	O1–O3	C2–C1–O3	C1–C2–O1	O1–C2–O2	C1–C2–O2	
Na$_4$[(VO)$_2$(d-C$_6$H$_6$O$_6$)(ℓ-C$_6$H$_6$O$_6$)]·12H$_2$O, sodium μ-d-dimethyltartrato(4-)-μ-ℓ-dimethyltartrato(4-)bis(oxovanadate(IV)) 12-hydrate											292
c	R	+23.7	1.554(4)	1.276(4)	1.234(4)	1.420(5)	2.574(3)	107.7(3)	114.5(3)	124.2(3)	121.3(3)	
c	S	+1.1	1.544(5)	1.285(5)	1.229(4)	1.434(4)	2.544(3)	108.4(3)	114.8(3)	122.7(3)	122.5(3)	
(NH$_4$)$_4$[(VO)$_2$(d-C$_4$H$_2$O$_6$)$_2$]·2H$_2$O, ammonium di-μ-d-tartrato(4-)bis(oxovanadate(IV)) dihydrate											294
c	(S)	-19.9	1.45(3)	1.28(4)	1.22(4)	1.46(3)	2.57(2)	110(2)	118(2)	122(2)	115(2)	
c	(S)	+5.8	1.51(3)	1.35(3)	1.37(3)	1.41(3)	2.55(2)	106(2)	118(2)	115(1)	124(2)	
(NH$_4$)$_2$[Sb$_2$(d-C$_4$H$_2$O$_6$)$_2$]·3H$_2$O, ammonium di-μ-d-tartrato(4-)bis(antimonate(III)) trihydrate[n]											295
c	R	-20.5	1.70	1.32	1.26	1.77	2.79	93	130	113	115	
c	R	+24.2	2.08	1.91	0.65	1.44	2.44	140	61	125	121	
c	R	-47.4	1.44	1.31	1.40	1.55	2.80	103	127	100	109	
c	R	+9.4	1.36	1.44	1.08	1.36	2.62	124	109	124	127	
dℓ-K$_2$[Sb$_2$(C$_4$H$_2$O$_6$)$_2$]·3H$_2$O, potassium di-μ-dℓ-tartrato(4-)bis(antimonate(III)) trihydrate											296
c	S	+2.8	1.57(5)	1.31(4)	1.20(5)	1.40(4)	2.68(3)	113(3)	114(3)	120(3)	124(3)	
c	S	-2.6	1.50(5)	1.34(4)	1.23(4)	1.39(4)	2.69(3)	116(3)	115(3)	118(3)	125(3)	
c	S	+5.2	1.51(5)	1.29(4)	1.31(5)	1.43(4)	2.68(4)	112(3)	118(3)	121(3)	119(3)	
c	S	+1.0	1.48(5)	1.29(4)	1.26(4)	1.47(5)	2.65(4)	112(3)	118(3)	120(3)	121(3)	
K$_2$[Sb$_2$(d-C$_4$H$_2$O$_6$)$_2$]·3H$_2$O, potassium di-μ-d-tartrato(4-)bis(antimonate(III)) trihydrate											298
c	(S)	+1.6	1.55(2)	1.26(2)	1.24(2)	1.39(2)	2.61(2)	113.2(12)	114.3(12)	125.2(13)	120.5(12)	

c	(S)	+2.0	1.50(2)	1.32(2)	1.19(2)	1.38(2)	2.64(2)	115.9(13)	113.7(14)	123.2(16)	123.1(17)	
c	(S)	-2.7	1.52(2)	1.24(2)	1.28(2)	1.45(2)	2.61(1)	108.6(12)	119.4(13)	125.0(13)	115.5(15)	
c	(S)	+2.2	1.53(2)	1.24(3)	1.22(2)	1.38(2)	2.62(2)	112.6(17)	116.7(17)	126.2(18)	117.1(21)	

[Pt(C₃H₁₀N₂)(C₄H₁₂N₂)][Sb₂(d-C₄H₂O₆)₂]·2H₂O, (+)-(R)-propylenediamine-N,N'-dimethylethylenediamineplatinum(II) di-μ-d-tartrato(4-)-bis(antimonate(III)) dihydrate   299

c	R	-0.6	1.55(2)	1.26(2)	1.24(2)	1.43(2)	2.62(2)	109(1)	118(1)	126(2)	117(1)	
c	R	+4.2	1.53(2)	1.32(2)	1.22(2)	1.40(2)	2.66(2)	113(1)	116(2)	124(2)	120(2)	
c	R	+1.6	1.51(3)	1.30(2)	1.22(2)	1.46(2)	2.64(2)	111(2)	118(2)	121(2)	121(2)	
c	R	-4.5	1.52(3)	1.28(2)	1.23(2)	1.42(2)	2.60(2)	112(1)	115(2)	124(2)	121(2)	

[Pt(NH₃)₂(C₅H₁₄N₂)][Sb₂(d-C₄H₂O₆)₂]·H₂O, (+)-diamine-N,N,N'-trimethylethylenediamineplatinum(II) di-μ-d-tartrato(4-)-bis(antimonate(III)) monohydrate   300

c	R	-2.1	1.50(2)	1.28(2)	1.24(2)	1.44(2)	2.63(1)	112(1)	118(1)	123(1)	119(1)	
c	R	-3.3	1.54(2)	1.30(2)	1.24(2)	1.44(2)	2.63(1)	110(1)	117(1)	122(1)	121(1)	
c	R	-6.5	1.55(2)	1.28(2)	1.23(2)	1.42(2)	2.66(1)	111(1)	117(1)	123(1)	120(1)	
c	R	+4.0	1.54(2)	1.27(2)	1.22(2)	1.43(2)	2.59(1)	110(1)	116(1)	123(1)	121(1)	

[Pt(NH₃)₂(C₄H₁₂N₂)][Sb₂(d-C₄H₂O₆)₂]·H₂O, (-)-diamine-N-methyl-(S)-propylenediamineplatinum(II) di-μ-d-tartrato(4-)-bis(antimonate(III)) monohydrate   301

c	R	-1.2	1.52(3)	1.30(3)	1.27(3)	1.39(3)	2.63(3)	112(2)	117(2)	123(2)	121(2)	
c	R	-5.4	1.54(3)	1.23(3)	1.24(3)	1.41(2)	2.59(3)	108(2)	119(2)	124(2)	117(2)	
c	R	-5.9	1.56(3)	1.31(3)	1.22(4)	1.41(3)	2.64(3)	112(2)	115(2)	125(2)	120(3)	
c	R	+3.9	1.52(3)	1.29(2)	1.22(2)	1.40(2)	2.60(3)	112(1)	116(2)	124(2)	120(1)	

[Fe(C₁₂H₈N₂)₃][Sb₂(d-C₄H₂O₆)₂]·8H₂O, (-)-tris(1,10-phenanthroline)iron(II) di-μ-d-tartrato(4-)-bis(antimonate(III)) 8-hydrate   302

c	R	-2.7	1.50(2)	1.28(2)	1.21(2)	1.44(2)	2.60(1)	111(2)	116(2)	127(2)	117(2)	

Table 4 (Continued)

Absolute config.b	Dihed. angle	Distances, Å					Angles, deg				Ref.
		C1-C2	C2-O1	C2-O2	C1-O3	O1-O3	C2-C1-O3	C1-C2-O1	O1-C2-O2	C1-C2-O2	
[Co(NCS)₂(C₃H₁₀N₂)₂]₂[Sb₂(d-C₄H₂O₆)₂]·4H₂O·(-)-dithiocyanatobis(1,3-diaminopropane)cobalt(III) di-μ-d-tartrato(4-)bis(antimonate(III)) tetrahydrate											303
c	R	-1.2	1.50(2)	1.27(2)	1.26(2)	1.41(2)	2.61(1)	111(2)	118(2)	123(2)	119(2)
c	R	+5.8	1.57(3)	1.27(2)	1.25(2)	1.40(2)	2.64(1)	112(1)	115(2)	126(2)	119(2)
c	R	+5.4	1.59(2)	1.27(2)	1.23(2)	1.38(2)	2.64(2)	112(1)	114(1)	127(2)	119(1)
[Cr(NCS)₂(C₃H₁₀N₂)₂]₂[Sb₂(d-C₄H₂O₆)₂]·4H₂O, (-)-dithiocyanatobis(1,3-diaminopropane)chromium(III) di-μ-d-tartrato(4-)bis(antimonate-(III)) tetrahydrate											304
c	R	+8.9	1.556(11)	1.280(9)	1.237(10)	1.393(9)	2.619(7)	112.4(6)	114.1(7)	125.7(7)	120.1(7)
c	R	+3.3	1.571(10)	1.256(9)	1.232(9)	1.383(9)	2.632(7)	111.2(6)	116.2(6)	125.6(8)	118.2(7)
dℓ-(NH₄)₂[Sb₂(C₄H₂O₆)₂]·4H₂O, ammonium di-μ-dℓ-tartrato(4-)bis(antimonate(III)) tetrahydrate											251
c	S	+9.3	1.43	1.42	1.17	1.48	2.68	118	112	117	130
c	S	-10.9	1.53	1.29	1.30	1.46	2.66	108	120	123	116
[Co(C₅H₇O₂)(C₃H₁₀N₂)₂][As₂(d-C₄H₂O₆)₂]·H₂O, acetylacetonato(1-)bis(1,3-diaminopropane)cobalt(III) di-μ-d-tartrato(4-)bis(arsenate(III)) hydrate											305
c	(S)	-6.0	1.58(5)	1.31(4)	1.23(4)	1.49(4)	2.60(3)	106(3)	117(3)	126(3)	117(3)
c	(S)	-2.7	1.55(4)	1.28(3)	1.23(3)	1.42(3)	2.63(3)	111(2)	117(4)	126(3)	118(2)
c	(S)	+3.3	1.54(5)	1.31(4)	1.22(4)	1.48(4)	2.60(3)	110(2)	115(3)	127(3)	119(3)
c	(S)	-6.6	1.53(4)	1.32(4)	1.23(4)	1.42(3)	2.60(3)	110(2)	115(3)	120(3)	125(3)
Na₄[Cu₂(d-C₄H₂O₆)(ℓ-C₄H₂O₆)]·10H₂O, sodium μ-d-tartrato(4-)-μ-ℓ-tartrato(4-)cuprate(II) 10-hydrate											307
c	R	+18.7	1.521(5)	1.298(4)	1.233(4)	1.411(4)	2.619(3)	110.4(3)	116.0(3)	123.1(4)	120.9(3)
c	R	+19.5	1.525(5)	1.299(5)	1.243(5)	1.414(4)	2.607(3)	110.3(3)	115.1(3)	123.5(3)	121.4(3)

^aUnderlined distances, angles, and esd's were calculated from reported coordinates.

^bAbsolute configuration of carbinol carbon. Parentheses denote that the absolute configuration calculated from the reported coordinates is opposite to that which the compound should have.

^cChelated group.

^dThe c cell dimension appears to be in error. A value of 5.79 rather than 5.97 was used.

^eDifficulty was encountered in reproducing the bond parameters using the reported cell dimensions and atom coordinates.

^fProtonated carboxyl.

^gFor the Λ absolute configuration complex.

^hFor the absolute configuration of the complex pictured in the text.

ⁱFor the Δ absolute configuration complex.

^jThe "absolute configuration" was established by setting the priority of the carboxylate carbon not in the α-hydroxycarboxylate group under consideration equal to that of a methyl group.

^kBonding parameters were calculated from a set of refined coordinates communicated by A.J.J. Sprenkels.

^ℓThe protonated carboxyl oxygen lies on the same side as the α-hydroxyl oxygen.

^mThe protonated carboxyl group could not be determined.

ⁿVery low precision structure.

Table 5 Geometries of α-Hydroxycarboxylate Chelate Rings[a]

Plane 1: M, O1, O3
Plane 2: O3, C1, C2, O1

			Distances, Å				Angles, deg				
M	Absol. config.[b]	Dihed. Angle	M-O3	M-O1	C1-plane 1	C2-plane 1	C1-O3-M	C2-O1-M	O3-M-O1	Plane 1/plane 2	Ref.
C	--	8.4	1.48(7)	1.44(5)	.371	.283	106(4)	111(3)	104(3)	13.9	336
Li(I)	--	7.8	2.357(3)	1.952(3)	.119	.227	107.2(10)	124.4(10)	76.2(10)	8.5	163
Li(I)	--	0.0	2.01(1)	2.10(2)	0.0	0.0	118.3(5)	114.9(5)	78.5(5)	0.0	164
K(I) e	--	6.2	2.778(8)	2.807(7)	.258	.224	121.5(4)	120.9(6)	58.7(2)	11.7	165
Rb(I) e	--	7.9	2.879(9)	3.009(9)	.229	.177	123.4(6)	120.2(8)	55.9(2)	9.9	166
Cu(II)	--	2.5	1.93(1)	1.91(1)	.052	.025	115.5(8)	116.0(9)	83.4(4)	1.7	167
Zn(II)	--	0.0	2.162(5)	2.056(5)	.010	.011	115.8(4)	120.3(4)	76.2(2)	0.5	168,169
Zn(II)	--	+6.2[f]	2.087(6)	2.104(5)	.142	.078	116.1(4)	116.8(4)	78.3(2)	5.0	
Er(III)	--	4.7	2.34(1)	2.30(1)	.162	.239	121.7(8)	122.2(9)	67.1(4)	9.8	173
Er(III)	--	0.2	2.29(1)	2.27(1)	.459	.531	119.7(7)	121.9(8)	67.8(3)	22.6	
Er(III)	--	7.4	2.34(1)	2.29(1)	.101	.212	120.8(8)	126.2(7)	66.5(3)	7.4	
Sc(III)	--	3.7	2.24(4)	2.18(4)	.262	.374	121(3)	119(3)	69.0(13)	15.6	174
Sc(III)	--	2.8	2.22(5)	2.17(4)	.495	.632	118(3)	119(3)	68.2(14)	26.0	
Sc(III)	--	7.1	2.27(4)	2.19(4)	.051	.149	120(3)	125(2)	67.4(13)	4.7	

Compounds (in row order):
$(C_6H_7Cl_3)C(OCH_2COO)$
$Li(HOCH_2COO)$
$Li(HOCH_2COO) \cdot H_2O$[c]
$K(HOCH_2COOH)(HOCH_2COO)$[d]
$Rb(HOCH_2COOH)(HOCH_2COO)$
$Cu(HOCH_2COO)_2$
$[Zn(HOCH_2COO)_2(H_2O)_2]$
$Er(HOCH_2COO)_3 \cdot 2H_2O$
$Sc(HOCH_2COO)_3 \cdot 2H_2O$

Compound	Ion											Ref
[Er(HOCH₂COO)(OCH₂COO)H₂O]·H₂O	Er(III)	--	10.1	2.476(12)	2.328(12)	.256	.417	115.0(11)	130.9(10)	63.5(4)	16.7	172
	g Er(III)	--	3.2	2.240(9)	2.372(11)	.058	.021	121.4(8)	121.4(8)	68.9(4)	1.8	
Gd(HOCH₂COO)₃	Gd(III)	--	+16.7ʰ	2.463(9)	2.497(11)	.471	.314	118.9(7)	118.7(11)	64.8(3)	18.4	175,177
	Gd(III)	--	+29.0ʰ	2.486(6)	2.417(5)	.121	.516	117.9(5)	123.3(4)	64.5(2)	16.8	
	Gd(III)	--	+30.8ʰ	2.467(8)	2.438(10)	.269	.736	116.9(7)	125.6(11)	63.6(3)	25.5	
La(HOCH₂COO)₃	La(III)	--	-17.7ʰ	2.569(7)	2.533(7)	.190	.411	117.4(4)	120.6(6)	63.8(2)	14.7	177
	La(III)	--	-13.6ʰ	2.592(4)	2.525(4)	-.107	+.049	121.6(4)	124.3(3)	61.9(1)	1.9	
	La(III)	--	-5.1ʰ	2.587(6)	2.594(7)	.515	.508	120.7(5)	124.0(6)	60.0(2)	24.2	
Eu(HOCH₂COO)₃	Eu(III)	--	0.1	2.62(2)	2.44(2)	.547	.596	117(2)	117(1)	65.6(6)	27.1	176
	Eu(III)	--	7.4	2.70(2)	2.53(2)	.349	.477	116(1)	122(1)	62.7(6)	19.7	
	Eu(III)	--	3.8	2.55(2)	2.41(2)	.371	.450	116(1)	122(1)	65.5(6)	19.5	
[Cu(HOCH(CH₃)COO)₂(C₆H₁₆N₂)]	Cu(II)	S	+0.7	2.310(7)	1.982(7)	.068	.087	112.1(5)	122.9(6)	75.7(3)	3.7	189
[Ni(HOCH(CH₃)COO)₂(C₆H₁₆N₂)]	Ni(II)	S	-0.5	2.080(5)	2.047(5)	.009	.016	115.2(4)	117.4(4)	79.0(2)	0.6	190
[Cu(HOCH(CH₃)COO)₂(C₁₀H₂₄N₂)]·½H₂O	Cu(II)	S	-0.6	2.28(1)	2.01(2)	.346	.404	110(1)	120(2)	75.8(7)	18.4	191
	Cu(II)	S	-6.6	2.26(2)	2.01(1)	.293	.408	111(1)	124(1)	75.4(6)	16.8	
[Zn(HOCH(CH₃)COO)₂(H₂O)₂]·H₂O	Zn(II)	S	-11.9ⁱ	2.194(9)	2.081(9)	.022	.178	112.6(7)	120.5(9)	77.3(3)	5.3	192
	Zn(II)	R	+9.8ⁱ	2.092(11)	2.154(9)	.096	.243	119.3(8)	118.0(8)	75.6(4)	8.5	
[Cu(HOCH(CH₃)COO)₂(H₂O)]	Cu(II)	S	+3.6	1.97(2)	1.93(2)	.181	.253	115.9(15)	116.8(15)	82.2(4)	10.3	167
	Cu(II)	R	+3.5	1.97(2)	1.93(2)	.148	.214	115.7(16)	116.8(15)	81.0(7)	8.7	
K[CrO(OC(CH₃)(C₂H₅)COO)₂]·H₂O	g Cr(V)	S	+8.9	1.767(12)	1.921(10)	.012	.139	117.9(10)	115.9(11)	84.0(5)	3.8	195
	g Cr(V)	S	+4.6	1.795(10)	1.902(11)	.237	.329	117.1(9)	114.7(10)	83.8(5)	13.0	

Table 5 (Continued)

Compound	M	Absol config.[b]	Dihed. Angle	M-O3	M-O1	C1-plane 1	C2-plane 1	C1-O3-M	C2-O1-M	O3-M-O1	Plane 1/plane 2	Ref.
				Distances, Å				Angles, deg				
Na(C₂H₅)₄N[VO(OC(C₆H₅)₂COO)₂]·2C₃H₇OH	g V(IV)	--	+12.4	1.900(8)	1.971(8)	.113	.308	119.1(6)	116.4(9)	81.4(4)	10.2	196
	g V(IV)	--	-11.3	1.933(9)	1.970(9)	.230	.093	117.1(7)	115.9(9)	81.6(4)	7.3	
[Cu(HOC(CH₃)₂COO)₂(H₂O)₂]	Cu(II)	--	0.0	2.01(2)	1.89(2)	.0	.0	117.8(18)	118.3(17)	82.7(8)	0.0	167
Ca(OH)(HOCH(C₆H₅)COO)·3CH₃COOH	Ca(II)	R	+19.1	2.428(9)	2.420(6)	-.017	+.230	120.4(7)	124.0(6)	65.8(2)	5.8	201
[Ge(OCH(C₆H₅)COO)₂(H₂O)₂]·2H₂O	g Ge(IV)	R	-1.8	1.81(1)	1.86(1)	.112	.146	113	113	88.8(2)	5.9	202
KH(OOCCH(OH)COO)(α)	K(I)	Sʲ	-23.0	3.028(2)	2.673(2)	-.111	+.172	117.7(2)	135.3(2)	55.1(1)	3.2	206,207
KH(OOCCH(OH)COO)(β)	K(I)	Rʲ	+5.4	2.967(3)	2.853(3)	-.010	+.052	121.8(2)	126.4(2)	56.0(1)	1.3	206,207
[Mg(OOCCH(OH)CH₂COO)(H₂O)₄]·H₂O	Mg(II)	S	+2.7	2.125(5)	2.070(5)	-.006	+.030	117.5(3)	121.8(4)	75.6(2)	0.7	213
[Mn(OOCCH(OH)CH₂COO)(H₂O)₂]·H₂O	Mn(II)	S	-1.2	2.192(3)	2.190(3)	.293	.318	117.1(2)	118.1(2)	74.2(1)	14.2	218
[Cu(OOCCH(OH)CH₂COOH)₂]·2H₂O	Cu(II)	S	+11.9	1.965(4)	1.905(4)	-.100	+.046	114.2(3)	115.2(4)	83.5(2)	1.7	219
	Cu(II)	S	+25.2	1.941(4)	1.939(4)	.310	.675	110.9(3)	114.1(4)	81.9(2)	24.3	
[Ni(OOCCH(OH)CH₂COOH)₂(H₂O)₂]·2H₂O	Ni(II)	S	-4.7	2.081(3)	2.014(3)	.085	.033	115.2(3)	119.3(3)	78.8(3)	2.6	220
[Ca(OOCCH(OH)CH₂COOH)₂(H₂O)₂]·2H₂O	Ca(II)	S	-8.4	2.523(1)	2.419(1)	.294	.220	119.8(1)	124.4(1)	64.3(1)	12.0	221
	Ca(II)	S	+14.7	2.434(1)	2.414(1)	-.002	+.201	124.6(1)	127.4(2)	63.0(1)	5.4	
K[B(OOCCH(O)CH₂COOH)₂]·H₂O	g B(III)	S	+10.1	1.429(3)	1.519(4)	.146	.312	110.1(3)	111.0(3)	103.7(3)	10.4	222
	g B(III)	S	+9.3	1.428(3)	1.517(3)	.066	.211	108.8(4)	109.0(5)	105.2(3)	6.5	

Compound	Ion											Ref
$[Zn(OOCCH(OH)CH_2COO)(H_2O)_2]\cdot H_2O$	Zn(II)	S	−26.8	2.170(2)	2.063(3)	.413	.779	106.8(2)	116.4(3)	76.1(1)	30.3	225
$[Co(OOCCH(OH)CH_2COO)(H_2O)_2]\cdot H_2O$	Co(II)	S	−27.3	2.136(3)	2.067(3)	.423	.793	106.3(2)	115.8(2)	76.82(12)	31.1	227
$Ca(OOCCH(OH)CH_2COO)\cdot 2H_2O$	Ca(II)	S	−15.8	2.465(7)	2.448(6)	.521	.757	110.5(4)	117.8(5)	65.8(2)	32.7	229
$(NH_4)_2[(MoO_2)_4O_3(OOCCH(O)CH_2COO)_2]\cdot 6H_2O$	g Mo(VI)	R	+1.6	2.00(4)	2.21(4)	.661	.736	109.7	117.6	75.4	33.5	230
$K(HOOCCH(OH)CH(COO)CH_2COOH)^e$	K(I)	R	−5.5	2.799(5)	2.814(5)	.817	.904	112.9(4)	109.5(4)	55.7(.2)	45.7	238
$Na(HOOCCH_2C(OH)(COO)CH_2COOH)$	Na(I)	--	16.9	2.451(3)	2.305(4)	.336	.590	113.8(2)	122.9(2)	67.3(1)	23.0	240
$Li(HOOCCH_2C(OH)(COO)CH_2COOH)$	Li(I)	--	19.4	2.22(2)	1.96(2)	.266	.556	109.6(7)	122.0(8)	76.2(7)	20.3	240
$RbH_2(OOCCH_2C(OH)(COO)CH_2COO)$	Rb(I)	--	15.1	3.00(7)	2.97(7)	.484	.689	118(5)	124(5)	54(2)	29.4	241
$[Mg(H_2O)_6][Mg(OOCCH_2C(OH)(COO)(COO)]$	Mg(II)	--	3.7	2.118(1)	2.081(1)	.708	.838	107.3(1)	111.3(1)	76.32(5)	38.5	242
$[Mn(H_2O)_6][Mn(OOCCH_2C(OH)(COO)CH_2COO)(H_2O)]_2\cdot 2H_2O$	Mn(II)	--	0.6	2.224(2)	2.194(2)	.740	.825	108.0(1)	110.2(1)	73.3(1)	39.4	243,244
$[Fe(H_2O)_6][Fe(OOCCH_2C(OH)(COO)CH_2COO)(H_2O)]_2\cdot 2H_2O$	Fe(III)	--	1.6	2.178(2)	2.130(2)	.732	.835	106.8(1)	110.5(2)	75.0(1)	39.2	245
$Ca(OOCCH_2C(OH)(COO)CH_2COOH)\cdot 3H_2O$	Ca(II)	--	16.3	2.490(4)	2.372(5)	−.167	+.033	122.0(2)	127.4(3)	63.9(1)	3.3	246
$[Co(OOCCH_2C(OH)(COO)CH_2COO)(C_6H_{18}N_4)]^k$	Co(III)	--	2.4	1.94	1.75	.094	.137	114	124	83	5.2	247a
	Co(III)	--	10.1	1.99	1.84	.105	.234	110	118	84	8.0	
$[Co(NH_3)_6][Sb(OOCCH_2C(O)(COO)CH_2COOH)_2]\cdot 5H_2O$	Sb(III)	--	3.4	2.00	2.16	.214	.281	118.4	114.5	77.6	11.7	247b
		--	1.4	1.99	2.16	.312	.325	117.7	113.7	77.9	14.9	

383

Table 5 (Continued)

Compound	M	Absol. config.[b]	Dihed. Angle	Distances, Å M-03	M-01	C1-plane 1	C2-plane 1	Angles, deg C1-03-M	C2-01-M	03-M-01	Plane 1/plane 2	Ref.
Cu₂(OOCCH₂C(O)(COO)CH₂COO)·2H₂O g	Cu(II)	--	18.9	1.925(8)	1.912(11)	.169	.437	110.9(7)	114.7(9)	84.7(4)	14.6	248
[N(CH₃)₄]₁₀[Ni₈(OOCCH₂C(O)(COO)CH₂COO)₆(OH)₂(H₂O)₂]·36H₂O g	Ni(II)	--	16.9	2.11(1)	2.01(1)	.198	.441	110.7(7)	116.4(9)	79.9(4)	16.0	245
	Ni(II)	--	30.9	2.05(1)	1.95(1)	.041	.438	107.2(6)	113.0(9)	84.3(4)	13.2	
g	Ni(II)	--	23.0	2.05(1)	2.01(1)	.284	.577	105.3(8)	111.3(8)	84.9(4)	21.9	
LiNH₄(OOCCH(OH)CH(OH)COO)·H₂O	Li(I)	R	- 1.6	2.199(4)	2.036(4)	.326	.386	113.2(1)	120.3(2)	75.3(1)	17.0	256
LiTℓ(OOCCH(OH)CH(OH)COO)·H₂O	Li(I)	R	- 2.9	2.190(10)	1.995(11)	.306	.382	111.6(2)	119.8(3)	77.0(3)	16.5	257
Na₂(OOCCH(OH)CH(OH)COO)·2H₂O	Na(I)	R	+18.0	2.82(3)	2.40(3)	.649	.523	109.3(1)	120.8(1)	62.0(1)	29.2	258
K(OOCCH(OH)CH(OH)COOH)	K(I)	S	- 8.1	2.766(1)	2.695(1)	.541	.532	116.3(1)	117.9(1)	59.97(2)	17.1	264
Rb(OOCCH(OH)CH(OH)COOH)	Rb(I)	R	+ 8.0	3.11	2.88	.339	.490	116	128	54	21.1	266
Cs(OOCCH(OH)CH(OH)COOH)	Cs(I)	R	+10.1	3.321(2)	3.008(2)	.332	.502	116.9(1)	130.4(1)	50.62(4)	21.4	267
Sr(OOCCH(OH)CH(OH)COO)·3H₂O	Sr(II)	R	+ 8.8	2.58(2)	2.66(2)	.225	.139	125	121	62	8.4	268
	Sr(II)	R	+12.7	2.55(2)	2.77(2)	.580	.532	117	121	60	26.8	
Ca(OOCCH(OH)CH(OH)COO)·4H₂O	Ca(II)	R	-25.6	2.54(1)	2.43(1)	.219	.557	115	126	63	19.6	268
	Ca(II)	R	- 1.4	2.51(1)	2.39(1)	.173	.169	119	125	65	7.9	
Na₅[Fe((d-C₄H₂O₆)(ℓ-C₄H₂O₆)]·14H₂O g	Fe(III)	R	+24.1	1.978(3)	2.149(3)	.771	1.062	100.3(1)	105.6(2)	77.5(1)	49.1	269

Compound												
Na₄[(VO)₂(d-C₄H₂O₆)(ℓ-C₄H₂O₆)]·12H₂O	g	V(IV)	R	+15.0	1.917(6)	2.004(6)	.271	.477	113.0(3)	113.7(3)	82.5(2)	17.8	290
	g	V(IV)	S	− 1.0	1.902(6)	1.994(6)	.234	.239	116.7(5)	116.7(6)	81.2(2)	10.9	
[N(C₂H₅)₄]₄[(VO)₂(d-C₄H₂O₆)(ℓ-C₄H₂O₆)]·8H₂O	g	V(IV)	S	−20.2	1.955(3)	2.021(3)	.308	.577	111.3(2)	114.2(2)	80.8(1)	21.7	·291
	g	V(IV)	R	+ 5.2	1.913(2)	2.026(2)	.306	.261	115.9(2)	114.1(2)	81.4(1)	13.3	
Na₄[(VO)₂(d-C₆H₆O₆)(ℓ-C₆H₆O₆)]·12H₂O	g	V(IV)	R	+23.7	1.964(3)	2.014(3)	.499	.824	106.8(1)	112.5(2)	80.6(1)	33.3	292
	g	V(IV)	S	+ 1.1	1.974(3)	1.981(2)	.526	.592	112.6(2)	114.8(2)	80.1(1)	26.3	
(NH₄)₄[(VO)₂(d-C₄H₂O₆)₂]·2H₂O	g	V(IV)	S	−19.9	1.93(2)	2.06(2)	-.170	+.069	115(1)	114(1)	81.0(6)	2.9	294
	g	V(IV)	S	+ 5.8	1.79(2)	2.01(2)	.230	.169	121(1)	109(1)	84.1(7)	9.0	
(NH₄)₂[Sb₂(d-C₄H₂O₆)₂]·3H₂O	g	Sb(III)	R	−20.5	2.02	2.12	.096	.450	117	110	85	12.7	295
	g	Sb(III)	R	+24.2	2.07	1.96	.146	.376	105	153	75	9.9	
	g	Sb(III)	R	−47.4	1.94	2.28	-.084	+.453	112	99	83	12.0	
	g	Sb(III)	R	+ 9.4	2.05	2.26	.121	.011	117	114	75	3.3	
dℓ-K₂[Sb₂(C₄H₂O₆)₂]·3H₂O	g	Sb(III)	S	+ 2.8	2.06(3)	2.25(3)	.039	.078	119(2)	115(2)	77(1)	2.8	296
	g	Sb(III)	S	− 2.6	2.01(3)	2.16(3)	.200	.173	114(2)	111(2)	81(1)	8.7	
	g	Sb(III)	S	+ 5.2	2.05(2)	2.17(3)	.141	.217	115(2)	113(2)	79(1)	8.5	
	g	Sb(III)	S	+ 1.0	1.91(3)	2.22(3)	.044	.036	118(2)	111(2)	80(1)	1.8	
K₂[Sb₂(d-C₄H₂O₆)₂]·3H₂O	g	Sb(III)	(S)	+ 1.6	2.00(1)	2.13(1)	.086	.112	116.7(8)	116.9(8)	78.6(4)	4.7	298
	g	Sb(III)	(S)	+ 2.0	2.01(1)	2.12(1)	.069	.094	115.4(9)	115.3(10)	79.5(5)	3.9	

Table 5 (Continued)

M	Absol. config.[b]	Dihed. Angle	M-O3	M-O1	C1-plane 1	C2-plane 1	C1-O3-M	C2-O1-M	O3-M-O1	Plane 1/plane 2	Ref.
			Distances, Å				Angles, deg.				
[Pt(C$_3$H$_{10}$N$_2$)(C$_4$H$_{12}$N$_2$)][Sb$_2$(d-C$_4$H$_2$O$_6$)$_2$]$_2$·2H$_2$O											299
g Sb(III)	(S)	− 2.7	1.97(1)	2.18(1)	.156	.143	119.3(7)	114.1(9)	77.8(4)	6.9	
g Sb(III)	(S)	+ 2.2	1.99(1)	2.21(1)	.108	.146	118.5(11)	114.5(12)	77.2(5)	6.2	
g Sb(III)	R	− 0.6	1.97(1)	2.14(1)	.112	.134	118.7(10)	115.0(11)	79.0(5)	5.7	
g Sb(III)	R	+ 4.2	1.96(1)	2.17(2)	.071	.022	118.4(10)	112.6(11)	80.0(5)	2.2	
g Sb(III)	R	+ 1.6	1.96(1)	2.18(1)	.036	.019	118.4(12)	113.8(12)	79.0(5)	1.2	
g Sb(III)	R	− 4.5	1.94(1)	2.15(2)	.069	.133	118.8(11)	115.2(13)	78.6(6)	4.8	
[Pt(NH$_3$)$_2$ C$_5$H$_{14}$N$_2$)][Sb$_2$(d-C$_4$H$_2$O$_6$)$_2$]·H$_2$O											300
g Sb(III)	R	− 2.1	1.96(1)	2.19(1)	.057	.035	118.5(8)	113.8(8)	78.4(4)	2.1	
g Sb(III)	R	− 3.3	1.97(1)	2.18(1)	.104	.157	120.1(8)	113.9(8)	78.5(4)	6.1	
g Sb(III)	R	− 6.5	1.99(1)	2.19(1)	.128	.223	117.7(8)	113.9(8)	78.8(4)	8.4	
g Sb(III)	R	+ 4.0	1.97(1)	2.12(1)	.052	.003	118.8(7)	116.5(8)	78.7(4)	1.3	
[Pt(NH$_3$)$_2$(C$_4$H$_{12}$N$_2$)][Sb$_2$(d-C$_4$H$_2$O$_6$)$_2$]·H$_2$O											301
g Sb(III)	R	− 1.2	1.99(2)	2.14(2)	.134	.157	117(1)	114(1)	79.3(7)	6.8	
g Sb(III)	R	− 5.4	1.93(2)	2.20(2)	−.058	+.005	122(1)	114(2)	77.2(7)	1.2	
g Sb(III)	R	− 5.9	1.98(2)	2.13(2)	.131	.215	117(1)	115(1)	79.9(7)	8.1	
g Sb(III)	R	+ 3.9	1.97(2)	2.19(2)	.112	.069	120(1)	115(1)	77.0(7)	4.2	

Formula	g	M	R/S										Ref
[Fe(C$_{12}$H$_8$N$_2$)$_3$][Sb$_2$(d-C$_4$H$_2$O$_6$)$_2$]·8H$_2$O	g	Sb(III)	R	− 2.7	1.94(1)	2.11(2)	.045	.084	117.3(7)	115.3(8)	79.7(6)	3.1	302
	g	Sb(III)	R	− 1.2	1.94(1)	2.16(2)	.050	.071	118.6(7)	113.0(7)	79.1(5)	2.9	
[Co(NCS)$_2$(C$_3$H$_{10}$N$_2$)$_2$]$_2$[Sb$_2$(d-C$_4$H$_2$O$_6$)$_2$]·4H$_2$O	g	Sb(III)	R	+ 5.8	2.02(1)	2.13(1)	.270	.216	116.8(9)	115.4(11)	78.9(4)	11.3	303
	g	Sb(III)	R	+ 5.4	2.00(1)	2.19(1)	.238	.182	119.1(10)	115.4(10)	77.7(4)	9.8	
[Cr(NCS)$_2$(C$_3$H$_{10}$N$_2$)$_2$]$_2$[Sb$_2$(d-C$_4$H$_2$O$_6$)$_2$]·4H$_2$O	g	Sb(III)	R	+ 8.9	1.997(5)	2.153(5)	.257	.157	118.0(4)	115.4(5)	78.2(2)	9.6	304
	g	Sb(III)	R	+ 3.3	1.989(5)	2.195(5)	.205	.180	119.3(4)	114.2(5)	77.8(2)	3.3	
d-α-(NH$_4$)$_2$[Sb$_2$(C$_4$H$_2$O$_6$)$_2$]·4H$_2$O	g	Sb(III)	S	+ 9.3	2.04	2.16	.135	.246	114	115	79	8.5	**251**
	g	Sb(III)	S	−10.9	2.04	2.14	− .023	+ .118	118	115	79	2.8	
[Co(C$_5$H$_7$O$_2$)(C$_3$H$_{10}$N$_2$)$_2$][As$_2$(d-C$_4$H$_2$O$_6$)$_2$]·H$_2$O	g	As(III)	(S)	− 6.0	1.79(2)	2.06(2)	.100	.024	121(2)	112(2)	85(1)	2.7	305
	g	As(III)	(S)	− 2.7	1.83(3)	2.11(2)	.284	.275	117(2)	110(2)	84(1)	13.0	
	g	As(III)	(S)	+ 3.3	1.82(2)	2.01(2)	.182	.247	115(2)	113(2)	85(1)	9.7	
	g	As(III)	(S)	− 6.6	1.77(2)	1.99(2)	.090	.008	117(2)	111(2)	87(1)	2.3	
Na$_4$[Cu(d-C$_4$H$_2$O$_6$)(α-C$_4$H$_2$O$_6$)]·10H$_2$O	g	Cu(II)	R	+18.7	1.915(3)	1.933(2)	.304	.555	108.4(2)	111.7(2)	85.8(1)	20.9	307
	g	Cu(II)	R	+19.5	1.890(2)	1.960(3)	.422	.683	107.3(2)	110.5(2)	85.2(1)	27.1	
Cu(d-C$_4$H$_4$O$_6$)·3H$_2$O	g	Cu(II)	R	− 4.3	2.40(1)	1.93(2)	.042	.101	108(1)	122(1)	77.2(6)	3.6	308
	g	Cu(II)	R	+ 9.7	2.00(2)	1.92(2)	.117	.002	114(1)	118(2)	82.1(7)	2.7	
	g	Cu(II)	R	+ 4.2	2.01(1)	1.92(2)	.060	.126	114(1)	119(1)	80.9(6)	4.4	

Table 5 (Continued)

	M	Absol. config.	Dihed.[b] Angle	Distances, Å				Angles, deg				Ref.
				M-O3	M-O1	C1-plane 1	C2-plane 1	C1-O3-M	C2-O1-M	O3-M-O1	Plane 1/plane 2	
Cu(ms-$C_4H_4O_6$)·$3H_2O$	Cu(II)	R	− 2.0	2.43(2)	1.89(2)	.039	.017	106(1)	126(1)	75.9(6)	1.3	308
	Cu(II)	R	− 6.9	1.955(2)	1.918(3)	.130	.235	112.5(2)	115.5(2)	84.3(1)	8.6	
	Cu(II)	S	+ 9.1	1.975(3)	1.920(3)	.249	.391	111.5(2)	116.3(3)	83.5(1)	15.0	
Ca(OOCCH(OH)CH_2OH)$_2$·$2H_2O$	Ca(II)	S	− 1.2	2.389(2)	2.355(2)	.351	.377	120.3(2)	122.9(2)	65.8(1)	17.2	317
	Ca(II)	S	+ 1.0	2.411(2)	2.371(2)	.104	.132	122.9(2)	125.4(2)	64.8(1)	5.5	
Cd(OOCCH(OH)CH_2OPO_3H)·$3H_2O$	Cd(II)	S	+ 3.2	2.38(2)	2.27(2)	.273	.327	121(1)	127(2)	66.4(6)	13.6	320
[Cu(OOCC(CH_3)(OH)(CH_2OH)$_2$(H_2O)$_2$]	Cu(II)	R	− 5.3	1.974(8)	1.890(8)	.080	.018	114.5(6)	116.9(7)	83.5(3)	2.2	321
K(OOCCH(OH)CH(OH)CH(OH)CH(OH)CH_2OH)·H_2O	K(I)	R	+ 5.3	3.003(6)	2.685(6)	.514	.513	114.9(2)	127.1(3)	55.7(1)	25.4	324
Mn(OOCCH(OH)CH(OH)CH(OH)CH(OH)CH(OH)CH_2OH)$_2$·$2H_2O$	Mn(II)	R	+ 3.9	2.258(6)	2.136(6)	.085	.150	117.9(5)	123.1(5)	72.1(2)	5.6	326
Na_3(OOCCH(OH)CH(OH)CH(OH)CH(OH)CH_2OPO_3)·$2H_2O$	Na(I)	R	+15.3	2.412(2)	2.423(3)	.728	.939	102.8(1)	109.5(2)	68.5(1)	44.8	329
Ca($C_{12}H_{21}O_{12}$)Br·$4H_2O$	Ca(II)	R	+12.0	2.480(5)	2.456(5)	.716	.905	106.9(4)	113.1(4)	65.4(2)	42.6	330
Ca(OOCCH(OH)CH(OH)CH(OH)CH(OH)COO)·$4H_2O$	Ca(II)	R	+ 0.7	2.445(10)	2.478(11)	.094	.095	123.3(6)	126.0(8)	64.1(3)	4.3	332
Ca(OOCCH(OH)CH(OH)CH(OH)CH_2OH)$_2$·$5H_2O$	Ca(II)	(R)	+24.6	2.45	2.47	.602	.922	107	115	65	41.3	333

333
334
335

Compound	Cation	Config.	Opt. Rot.	d	d	r	r	Angle	Angle	Angle	Angle	Ref
$Sr(OOCCH(OH)CH(OH)CH(OH)CH_2OH)_2 \cdot 5H_2O$	Sr(II)	(R)	+16.1	2.60	2.59	.653	.926	111	111	61	43.7	333
$Ca(OOCC(OH)(CH_2OH)CH_2CH(OH)CH_2OH)_2$	Ca(II)	S	+22.2	2.44(1)	2.38(2)	.373	.151	121.4(9)	122.1(15)	65.3(5)	11.6	334
$Sr(OOCC(OH)(CH_2OH)CH_2CH(OH)CH_2OH)_2$	Sr(II)	S	+13.8	2.555(7)	2.493(8)	.315	.178	122.9(8)	125.0(9)	62.1(3)	11.2	335

Average Values[m]

Unionized Hydroxyl Group

Compound	d	d	r	r	Angle	Angle	Angle	Angle
Mn(II), 3	2.22(2)	2.17(2)	.37(24)	.43(26)	114.3(42)	117.1(46)	73.2(7)	19.7(13)
Fe(II), 1	2.178	2.130	.73	.84	106.8	110.5	75.0	39.2
Co(II), 1	2.136	2.067	.42	.79	106.3	115.8	76.8	31.1
Co(III), 2	1.96(2)	2.03(2)	.10(1)	.19(5)	112(2)	121(3)	83.5(5)	6.6(14)
Ni(II), 2	2.08(0)	1.92(1)	.05(4)	.02(1)	115.2(0)	118.3(10)	78.9(1)	3.2(16)
[n]Cu(II), 11	1.97(2)	1.96(4)	.13(7)	.17(16)	114.2(15)	116.6(11)	82.6(9)	7.2(56)
[n]Cu(II), 5	2.34(6)	2.01(5)	.16(13)	.20(16)	109.4(19)	123.0(16)	76.0(5)	8.8(71)
Li(I), 5	2.20(8)	2.38(5)	.20(12)	.31(16)	112.0(30)	120.3(23)	76.6(9)	12.5(66)
Na(I), 3	2.56(17)	2.75(7)	.57(16)	.68(17)	108.6(39)	117.7(55)	65.9(26)	32.3(83)
K(I), 6	2.89(11)	2.95(5)	.33(29)	.40(25)	117.5(28)	122.8(67)	56.9(16)	17.4(121)
Rb(I), 3	3.00(8)	3.01	.35(9)	.45(18)	119.1(28)	124.1(26)	54.6(8)	20.1(68)
Cs(I), 1	3.32	2.08(1)	.33	.50	116.9	130.4	50.6	21.4
Mg(II), 2	2.12(0)	2.42(3)	.35(36)	.43(40)	112.4(51)	116.6(52)	76.0(3)	19.6(189)
Ca(II), 13	2.46(4)	2.63(9)	.28(18)	.36(26)	117.9(49)	122.0(42)	64.7(8)	16.1(112)
Sr(II), 4	2.57(2)		.44(17)	.44(28)	119.0(50)	119.5(42)	61.3(8)	22.5(127)

Table 5 (Continued)

M	Absol. config.b	Dihed. Angle	Distances, Å				Angles, deg.				
			M-O3	M-O1	C1-plane 1	C2-plane 1	C1-O3-M	C2-O1-M	O3-M-O1	Plane 1/plane 2	Ref.
Zn(II), 5			2.14(4)	2.09(3)	.14(11)	.26(21)	114.1(35)	118.4(16)	76.7(8)	9.9	(81)
Cd(II), 1			2.38	2.27	.27	.33	121	127	66.4	13.6	
Sc(III), 3			2.24(2)	2.18(1)	.27(15)	.38(16)	119.7(11)	121.0(27)	68.2(5)	15.4	(72)
La(III), 3			2.58(1)	2.55(3)	.27(16)	.29(23)	119.9(17)	123.0(16)	61.9(13)	13.6	(78)
Eu(III), 3			2.62(5)	2.46(5)	.42(8)	.51(6)	116.3(4)	120.3(22)	64.6(13)	22.1	(33)
Gd(III), 3			2.47(1)	2.45(3)	.29(12)	.52(14)	117.9(7)	122.5(25)	64.3(5)	20.2	(35)
Er(III), 4			2.36(6)	2.30(2)	.24(11)	.35(12)	119.3(22)	125.3(32)	66.2(14)	14.1	(55)
Deprotonated Hydroxyl Group											
V(IV), 10			1.92(3)	2.01(2)	.29(10)	.35(22)	114.8(31)	114.1(15)	81.5(8)	15.3	(76)
Cr(V), 2			1.78(1)	1.91(1)	.12(11)	.23(10)	117.5(4)	115.3(6)	83.9(1)	8.4	(46)
Fe(III), 1			1.98	2.15	.77	1.01	100.3	105.6	77.5	49.1	
Ni(II), 3			2.07(3)	1.99(3)	.17(9)	.48(6)	107.7(2)	113.6(1.9)	83.0(21)	17.0	(32)
Cu(II), 3			1.91(1)	1.94(2)	.30(9)	.56(8)	108.9(14)	112.3(16)	85.2(4)	20.9	(42)
Mo(VI), 1			2.00	2.21	.66	.74	109.7	117.6	75.4	33.5	
Er(III), 1			2.24	2.37	.06	.02	121.4	121.4	68.9	1.8	
As(III), 4			1.80(2)	2.04(4)	.16(7)	.13(12)	117.5(18)	111.5(10)	85.2(9)	6.9	(44)

Sb(III), 34	1.99(3)	2.16(4)	.11(6)	.16(9)	117.3(18)	114.9(26)	78.7(12)	6.2(32)
B(III), 2	1.43(0)	1.52(0)	.11(4)	.26(5)	109.4(6)	110.0(10)	104.4(8)	8.4(20)
C, 1	1.48	1.44	.37	.28	106	111	104	13.9
Ge, 1	1.81	1.86	.11	.15	113	113	88.8	5.9

[a]Underlined distances, angles, and esd's were calculated from reported coordinates.

[b]Absolute configuration of carbinol carbon. Parentheses denote that the absolute configuration calculated from the reported coordinates is opposite to that which the compound should have.

[c]The c cell dimension appears to be in error. A value of 5.79 rather than 5.97 was used.

[d]Difficulty was encountered in reproducing the bond parameters using the reported cell dimensions and atom coordinates.

[e]Protonated carboxyl group.

[f]For Λ absolute configuration complex.

[g]Ionized hydroxyl group.

[h]For the absolute configuration of the complex pictured in the text.

[i]For the Δ absolute configuration complex.

[j]The "absolute configuration" was established by setting the priority of the carboxyl carbon not in the α-hydroxylcarboxylate group under consideration equal to that of a methyl group.

[k]Structure is still being refined. There are two molecules per unit cell.

[l]Very low precision structure.

[m]The number of values averaged is given after the symbol for each ion. The values in parentheses are the average absolute deviations where two or more values are averaged.

[n]The first values listed for Cu(II) are for these structures where neither coordinated oxygen atom of the α-hydroxycarboxylate is in an axial site of an axially elongated octahedron. The second set of values is for structures where an oxygen atom (always a hydroxyl oxygen) is located in such a site.

Table 6 Geometries of α-Alkoxycarboxylates and Related Compounds[a]

Structure: R–O3 / C1 / C2 with O1 and O2(H,R)

Absolute config.[b]	Dihed. angle	Distances, Å					Angles, deg				Ref.
		C1-C2	C2-O1	C2-O2	C1-O3	O1-O3	C2-C1-O3	C1-C2-O1	O1-C2-O2	C1-C2-O2	
						Uncoordinated Groups					
CH3OC6H4OCH(CH3)COOH, (+)-m-methoxyphenoxypropionic acid											
S	-25.7	1.51(2)	1.20(2)	1.34(2)	1.42(2)	2.75(2)	108(3)	127(3)	119(2)	113(4)	339
BrC6H4OCH(CH3)COOH, (-)-m-bromophenoxypropionic acid											
R	+27.4	1.50(4)	1.27(6)	1.31(4)	1.45(4)	2.76(2)	108(3)	124(3)	120(2)	116(3)	339
Cl2C6H3OCH2COOH, 2,5-dichlorophenoxyacetic acid											
--	4.0	1.491(6)	1.208(5)	1.325(4)	1.401(4)	2.710(3)	111.9(2)	124.1(2)	123.8(3)	112.0(2)	340
Cl2C6H3OCH2COOH, 2,4-dichlorophenoxyacetic acid											
--	3.7	1.520(6)	1.217(6)	1.304(5)	1.423(5)	2.717(2)	111.1(3)	123.2(3)	124.1(2)	112.2(2)	341
Cl3C6H2OCH(CH3)COOH, (±)-2-(2,4,5-trichlorophenoxy)propionic acid											
S	-16.3	1.510(4)	1.213(4)	1.311(4)	1.438(4)	2.730(3)	110.4(3)	123.9(3)	124.1(3)	111.9(3)	342
Cl3C6H2OCH2COOH, 2,4,6-trichlorophenoxyacetic acid											
--	32.2	1.488(6)	1.235(5)	1.262(5)	1.435(5)	2.683(6)	109.3(3)	120.9(4)	123.3(5)	115.7(4)	343a
Cl2C6H3OCH(CH3)COOH, (±)-2-(3,5-dichlorophenoxy)propionic acid											
R	33.4	1.504(8)	1.237(7)	1.276(7)	1.435(6)	2.745	110.7(4)	120.7(6)	124.0(6)	115.3(5)	343b

Cl_3C_6H_2OCH_2COOH, 2,4,5-trichlorophenoxyacetic acid

--	0.1	1.517(7)	1.213(7)	1.301(8)	1.417(8)	2.605(7)	106.7(3)	122.9(4)	124.4(4)	112.7(3)	344

$C_6H_5OC_6H_4OC(CH_3)_2COOH$, 2-[2-(phenoxy)phenoxy]-2-methylpropionic acid

d	--	35.2	1.519(3)	1.286(3)	1.239(2)	1.440(2)	2.708(3)	110.5(2)	116.3(2)	123.5(2)	120.0(2)	345

$C_6H_4OC_6H_3OC(CH_3)_2COOH$, 2-(4-dibenzofuranyloxy)-2-methylpropionic acid

d	--	35.9	1.510(4)	1.309(3)	1.224(4)	1.446(4)	2.718(3)	111.5(3)	115.3(3)	123.3(2)	121.3(3)	346

$HOOCCH_2OCH_2COOH$, oxydiacetic acid

--	4.3	1.510(2)	1.226(2)	1.300(3)	1.413(3)	2.723(1)	112.4(2)	122.6(2)	125.0(1)	112.5(2)	347

Cyclic Ethers

$(C_{10}H_{17}O_8)$-0-CHCH(OH)CH(OH)CH(OCH_3)CH(COOH)·3H_2O, an aldotriuronic acid trihydrate

S	-32.7	1.528(7)	1.202(b)	1.331(6)	1.430(6)	2.657(6)	104.9(4)	122.7(4)	125.2(4)	112.1(4)	348

$(C_7H_{13}NO_4)$-0-CHCH(OH)CH(OH)CH(OH)CH(COO)·H_2O, 2-amino-2-deoxy-3-0-β-D-glucopyronosyl-D-galactose monohydrate

S	+13.7	1.50(2)	1.26(1)	1.27(1)	1.45(1)	2.55(1)	105.8(9)	119.9(9)	125.6(10)	114.5(9)	349

$(C_4H_2N_3O_4)$CHCH(OH)CH(OH)CH(OH)CH(COOH)·H_2O, 5-nitro-1-(β-D-ribosyluronic acid)-uracil monohydrate

d	S	-20.4	1.531(6)	1.302(6)	1.209(6)	1.451(6)	2.678(5)	111.8(4)	115.8(4)	124.9(4)	119.4(4)	350

$(C_4N_4N_3O)$CHCH(OH)CH(OH)CH(NH_2)CH(COOH)·H_2O, 4-amino-1-[4-amino-2-oxo-1(2H)-pyrimidinyl]-1,4-dideoxy-β-D-glucopyranuronic acid monohydrate

S	+0.9	1.540(6)	1.237(6)	1.256(6)	1.428(6)	2.702(5)	111.4(4)	120.5(4)	126.1(4)	113.4(4)	351
S	+54.8	1.535(6)	1.253(6)	1.256(6)	1.427(6)	2.698(6)	104.0(4)	115.9(4)	126.2(4)	118.0(4)	

Table 6 (Continued)

Absolute config.b	Dihed. angle	Distances, Å					Angles, deg				Ref.
		C1-C2	C2-O1	C2-O2	C1-O3	O1-O3	C2-C1-O3	C1-C2-O1	O1-C2-O2	C1-C2-O2	
HOCH₂CH(OH)CH(OH)CHCH(NHCOCH₃)CH(OH)CH₂C(OH)COOH·2H₂O, β-D-N-acetylneuramic acid dihydrate[c]											
d S	-51.7	1.531(6)	1.290(7)	1.198(7)	1.420(6)	2.699(4)	106.0(4)	114.4(4)	122.7(5)	122.9(5)	81
CH₂CH(OH)CH(OH)CH(OH)C(OH)COOH·H₂O, 2-keto-L-gulonic acid monohydrate[c]											
d R	+52.7	1.539(3)	1.308(3)	1.197(2)	1.427(3)	2.685(2)	105.8(2)	112.3(2)	125.0(2)	122.7(2)	80
Esterified Carboxyl Group											
CH₃OOCC(OH)CH₂C(OH)CH₂CH(OH)CH(HNC(O)CH₃)CH(C₃H₇O₃)·H₂O, N-acetylneuramic acid methyl ester monohydrate											
d S	-87.0	1.541(3)	1.317(3)	1.192(3)	1.421(3)	2.970(3)	102.5(2)	113.3(2)	124.2(2)	122.5(2)	100
CH₃OOCHCH(OOCCH₃)CH(OOCCH₃)CH(OOCCH₃)CH-O-(C₂₃H₂₉O₅), 18R-α-D-glucopyranosiduronic acid tetraacetate methyl ester											
S	-76.3	1.518(5)	1.178(5)	1.304(5)	1.419(4)	2.966(5)	105.4(3)	125.2(4)	125.2(4)	109.5(2)	352
CH₃OOCCHCH(OOCCH₃)CH(OOCCH₃)CH(OOCCH₃)CH(OOCCH₃), methyl 1,2,3,4-tetra-0-acetyl-β-D-galactropyranuronate											
S	+5.4	1.514(3)	1.198(3)	1.325(2)	1.429(2)	2.632(1)	106.2(2)	125.0(2)	124.7(2)	110.4(2)	353
CH₃OOCC(CH₂COCH₃)C(C₂H₅)C(OCH₃)CO, methyl-2-acetonyl-3-ethyl-4-methoxy-5-oxo-dihydro-2H-furan-2-carboxylate											
S	-29.6	1.536(3)	1.196(3)	1.326(3)	1.443(2)	2.711(2)	106.8(2)	123.7(2)	125.7(2)	110.4(2)	354

$(C_9H_{11}N_2O_6)$-O-$CH_2COOCH_3 \cdot H_2O$, uridine-5-oxyacetic acid methyl ester monohydrate

$CH_3OOCC(CH_2COOCH_3)N(CH_3)CH(CH_3)CH(C_6H_5)$, 2-methoxycarbonyl-3,4-dimethyl-5-phenyl-1,3-oxazolidine-2-acetate

Metal Derivatives

$[Cu(CH_3OCH_2COO)_2(H_2O)_2]$, diaquobis(methoxyacetato)copper(II)

$[Ni(CH_3OCH_2COO)_2(H_2O)_2]$, diaquobis(methoxyacetato)nickel(II)

$[Cu(CH_3OCH_2COO)_2(C_5H_5N)_2] \cdot 4H_2O$, bis(methoxyacetato)bis(pyridine)copper(II) tetrahydrate

$[Cu(CH_3OCH_2COO)_2(C_3N_2H_4)_4]$, bis(methoxyacetato)tetrakis(imidazole)copper(II)

$[Cu(C_2H_5OCH_2COO)_2(H_2O)_2]$, diaquobis(ethoxyacetato)copper(II)

$Na(C_6H_5OCH_2COO) \cdot \tfrac{1}{2} H_2O$, sodium phenoxyacetate hemihydrate

$[Cu(C_6H_5OCH_2COO)_2(H_2O)_2]$, diaquobis(phenoxyacetato)copper(II)

											Ref.
--	4.3	1.514(7)	1.181(7)	1.329(7)	1.428(6)	2.656(5)	106.9(4)	126.0(5)	125.8(5)	108.2(5)	355
R	-27.7	1.544(6)	1.187(5)	1.339(5)	1.411(5)	2.751(4)	108.2(3)	125.2(4)	124.1(4)	110.6(3)	356
e	0.2	1.53(2)	1.24(2)	1.28(2)	1.44(2)	2.61(2)	107.1(11)	121.2(12)	124.7(12)	114.0(13)	167
e	0.1	1.49(2)	1.21(2)	1.24(2)	1.46(2)	2.60(1)	109.4(13)	119.3(14)	125.2(14)	115.5(13)	361
e	15.7	1.50(2)	1.24(1)	1.25(1)	1.41(1)	2.66(1)	110(1)	121(1)	123(1)	116(1)	362
e	6.8	1.57(1)	1.21(1)	1.26(1)	1.42(2)	2.69(1)	114.8(10)	114.4(8)	126.5(9)	119.1(8)	363
e	8.8	1.51(2)	1.27(1)	1.26(1)	1.44(1)	2.68(1)	111.1(8)	120.2(9)	123.7(10)	116.1(9)	364
e	2.4	1.542(11)	1.276(10)	1.240(10)	1.430(10)	2.657(8)	111.2(6)	117.5(7)	128.3(8)	114.2(7)	365
e	9.3	1.52(2)	1.28(2)	1.26(2)	1.43(2)	2.70(1)	110.2(11)	121.5(12)	120.9(12)	117.6(12)	167
e	9.8	1.54(2)	1.27(2)	1.24(2)	1.44(2)	2.76(3)	110.1(11)	122.4(12)	121.1(12)	116.4(12)	
e	5.9	1.51(2)	1.29(2)	1.26(2)	1.42(2)	2.70(3)	110.6(12)	121.3(12)	121.4(12)	117.3(12)	

Table 6 (Continued)

Absolute config.[b]	Dihed. angle	Distances, Å					Angles, deg				Ref.
		C1-C2	C2-O1	C2-O2	C1-O3	O1-O3	C2-C1-O3	C1-C2-O1	O1-C2-O2	C1-C2-O2	
[Cu(CH₃OC₆H₄OCH₂COO)₂(H₂O)₂], diaquobis(p-methoxyphenoxyacetato)copper(II)											
e	1.7	1.512(5)	1.264(5)	1.237(5)	1.421(4)	2.691(4)	111.2(3)	121.1(3)	124.6(3)	114.3(3)	366
Cu(C₆H₅OCH₂COO)₂, copper(II) phenoxyacetate											
--	5.5	1.54(2)	1.27(2)	1.24(2)	1.46(2)	2.70(1)	112	117	126	116	367
e	17.4	1.51(2)	1.21(2)	1.27(2)	1.43(2)	2.63(1)	108	121	124	113	
Cu₂(Cl₂C₆H₃OCH₂COO)₄(C₄H₈O₂)₂, tetra-μ-(2,4-dichlorophenoxyacetato)bis(dioxanecopper(II))											
--	0.6	1.52(1)	1.25(1)	1.24(1)	1.41(1)	2.675(6)	113(1)	118(1)	127(1)	115(1)	368
--	5.0	1.49(1)	1.27(1)	1.25(1)	1.43(1)	2.621(6)	109(1)	121(1)	126(1)	113(1)	
[Cu(CH₃OC₆H₄OCH₂COO)₂(H₂O)₂]·2H₂O, diaquobis(p-methoxyphenoxyacetato)copper(II) dihydrate											
--	21.5	1.52(1)	1.30(2)	1.22(2)	1.42(2)	2.62(1)	109(1)	117(1)	126(1)	117(1)	366
[Cu(C₆H₅OCH₂COO)₂(H₂O)₂]₃, triaquobis(phenoxyacetato)copper(II)											
--	3.7	1.525(25)	1.225(21)	1.250(21)	1.432(22)	2.606(20)	107.9(14)	120.4(16)	124.2(18)	115.4(15)	369
--	7.7	1.541(26)	1.200(20)	1.331(20)	1.450(22)	2.648(20)	105.8(15)	124.4(16)	123.0(17)	112.6(15)	
--	3.5	1.570(25)	1.231(21)	1.249(20)	1.447(22)	2.660(21)	106.7(14)	122.0(16)	124.8(18)	113.3(15)	
--	3.6	1.562(26)	1.221(21)	1.313(21)	1.416(22)	2.680(19)	107.7(15)	123.1(16)	124.3(17)	112.5(16)	
[Cu(C₆H₅OCH₂COO)₂(C₅H₅N)₂(H₂O)], aquobis(phenoxyacetato)bis(pyridine)copper(II)											
--	11.9	1.52(3)	1.21(2)	1.32(2)	1.44(2)	2.63(2)	107(2)	123(2)	128(2)	109(2)	362

Na(HOOCCH$_2$OCH$_2$COO), sodium oxydiacetate(1-)											
--	0.6	1.52(3)	1.24(3)	1.28(2)	1.52(3)	2.72(2)	110(2)	123(2)	128(2)	109(2)	
e	5.2	1.506(6)	1.235(4)	1.277(4)	1.415(4)	2.643(3)	110.1(3)	120.9(3)	126.4(3)	112.7(3)	372
e,i	14.1	1.513(4)	1.223(4)	1.286(4)	1.414(4)	2.659(3)	109.7(3)	121.5(3)	126.1(3)	112.3(3)	
K(HOOCCH$_2$OCH$_2$COO), potassium oxydiacetate(1-)											
e	2.3	1.514(3)	1.226(3)	1.279(3)	1.416(3)	2.660(2)	110.2(2)	121.5(2)	126.0(2)	112.5(2)	372
e,i	9.9	1.507(3)	1.217(3)	1.294(3)	1.415(3)	2.656(2)	109.7(2)	122.1(2)	125.0(2)	112.9(2)	
Li(HOOCCH$_2$OCH$_2$COO), lithium oxydiacetate(1-)											
i	5.8	1.510(2)	1.203(2)	1.304(2)	1.410(2)	2.775(2)	114.1(2)	123.9(2)	125.6(2)	110.5(1)	373
e	8.9	1.508(2)	1.237(2)	1.260(2)	1.417(2)	2.572(2)	109.1(1)	118.1(2)	125.6(2)	116.4(1)	
Rb(HOOCCH$_2$OCH$_2$COO), rubidium oxydiacetate(1-)											
--	1.5	1.510(2)	1.219(2)	1.289(2)	1.410(2)	2.659(2)	110.18(12)	122.02(11)	125.80(13)	112.17(12)	374
Cd(OOCCH$_2$OCH$_2$COO)·3H$_2$O, cadmium(II) oxydiacetate(2-) trihydrate (orthorhombic)											
e	4.4	1.48(2)	1.24(2)	1.27(2)	1.47(2)	2.61(2)	105.6(14)	124.5(16)	122.1(17)	113.2(15)	376
--	1.5	1.51(3)	1.22(2)	1.27(3)	1.45(2)	2.76(2)	112.4(14)	123.4(19)	122.1(19)	114.3(17)	
[Cd$_2${OOCCH$_2$OCH$_2$COO)$_2$(H$_2$O)$_6$], bis(triaquooxydiacetato(2-)cadmium(II))											
e	1.4	1.44(1)	1.26(1)	1.29(1)	1.43(1)	2.61(1)	110.0(9)	121.6(8)	123.8(9)	114.6(9)	378
e	6.0	1.51(1)	1.24(1)	1.25(1)	1.38(1)	2.61(1)	110.8(8)	119.0(9)	125.4(9)	115.7(9)	
Cd(OOCCH$_2$OCH$_2$COO)·3$\frac{1}{2}$H$_2$O, cadmium(II) oxydiacetate(2-) 3$\frac{1}{2}$-hydrate											
e	10.4	1.51(3)	1.20(3)	1.27(2)	1.36(3)	2.59(2)	112(2)	118(2)	125(2)	117(2)	379

Table 6 (Continued)

Absolute config.b	Dihed. angle	Distances, Å					Angles, deg				Ref.
		C1-C2	C2-O1	C2-O2	C1-O3	O1-O3	C2-C1-O3	C1-C2-O1	O1-C2-O2	C1-C2-O2	
e	--	0.3	1.45(3)	1.33(3)	1.25(2)	1.40(2)	2.59(2)	107(2)	123(2)	117(2)	121(2)	
e	--	0.6	1.49(2)	1.23(3)	1.31(2)	1.45(2)	2.58(2)	106(2)	123(2)	126(2)	111(2)	380
e	--	4.1	1.49(3)	1.26(3)	1.18(2)	1.48(2)	2.58(2)	108(2)	119(2)	123(2)	118(2)	
[Ca(OOCCH2OCH2COO)(H2O)5]·H2O, pentaquooxydiacetato(2-)calcium(II) monohydrate											
e	--	0.8	1.513(5)	1.254(5)	1.268(6)	1.403(5)	2.602(4)	109.0(3)	119.7(3)	126.0(4)	114.3(3)	381
e	--	2.1	1.505(6)	1.267(5)	1.250(6)	1.447(7)	2.592(3)	109.2(3)	118.2(3)	125.5(3)	116.3(3)	
Na2[UO2(OOCCH2OCH2COO)2]·2H2O, sodium bis(oxyacetato(2-),dioxouranium(VI) dihydrate[f]											
e	--	--	--	1.20(3)	1.26(3)	1.49(3)	--	104(2)	--	--	--	
e	--	--	--	1.22(3)	1.29(3)	1.44(3)	--	107(2)	--	--	--	
UO2(OOCCH2OCH2COO), uranyl(VI) oxydiacetate(2-)											
e	--	6.3	1.47(3)	1.29(2)	1.23(2)	1.42(2)	2.54(2)	109(1)	118(2)	120(2)	121(2)	382
Cu(OOCCH2OCH2COO)·1/2 H2O, copper(II) oxydiacetate(2-) hemihydrate											
e	--	10.1	1.50(2)	1.25(2)	1.26(1)	1.39(2)	2.70(1)	112.8(10)	120.8(11)	122.4(11)	116.7(11)	383
e	--	11.9	1.48(2)	1.23(1)	1.26(1)	1.45(1)	2.73(1)	112.7(9)	122.2(10)	122.7(10)	115.1(10)	
Na3[Gd(OOCCH2OCH2COO)3]·2NaClO4·6H2O, sodium tris(oxydiacetato(2-)gadolinate(III) di(sodium perchlorate) hexahydrate											
e	--	6.0	1.51(3)	1.26(3)	1.24(3)	1.40(2)	2.58(2)	110(2)	118(2)	122(2)	121(2)	384

Na₃[Nd(OOCCH₂OCH₂COO)₃]·2NaClO₄·6H₂O, sodium tris(oxydiacetato(2-))neodynate(III) di(sodium perchlorate) hexahydrate

e	1.8	1.56(2)	1.23(2)	1.28(2)	1.45(1)	2.58(1)	105.5(10)	119.8(12)	125.9(13)	113.7(15)	385

Na₃[Yb(OOCCH₂OCH₂COO)₃]·2NaClO₄·6H₂O, sodium tris(oxydiacetato(2-))ytterbate(III) di(sodium perchlorate) hexahydrate

e	5.1	1.503(16)	1.264(16)	1.249(16)	1.419(11)	2.546(8)	108.3(9)	117.5(9)	124.3(10)	118.1(12)	385

Na₃[Ce(OOCCH₂OCH₂COO)₃]·2NaClO₄·6H₂O, sodium tris(oxydiacetato(2-))cerate(III) di(sodium perchlorate) hexahydrate

e	4.2	1.541(11)	1.241(7)	1.234(6)	1.423(7)	2.602(5)	110.3(6)	116.5(5)	125.3(8)	117.9(6)	386

Er[Er(OOCCH₂OCH₂COO)₃]·6H₂O, erbium(III) tris(oxydiacetato(2-))erbate(III) hexahydrate

e	8.6	1.49(5)	1.17(3)	1.28(3)	1.48(6)	2.46(3)	106(4)	117(3)	130(3)	112(3)	388
e	2.6	1.43(5)	1.25(4)	1.40(5)	1.42(4)	2.57(3)	114(3)	117(3)	118(3)	120(3)	
e	1.9	1.54(5)	1.25(4)	1.22(4)	1.32(4)	2.54(2)	108(3)	119(3)	126(3)	116(3)	

Na₃[Ce(OOCCH₂OCH₂COO)₃]·9H₂O, sodium tris(oxydiacetato(2-))cerate(III) 9-hydrate

e	13.2	1.506(5)	1.271(5)	1.242(5)	1.413(5)	2.612(5)	111.2(3)	116.9(3)	125.1(4)	118.0(3)	389
e	10.6	1.526(5)	1.250(5)	1.243(5)	1.409(5)	2.608(5)	109.7(3)	118.1(4)	125.6(4)	116.4(3)	
e	0.2	1.520(6)	1.270(5)	1.225(6)	1.406(6)	2.587(6)	110.3(4)	116.8(4)	126.0(5)	117.2(4)	
e	3.8	1.528(5)	1.254(6)	1.240(5)	1.408(6)	2.589(6)	110.1(3)	117.0(3)	126.5(4)	116.5(4)	
e	6.7	1.517(5)	1.270(5)	1.229(5)	1.429(5)	2.585(5)	109.2(3)	117.4(3)	125.0(4)	117.6(3)	
e	13.2	1.511(5)	1.269(5)	1.240(5)	1.414(5)	2.579(6)	109.9(3)	116.4(3)	125.7(4)	118.0(3)	

K(OOCCHCH(OH)CH(OH)CH(OH)CHOH)·2H₂O, potassium D-glucuronate(1-) dihydrate

S	-35.3	1.51(2)	1.28(1)	1.24(1)	1.47(1)	2.72	108(1)	119(1)	125(1)	115(1)	391

Rb(OOCCHCH(OH)CH(OH)CH(OH)CHOH)·2H₂O, rubidium D-glucuronate(1-) dihydrate

Table 6 (Continued)

Absolute config.[b]	Dihed. angle	Distances, Å					Angles, deg				Ref.
		C1-C2	C2-O1	C2-O2	C1-O3	O1-O3	C2-C1-O3	C1-C2-O1	O1-C2-O2	C1-C2-O2	
S	-34.8	1.49	1.31	1.22	1.43	2.64	111	113	128	116	392

Ca(OOCCHCH(OH)(OH)CH(OH)CHOH)Br·3H$_2$O, calcium D-glucuronate(1-) bromide trihydrate

e S	+2.9	1.524(5)	1.245(5)	1.262(5)	1.425(5)	2.603(4)	109.2(3)	118.6(4)	125.6(4)	115.8(4)	393

CaNa(OOCCHCH(OH)(OH)CH(OH)CH(OH)CHOH)$_3$·6H$_2$O, calcium sodium galacturonate(1-) hexahydrate

e S	-18.3	1.529(8)	1.265(10)	1.232(10)	1.431(9)	2.616(5)	108.8(4)	117.6(5)	126.7(5)	115.6(4)	394

SrNA(OOCCHCH(OH)(OH)CH(OH)CH(OH)CHOH)$_3$·6H$_2$O, strontium sodium galacturonate(1-) hexahydrate

e S	-15.4	1.56(2)	1.25(2)	1.23(2)	1.42(2)	2.64(1)	110(1)	117(1)	128(1)	115(1)	395

Sr(C$_{12}$H$_{14}$O$_{12}$)$_2$·4$\frac{1}{2}$H$_2$O, strontium 4-0-(4-deoxy-β-L-threo-hex-4-enosyl)-α-D-galacturonate(2-) 4$\frac{1}{2}$-hydrate

e S	-9.6	1.59(5)	1.22(4)	1.23(3)	1.45(3)	2.46(4)	103(3)	115(2)	129(3)	116(3)	396
g	3.0	1.45(5)	1.29(2)	1.22(4)	1.44(4)	2.63(4)	113(2)	118(3)	123(3)	118(2)	
e S	+2.2	1.57(5)	1.33(4)	1.26(3)	1.45(3)	2.63(4)	103(3)	123(2)	120(3)	117(3)	
g	34.11	1.56(5)	1.29(3)	1.26(3)	1.39(3)	2.77(4)	111(2)	119(2)	126(3)	114(2)	

(-)-Ca(OOCCHC(OH)(COO)CH₂CO)·4H₂O, calcium (-)-hydroxycitrato(2-) lactone tetrahydrate

e,h	S	+7.6	1.538(7)	1.245(6)	1.242(7)	1.422(7)	2.569(6)	108.4(4)	117.1(5)	127.5(5)	115.3(5)	399
	S	+23.0	1.532(8)	1.248(7)	1.257(6)	1.452(6)	2.747(6)	111.4(4)	120.2(5)	126.0(5)	113.7(5)	

(+)-Ca(OOCCHC(OH)(COO)CH₂CO)·4H₂O, calcium (+)-allo-hydroxycitrate(2-) lactone tetrahydrate

e,h	R	-10.2	1.529(8)	1.256(7)	1.252(7)	1.412(7)	2.550(6)	107.4(4)	117.4(5)	123.3(5)	119.3(5)	400
	S	+0.2	1.520(8)	1.237(8)	1.261(7)	1.452(7)	2.722(7)	112.0(5)	121.3(5)	123.9(6)	115.0(5)	

K(OOCCHCH(COOH)CH₂CO), potassium isocitrate(1-) lactone

e	R	-23.4	1.510(5)	1.239(5)	1.279(5)	1.461(4)	2.695(4)	110.2(3)	120.0(3)	125.5(4)	114.5(3)	401

Ca(C₁₅H₆O₉)·9H₂O, calcium 4,10-dioxo-5-methoxy-2,8-dicarboxy-4H,10H-benzodipyran(2-) 9-hydrate

g	--	11.8	1.46(3)	1.28(2)	1.26(2)	1.37(2)	2.67(2)	112.4(15)	121.9(17)	121.7(18)	116.4(17)	402
g	--	12.8	1.47(2)	1.21(2)	1.30(2)	1.40(2)	2.58(2)	108.3(14)	121.7(16)	125.5(16)	112.6(15)	

Na(C₁₈H₁₉O₅)·H₂O, sodium 8-allyl-5-(3-methylbutoxy)-4-oxo-4H-1-benzopyran-2-carboxylate(1-) monohydrate

g	--	13.3	1.541(5)	1.234(4)	1.226(4)	1.347(4)	2.631(5)	112.7(2)	116.6(3)	129.3(3)	114.1(3)	403

Ca(OOCCHCH(OH)CH(OH)C(OH)CH₂OH)₂·2H₂O, calcium 5-keto-D-gluconate(1-) lactol dihydrate

	R	+1.4	1.49(3)	1.23(3)	1.28(3)	1.45(3)	2.63(3)	110(4)	121(4)	125(4)	115(4)	404

a Underlined distances, angles, and esd's were calculated from reported coordinates.

b Absolute configuration of carbon α to carboxyl group.

Table 6 (Continued)

[c] α-Hydroxycarboxylate grouping parameters are given in Table 3.

[d] The protonated or esterified carboxyl oxygen lies proximal to the alkoxy group.

[e] Chelated group.

[f] Coordinates not given so that some parameters could not be calculated.

[g] α-Carbon atom is sp^2.

[h] α-Hydroxycarboxylate group.

[i] Protonated carboxyl group.

Table 7 Geometries of Chelate Rings in α-Alkoxycarboxylates[a]

Plane 1: M, O1, O3
Plane 2: O3, C1, C2, O1

Compound	M	Absolute config.[b] O3	Absolute config.[b] C1	Dihed. angle	Distances, Å M-O3	Distances, Å M-O1	Distances, Å C1-plane 1	Distances, Å C2-plane 1	Angles, deg C1-O3-M	Angles, deg C2-O1-M	Angles, deg O3-M-O1	Plane 1/ plane 2	Ref.
[Cu(CH₃OCH₂COO)₂(H₂O)₂]	Cu(II)	R	--	+ 0.2	2.13(1)	1.93(1)	.056	.068	112.7(8)	119.0(9)	79.9(4)	2.9	167
[Ni(CH₂OCH₂COO)₂(H₂O)₂]	Ni(II)	R	--	+ 0.1	1.99(1)	2.05(1)	.043	.051	114.7(9)	117.1(10)	79.5(4)	2.2	361
[Cu(CH₃OCH₂COO)₂(C₅H₅N)₂]·4H₂O	Cu(II)	S	--	-15.7	2.36(1)	1.94(1)	.173	.386	107(1)	123(1)	75.5(3)	14.2	362
[Cu(C₂H₅OCH₂COO)₂(H₂O)₂]	Cu(II)	S	--	- 8.8	2.38(1)	1.97(1)	.042	.158	108.4(6)	124.3(7)	75.2(3)	5.1	364
Na(C₆H₅OCH₂COO)·½ H₂O	Na(I)	R	--	- 2.4	2.658(6)	2.492(6)	.728	.774	111.6(3)	115.1(4)	63.3(2)	37.5	365
[Cu(C₆H₅OCH₂COO)₂(H₂O)₂]	Cu(II)	S	--	- 9.3	2.50(1)	1.94(1)	.256	.400	106.2(7)	124.6(9)	73.6(3)	16.0	167
	Cu(II)	R	--	+ 9.8	2.50(1)	1.95(1)	.230	.383	106.0(7)	124.0(9)	74.3(3)	15.1	
	Cu(II)	S	--	- 5.9	2.44(1)	1.94(1)	.202	.297	107.3(7)	123.6(9)	75.1(3)	12.0	
[Cu(CH₃OC₆H₄OCH₂COO)₂(H₂O)₂]	Cu(II)	R	--	- 1.7	2.432(3)	1.955(2)	.351	.373	106.5(2)	122.2(2)	74.8(1)	17.4	366
Cu(C₆H₅OCH₂COO)₂	Cu(II)	S	--	-17.4	2.47(1)	1.96(1)	.420	.285	108	123	71.8	16.6	367
Na(HOOCCH₂OCH₂COO)	Na(I)	R	--	- 5.2	2.398(3)	2.742(3)	.325	.308	127.9(2)	115.7(2)	61.5(1)	15.2	372
f	Na(I)	S	--	-14.1	2.398(3)	2.468(3)	.123	.314	120.1(2)	119.4(2)	66.2(1)	11.2	

Table 7 (Continued)

M	Absolute config.[b] O3	C1	Dihed. angle	M-O3	M-O1	C1-plane 1	C2-plane 1	C1-O3-M	C2-O1-M	O3-M-O1	Plane 1/plane 2	Ref.
K(HOOCH₂CH₂COO) K(I)	R	--	- 2.3	2.740(2)	2.805(2)	.274	.285	124.8(1)	123.1(1)	57.3(1)	13.5	372
f	S	--	- 9.9	2.740(2)	2.708(2)	.202	.353	121.6(1)	124.6(1)	58.4(1)	14.0	
Li(HOOCH₂CH₂COO) Li(I)	R	--	+ 8.9	2.122(3)	2.052(3)	.242	.158	115.6(1)	119.3(1)	76.1(1)	9.3	373
Cd(OOCCH₂CH₂COO)·3H₂O (orthorhombic) Cd(II)	S	--	+ 4.4	2.490(11)	2.296(13)	.200	.183	118.7(9)	123.8(12)	66.0(4)	8.8	376
[Cd₂(OOCCH₂CH₂COO)₂(H₂O)₆] Cd(II)	R	--	+ 1.4	2.356(7)	2.352(8)	.224	.272	119.1(6)	120.0(6)	67.2(3)	11.9	378
Cd(II)	S	--	- 6.0	2.356(7)	2.492(8)	.298	.401	121.8(6)	118.6(6)	65.1(2)	17.3	
Cd(OOCCH₂CH₂COO)·3½H₂O Cd(II)	R	--	+10.4	2.412(14)	2.433(18)	.171	.062	121.0(12)	123.2(12)	64.7(5)	5.6	379
Cd(II)	S	--	- 0.3	2.412(14)	2.408(13)	.301	.335	123.6(11)	118.0(12)	65.1(5)	14.9	
Cd(II)	R	--	+ 0.6	2.278(15)	2.310(17)	.086	.110	122.2(9)	120.1(10)	68.6(5)	4.6	
Cd(II)	S	--	- 4.1	2.278(15)	2.310(14)	.034	.093	121.4(12)	122.9(15)	68.4(5)	3.1	
[Ca(OOCCH₂CH₂COO)(H₂O)₅]·H₂O Ca(II)	R	--	- 0.8	2.431(3)	2.446(3)	.110	.114	123.6(2)	122.8(2)	64.5(1)	5.3	380
Ca(II)	S	--	+ 2.1	2.431(3)	2.472(3)	.049	.082	124.0(2)	124.6(2)	63.8(1)	3.1	
Na₂[UO₂(OOCCH₂CH₂COO)₂]·2H₂O [c] U(VI)	--	--	--	2.63(2)	2.43(1)	--	--	122(2)	132(2)	58(1)	--	381
U(VI)	--	--	--	2.65(2)	2.46(1)	--	--	120(2)	131(2)	58(1)	--	
UO₂(OOCCH₂CH₂COO) U(VI)	d	--	6.3	2.55(1)	2.37(1)	.173	.109	121(1)	129(1)	62.0(4)	6.4	382

Compound	Ion												Ref.
Cu(OOCCH₂OCH₂COO)·½H₂O	Cu(II)	R	--	-10.1	2.488(7)	1.948(9)	.117	.249	105.6(7)	125.3(8)	73.8(3)	9.3	383
	Cu(II)	S	--	+11.9	2.488(7)	1.946(8)	.163	.325	103.1(6)	124.6(7)	74.7(3)	11.9	
Na₃[Gd(OOCCH₂OCH₂COO)₃]·2NaClO₄·6H₂O	Gd(III)	d	--	6.0	2.49(3)	2.41(1)	-.012	+.064	122(1)	126(1)	64(1)	1.5	384
Na₃[Nd(OOCCH₂OCH₂COO)₃]·2NaClO₄·6H₂O	Nd(III)	d	--	1.8	2.523(10)	2.428(6)	.011	.038	123.9(6)	127.9(7)	62.8(2)	1.2	385
Na₃[Yb(OOCCH₂OCH₂COO)₃]·2NaClO₄·6H₂O	Yb(III)	d	--	5.1	2.431(9)	2.339(6)	-.018	+.048	122.5(5)	127.0(6)	64.5(2)	1.0	385
Na₃[Ce(OOCCH₂OCH₂COO)₃]·2NaClO₄·6H₂O	Ce(III)	d	--	4.2	2.564(6)	2.471(4)	-.020	+.034	121.8(4)	129.0(3)	62.2(1)	0.7	386
Er[Er(OOCCH₂OCH₂COO)₃]·6H₂O	Er(III)	d	--	8.6	2.50(3)	2.33(2)	.132	.039	121(2)	134(2)	61.2(6)	3.7	388
	Er(III)	R	--	+2.6	2.49(2)	2.36(2)	.318	.377	116(2)	126(2)	63.9(6)	16.9	
	Er(III)	S	--	+1.9	2.49(2)	2.36(2)	.011	.036	124(2)	126(2)	63.2(6)	1.2	
Na₃[Ce(OOCCH₂OCH₂COO)₃]·9H₂O	Ce(III)	S	--	-13.2	2.593(5)	2.452(5)	.123	.301	119.1(2)	127.8(2)	62.3(1)	10.5	389
	Ce(III)	R	--	+10.6	2.593(5)	2.524(6)	.219	.377	120.8(2)	126.1(2)	61.3(1)	14.6	
	Ce(III)	S	--	+0.2	2.577(6)	2.488(6)	.054	.061	123.2(3)	128.2(3)	61.4(1)	2.7	
	Ce(III)	R	--	-3.8	2.577(6)	2.480(6)	.132	.095	122.5(2)	128.2(3)	61.6(1)	5.3	
	Ce(III)	R	--	+6.7	2.606(5)	2.486(6)	.035	.128	122.7(2)	129.2(3)	60.9(2)	4.0	
	Ce(III)	S	--	-13.2	2.606(5)	2.480(5)	.116	.299	121.0(3)	129.0(3)	60.9(2)	10.2	
Ca(OOCCHCH(OH)CH(OH)CH(OH)CHOH)Br·3H₂O	Ca(II)	S	S	+2.9	2.503(3)	2.468(3)	.156	.214	122.2(2)	125.5(3)	63.1(1)	8.8	393

Table 7 (Continued)

M	Absolute config. O3 C1	Dihed. angle	Distances, Å				Angles, deg				Ref.
			M-O3	M-O1	C1-plane 1	C2-plane 1	C1-O3-M	C2-O1-M	O3-M-O1	Plane 1/plane 2	
CaNa(OOCCHCH(OH)CH(OH)CHOH)$_3$											
·6H$_2$O	Ca(II)	S S	-18.3	2.831(2)	2.398(5)	-.038	+.198	116.6(2)	134.8(4)	59.4(1)	4.8	394
SrNa(OOCCHCH(OH)CH(OH)CHOH)$_3$											
·6H$_2$O	Sr(II)	S S	-15.4	2.87(1)	2.52(1)	.006	.209	118(1)	135(1)	58.1(3)	5.8	395
Sr(C$_{12}$H$_{14}$O$_{12}$)$_2$·4$\frac{1}{2}$ H$_2$O	Sr(II)	S S	- 9.6	2.70(2)	2.48(2)	.045	.198	126(2)	138(2)	56.4(8)	6.1	396
	Sr(II)	S S	+ 2.2	2.58(3)	2.55(2)	.190	.187	128(2)	123(2)	61.9(8)	8.4	
(-)-Ca(OOCCH(OH)(COO)CH$_2$CO)											
·4H$_2$O^e	Ca(II)	-- S	+ 7.6	2.481(4)	2.419(4)	.022	.127	123.0(3)	127.7(4)	63.2(2)	3.8	399

(+)-Ca(OOCCH(OH)(COO)CH₂CO) [e]
 ·4H₂O
 (O bridge)

K(OOCCHCH(COOH)CH₂CO)
 (O bridges)

Ca(OOCCHCH(OH)CH(OH)C(OH)
 CH₂OH)₂·2H₂O
 (O bridge)

Ca(II)	--	R	-10.2	2.449(5)	2.411(4)	.162	.317	123.1(3)	126.0(4)	63.3(1)	11.6	400
K(I)	R	R	-23.4	3.017(3)	2.772(3)	.871	.793	109.6(2)	110.8(2)	55.3(1)	43.2	401
Ca(II)	R	R	+ 1.4	2.47(2)	2.45(2)	.067	.095	120.7	123.8	64.8	3.9	404

[a] Underlined distances, angles, and esd's were calculated from reported coordinates.

[b] Absolute configuration about O3 are for priority ordering $M > C_1 > R >$ lone pair in all cases. In the oxydiacetates, R is the $CH_2COO(H)$ group not under consideration.

[c] Some parameters could not be calculated as coordinates were not published.

[d] Planar by symmetry.

[e] α-Hydroxycarboxylate group.

[f] Protonated carboxyl group.

Table 8 Geometries of α-Thiocarboxylate Groups[a]

Structure: $(H,R)S\,$—$\,C1$—$C2$ with $O1$ and $O2(H,R)$

	Distances, Å					Angles, deg.				Ref.
Dihed. angle	C1-C2	C2-O1	C2-O2	C1-S	O1-S	C2-C1-S	C1-C2-O1	O1-C2-O2	C1-C2-O2	
α-Mercaptocarboxylates										
$[Co(C_2H_8N_2)_2(SCH_2COO)]Cl·H_2O$, mercaptoacetato(2-)bis(ethylenediamine)cobalt(III) chloride monohydrate										405
-5.6[b]	1.507(7)	1.287(6)	1.336(6)	1.810(6)	2.917(3)	113.8(4)	119.2(4)	121.4(4)	119.5(5)	
$[Cr(C_2H_8N_2)_2(SCH_2COO)]ClO_4$, mercaptoacetato(2-)bis(ethylenediamine)chromium(III) perchlorate										405
+7.8[b]	1.507(8)	1.300(6)	1.229(6)	1.812(6)	2.942(3)	115.7(4)	117.6(4)	122.8(4)	119.6(5)	
$Fe(SCH_2COO)·H_2O$, iron(II) mercaptoacetate(2-) monohydrate										406
15.7	1.542(14)	1.279(12)	1.234(11)	1.831(9)	3.011(8)	114.7(7)	119.7(7)	125.5(9)	115.0(9)	
$H[Sb(SCH_2COO)_2]$, hydrogen bis(mercaptoacetato(2-))antimonate(III)										407
+9.6[b]	1.53(5)	1.25(4)	1.27(4)	1.84(4)	3.07(2)	114(2)	125(3)	120(3)	115(2)	
-33.6[b]	1.54(4)	1.29(3)	1.27(3)	1.76(3)	2.98(2)	113(2)	118(2)	117(2)	124(2)	
$(C_4H_9)_4N[TcO(SCH_2COS)_2]$, tetrabutylammonium bis(thiomercaptoacetato(2-))oxotechnetate(V)[d]										408
25.5	1.46(2)	1.741(12)	1.26(2)	1.762(13)	3.139(4)	115.6(8)	118.9(9)	117(1)	124(1)	
$Tℓ_5[Cu^{II}_6Cu^{I}_8(SC(CH_3)_2COO)_{12}Cl]·12H_2O$, pentathallium(I) μ₈-chlorododecakis(α-mercaptoisobutyrato(2-))octacuprate(I)hexacuprate(II) hydrate										409
10.7	1.27(10)	1.53(7)	1.19(10)	1.83(7)	2.88(5)	122(5)	114(6)	105(5)	140(6)	
7.0	1.49(8)	1.25(6)	1.30(8)	1.81(5)	2.85(4)	109(3)	122(4)	121(5)	117(4)	

c	14.9	1.28(11)	1.32(13)	1.32(3)	1.80(6)	2.84(5)	113(6)	126(9)	91(7)	132(9)	410
c	30.3	1.40(8)	1.18(8)	1.38(7)	1.82(5)	2.77(4)	104(4)	127(6)	118(6)	115(6)	
c	39.7	1.38(10)	1.33(9)	1.28(11)	1.85(5)	2.85(4)	100(4)	132(7)	105(7)	119(7)	
c	29.9	1.44(7)	1.24(7)	1.34(6)	1.89(5)	2.90(5)	107(3)	125(4)	119(5)	115(4)	

α-Thioethercarboxylates

HOOCCH$_2$SCH$_2$COOH, thiodiacetic acid

c	12.6	1.50(2)	1.22(2)	1.30(2)	1.80(1)	2.918(4)	111.9(7)	122.9(11)	124.3(9)	112.6(11)	411

HOOCH(CH$_3$)CH(CH$_3$)COOH, thiodilactic acid

	80.0	1.52(2)	1.28(1)	1.25(1)	1.81(1)	3.28(1)	107.9(3)	118.1(4)	122.2(5)	119.7(4)	412
	81.3	1.50(2)	1.28(2)	1.24(2)	1.80(1)	3.22(1)	107.4(4)	115.0(5)	122.2(5)	122.7(6)	

[Zn(OOCCH$_2$SCH$_2$COO)$_3$]·H$_2$O, triaquothiodiacetato(2-)zinc(II) monohydrate

c	10.6	1.520(9)	1.243(8)	1.264(8)	1.807(7)	3.046(5)	116.5(4)	122.0(6)	123.7(6)	114.3(6)	413
c	9.7	1.526(9)	1.248(7)	1.263(9)	1.790(7)	2.990(5)	115.5(4)	120.6(5)	123.8(6)	115.6(5)	

Cd(OOCCH$_2$SCH$_2$COO)·H$_2$O, cadmium(II) thiodiacetate(2-) monohydrate

c	0.2	1.532(10)	1.248(9)	1.257(10)	1.776(9)	3.106(5)	119.6(6)	121.8(7)	124.3(6)	113.9(6)	414
c	6.8	1.545(11)	1.234(7)	1.251(8)	1.776(7)	3.113(5)	119.7(4)	121.5(6)	126.6(8)	111.9(5)	

Nd(OOCCH$_2$SCH$_2$COO)Cl·4H$_2$O, neodynium(III) thiodiacetate(2-) chloride tetrahydrate

c	11.9	1.46(12)	1.34(8)	1.13(7)	1.88(8)	2.94(6)	114(6)	119(5)	122(7)	118(7)	415

[Ni(OOCCH$_2$SCH$_2$CH$_2$SCH$_2$COO)(H$_2$O)$_2$], diaquoethylenedithiodiacetato(2-)nickel(II)

c	24.1	1.50	1.22	1.21	1.85	2.97	114	119	123	117	

[a] Underlined distances, angles, and esd's were calculated from reported coordinates. [b] For Δ absolute configuration. [c] Chelated group. [d] Thiomercaptocarboxylate ligand.

Table 9 Geometries of α-Thiocarboxylate Chelate Rings[a]

Plane 1: M, O1, S
Plane 2: S, C1, C2, O1

		Distances, Å				Angles, deg				
M	Dihed. angle	M-S	M-O1	C1-plane 1	C2-plane 1	C1-S-M	C2-O1-M	S-M-O1	Plane 1/Plane 2	Ref.

α-Mercaptocarboxylates

Compound	M	Dihed. angle	M-S	M-O1	C1-plane 1	C2-plane 1	C1-S-M	C2-O1-M	S-M-O1	Plane 1/Plane 2	Ref.
$[Co(C_2H_8N_2)_2(SCH_2COO)]Cl \cdot H_2O$	Co(III)	- 5.6[b]	2.243(2)	1.918(3)	.088	.031	96.7(2)	121.4(3)	88.7(1)	2.2	405
$[Cr(C_2H_8N_2)_2(SCH_2COO)]ClO_4$	Cr(III)	+ 7.8[b]	2.337(2)	1.966(3)	.040	.155	96.3(2)	124.0(3)	85.8(1)	4.7	405
$Fe(SCH_2COO) \cdot H_2O$	Fe(II)	15.7	2.436(7)	2.207(7)	.371	.658	96.6(4)	120.6(6)	80.9(2)	23.8	406
$H[Sb(SCH_2COO)_2]$	Sb(III)	+ 9.6[b]	2.431(8)	2.304(22)	.056	.196	100.7(12)	118.5(19)	80.9(6)	6.2	407
	Sb(III)	-33.6[b]	2.426(8)	2.255(19)	.287	.760	97.2(11)	119.1(16)	79.1(5)	26.1	
$(C_4H_9)_4N[TcO(SCH_2COS)_2]$[c]	Tc(V)	25.5	2.303(3)	2.336(3)	.274	.588	105.9(4)	106.5(5)	85.1(1)	16.8	408
$T\ell_5[Cu^{II}_6Cu^I_8(SC(CH_3)_2COO)_{12}Cl]$ $\sim 12H_2O$	Cu(II)	10.7	2.25(2)	1.95(4)	.257	.392	96(2)	119(4)	86(1)	13.1	409
	Cu(II)	7.0	2.24(1)	1.87(4)	.106	.234	98(2)	122(4)	87(1)	7.8	
	Cu(II)	14.9	2.23(2)	1.99(5)	.201	.075	99(2)	116(5)	84(2)	5.3	
	Cu(II)	30.3	2.24(1)	1.90(4)	.053	.461	98(2)	120(4)	84(1)	14.2	
	Cu(II)	39.7	2.25(2)	1.89(4)	.032	.523	99(2)	112(4)	87(1)	14.7	

Compound	Metal										Ref.
	Cu(II)	29.9	2.26(2)	1.96(4)	.018	.416	96(2)	118(4)	86(1)	11.9	
α-Thioethercarboxylates											
[Zn(OOCCH₂SCH₂COO)(H₂O)₃]·H₂O	Zn(II)	10.6	2.601(2)	2.029(5)	.254	.447	92.3(3)	124.6(5)	81.2(2)	16.4	412
	Zn(II)	9.7	2.601(2)	2.102(4)	.401	.613	94.3(4)	124.4(5)	78.2(1)	23.5	
Cd(OOCCH₂SCH₂COO)·H₂O	Cd(II)	0.2	2.663(2)	2.276(6)	.144	.174	96.8(2)	123.5(4)	77.5(1)	7.3	413
	Cd(II)	6.8	2.663(2)	2.277(5)	.255	.227	96.3(2)	122.8(4)	77.7(1)	10.6	
Nd(OOCCH₂SCH₂COO)Cl·4H₂O	Nd(II)	11.9	3.15(3)	2.44(4)	.358	.277	102(3)	139(4)	62(1)	12.4	414
[Ni(OOCCH₂SCH₂CH₂SCH₂COO)(H₂O)₂]	Ni(II)	24.1	2.46	2.01	.262	.009	95	125	83	4.7	415

ᵃUnderlined distances, angles, and esd's were calculated from reported coordinates.

ᵇFor Δ absolute configuration.

ᶜThiomercaptocarboxylate ligand.

REFERENCES

1. G. R. Choppin and J. A. Chopoorian, *J. Inorg. Nucl. Chem.*, *22*: 97 (1961).

2. D. T. Sawyer, *Chem. Rev.*, *64*:633 (1964).

3. C. H. Holten, *Lactic Acid*, Verlag Chemie, Weinheim, 1971, chap. 10.

4. R. E. Tapscott, R. L. Belford, and I. C. Paul, *Coord. Chem. Rev.*, 4:323 (1969).

5. I. V. Pyatniskii, *Russ. Chem. Rev.*, *32*:44 (1963).

6. S. Herzog and W. Kalies, *Z. Chem.*, *8*:81 (1968).

7. U. Casellato, P. A. Vigato, and M. Vidali, *Coord. Chem. Rev.*, *26*:85 (1978).

8. W. J. Cook and C. E. Bugg, in *Metal-Ligand Interactions in Organic Chemistry and Biochemistry* (B. Pullman and N. Goldblum, eds.), Part 2, Reidel, Dordrecht-Holland, 1977, pp. 231-256.

9. H. Einspahr and C. E. Bugg, in *Calcium Binding Proteins and Calcium Function* (R. H. Wasserman, R. A. Corradino, E. Carafoli, R. H. Kretsinger, D. H. MacLennan, and F. L. Siegel, eds.), North-Holland, New York, 1977, pp. 13-20.

10. S. M. Roberts, in *Comprehensive Organic Chemistry* (I. O. Sutherland, ed.), Vol. 2, Pergamon, New York, 1979, pp. 739-778.

11. S. M. Martin, in *Encyclopedia of Plant Physiologie* (W. Ruhland, ed.), Vol. 12, Part 2, Springer-Verlag, Berlin, 1969, pp. 605-640.

12. G. E. Fogg and J. W. Millbank, in *Encyclopedia of Plant Physiologie* (W. Ruhland, ed.), Vol. 12, Part 2, Springer-Verlag, Berlin, 1960, pp. 640-662.

13. D. Wang, in *Phytochemistry* (L. P. Miller, ed.), Vol. 3, Van Nostrand Reinhold, New York, 1973, pp. 74-111.

14. D. T. Downing, *Rev. Pure Appl. Chem.*, *11*:196 (1961).

15. E. T. Peltzer and J. L. Bada, *Nature*, *272*:443 (1978).

16. S. L. Miller, *Biochim. Biophys. Acta*, *23*:480 (1957).

17. R. S. Cahn, C. K. Ingold, and V. Prelog, *Experientia*, *12*:81 (1956).

18. W. Klyne and J. Buckingham, *Atlas of Stereochemistry*, Vol. 1, 2nd ed., Chapman and Hall, London, 1974.

19. R. Bentley, *Molecular Asymmetry in Biology*, Vol. 2, Academic Press, New York, 1970, pp. 374-403.

20. H. E. Whipple, ed., *Chemistry and Metabolism of L- and D-Lactic Acids*, Ann. N.Y. Acad. Sci., Vol. 119, 1965, pp. 851-1165.

21. D. Rehbinder, in *Lactic Acid* (by C. H. Holten), Verlag Chemie, Weinheim, 1971, pp. 413-460.

22. G. Wagner, J. C. Yang, and F. A. Loewus, *Plant Physiol.*, *55*:1071 (1975).

23. T. W. Goodwin, ed., *The Metabolic Roles of Citrate*, Academic Press, New York, 1968.

24. J. W. Green, in *The Carbohydrates* (W. Pigman, ed.), Academic Press, New York, 1957, pp. 299-366.

25. L. B. Lockwood, D. E. Yoder, and M. Zienty, *Ann. N.Y. Acad. Sci.*, *119*:854 (1965).

26. J. J. Christenson, R. M. Izatt, and L. D. Hansen, *J. Amer. Chem. Soc.*, *89*:213 (1967).

27. H. S. Harned and R. W. Ehlers, *J. Amer. Chem. Soc.*, *55*:652 (1933).

28. P. B. Davies and C. B. Monk, *Trans. Faraday Soc.*, *50*:128 (1954).

29. E. J. King, *J. Amer. Chem. Soc.*, *82*:3575 (1960).

30. H. S. Harned and R. W. Ehlers, *J. Amer. Chem. Soc.*, *55*:2379 (1933).

31. L. F. Nims and P. K. Smith, *J. Biol. Chem.*, *113*:145 (1936).

32. G. D. Pinching and R. G. Bates, *J. Res. Nat. Bur. Stand.*, *45*: 448 (1950).

33. G. D. Pinching and R. G. Bates, *J. Res. Nat. Bur. Stand.*, *45*: 327 (1950).

34. M. Eden and R. G. Bates, *J. Res. Nat. Bur. Stand.*, *62*:161 (1959).

35. R. G. Bates and R. G. Canham, *J. Res. Nat. Bur. Stand.*, 47:343 (1951).

36. R. B. Martin, *J. Amer. Chem. Soc.*, *65*:2053 (1961).

37. W. Versichel, W. Van de Meiroop, and A. T. H. Lenstra, *Acta Crystallogr.*, *B34*:2643 (1978).

38. J. Sicher, M. Tichý, F. Sipos, M. Svoboda, and J. Jonás, *Collect. Czech. Chem. Commun.*, *29*:1561 (1964).

39. I. M. Kolthoff and M. K. Chantooni, Jr., *J. Amer. Chem. Soc.*, *98*:5063 (1976).

40. R. A. Auerbach and C. A. Kingsbury, *Tetrahedron*, *27*:2069 (1971).

41. J. Murto, *Acta Chem. Scand.*, *18*:1043 (1964).

42. B. Csiszár, M. Halmos, M. T. Beck, and P. Szarvas, *Magy. Kem. Folyoirat*, *70*:214 (1964).

43. P. Ballinger and F. A. Long, *J. Amer. Chem. Soc.*, *82*:795 (1960).

44. M. T. Beck, B. Csiszár, and P. Szarvas, *Magy. Kem. Folyoirat*, *70*:217 (1964).

45. M. M. Petit-Ramel and C. M. Blanc, *J. Inorg. Nucl. Chem.*, *34*: 1233 (1972).

46. R. J. Irving, L. Nelander, and I. Wadsö, *Acta Chem. Scand.*, *18*: 769 (1964).

47. R. P. Bell, *The Proton in Chemistry*, 2nd ed., Cornell University Press, Ithaca, N. Y., 1973, pp. 96-97.

48. J. P. Danehy and K. N. Parameswaran, *J. Chem. Eng. Data*, *13*:386 (1968).

49. R. D. Ellison, C. K. Johnson, and H. A. Levy, *Acta Crystallogr.*, *B27*:333 (1971).

50. W. P. J. Gaykema, J. A. Kanters, and G. Roelofsen, *Crystal. Struct. Commun.*, *7*:463 (1978).

51. B. Dahlén, B.-M. Lundén, and I. Pascher, *Acta Crystallogr.*, *B32*: 2059 (1976).

52. M. Cesario and J. Guilhem, *Crystal Struct. Commun.*, *4*:245 (1975).

53. K.-T. Wei and D. L. Ward, *Acta Crystallogr.*, *B33*:797 (1977).

54. O. Korver, S. De Jong and T. C. van Soest, *Tetrahedron*, *32*:1225 (1976).

55. T. C. van Soest, Unilever Research, Vlaardingern, The Netherlands, private communication.

56. M. Cesario and J. Guilhem, *Crystal Struct. Commun.*, *4*:197 (1975).

57. M. Cesario and J. Guilhem, *Crystal Struct. Commun.*, *4*:193 (1975).

58. G. Roelofsen, J. A. Kanters, J. Kroon, H. M. Doesburg, and T. Koops, *Acta Crystallogr.*, *B34*:2565 (1978).

59. G. S. Parry, *Acta Crystallogr.*, *4*:131 (1951).

60. H. Hope and U. H. de la Camp, *Acta Crystallogr.*, *A28*:201 (1972).

61. G. A. Bootsma and J. C. Schoone, *Acta Crystallogr.*, *22*:522 (1967).

62. A. J. J. Sprenkels, Ph.D. dissertation, Der Rijksuniversiteit, Utrecht, The Netherlands, 1956.

63. A. J. van Bommel and J. M. Bijvoet, *Acta Crystallogr.*, *11*:61 (1958).

64. D. Carlström, *Acta Crystallogr.*, *B29*:161 (1973).

65. G. Hite and J. R. Soares, *Acta Crystallogr.*, *B29*:2935 (1973).

66. J. R. Ruble, G. Hite, and J. R. Soares, *Acta Crystallogr.*, *B32*: 136 (1976).

67. E. Bye, *Acta Chem. Scand.*, *B29*:22 (1975).

68. B. Jensen, European Crystallographic Meeting, 268 (1977).

69. S. Perez, *Acta Crystallogr.*, *B33*:1083 (1977).

70. M. Kuramoto, Y. Kushi, and H. Yoneda, *Bull. Chem. Soc. Jap.*, *51*: 3251 (1978).

71. M. Kuramoto, Y. Kushi, and H. Yoneda, *Bull. Chem. Soc. Jap.*, *53*: 125 (1980).

72. M. Kuramoto, *Bull. Chem. Soc. Jap.*, *52*:3702 (1979).

73. J. P. Glusker, J. A. Minkin, and A. L. Patterson, *Acta Crystallogr.*, *B25*:1066 (1969).

74. G. Roelofsen and J. A. Kanters, *Crystal Struct. Commun.*, *1*:23 (1972).

75. W. C. Stallings, J. F. Blount, P. A. Srere, and J. P. Glusker, *Arch. Biochem. Biophys.*, *193*:431 (1979).

76. M. E. Gress and G. A. Jeffrey, *Carbohyd. Res.*, *50*:159 (1976).

77. M. Schiffer, Ph.D. dissertation, Columbia University, 1965.

78. L. Brehm and T. Honore, *Acta Crystallogr.*, *B34*:2359 (1978).

79. L. Brehm and P. Krogsgaard-Larsen, *Acta Chem. Scand.*, *B33*:52 (1979).

80. J. Hvoslef and B. Bergen, *Acta Crystallogr.*, *B31*:697 (1975).

81. J. L. Flippen, *Acta Crystallogr.*, *B29*:1881 (1973).

82. J. Beintema, *Acta Crystallogr.*, *B32*:1631 (1976).

83. M. P. Gupta and N. P. Gupta, *Acta Crystallogr.*, *B24*:631 (1968).

84. L.-K. Liu and R. E. Davis, *Acta Crystallogr.*, *B36*:171 (1980).

85. M.-C. Brianso, M. Leclercq, and J. Jacques, *Acta Crystallogr.*, *B35*:2751 (1979).

86. V. S. Yadava and V. M. Padmanabhan, *Acta Crystallogr.*, *B29*:493 (1973).

87. B. Jensen, *Acta Chem. Scand.*, *B30*:5 (1976).

88. C. K. Fair and E. O. Schlemper, *Acta Crystallogr.*, *B33*:1337 (1977).

89. D. H. Templeton, A. Zalkin, H. W. Ruben, and L. K. Templeton, *Acta Crystallogr.*, *B35*:1608 (1979).

90. R. J. Geue and M. R. Snow, *Inorg. Chem.*, *16*:231 (1977).

91. M. Ahlgrén and R. Hämäläinen, *Finn. Chem. Lett.*, 211 (1975).

92. A. Wada, C. Katayama, and J. Tanaka, *Acta Crystallogr.*, *B32*: 3194 (1976).

93. W. A. Freeman, *Inorg. Chem.*, *15*:2235 (1976).

94. R. Hämäläinen and A. Pajunen, *Finn. Chem. Lett.*, 150 (1974).

95. J. Kansikas and R. Hämäläinen, *Finn. Chem. Lett.*, 54 (1978).

96. N. Gavrushenko, H. L. Carrell, W. C. Stallings, and J. P. Glusker, *Acta Crystallogr.*, *B33*:3936 (1977).

97. N. C. Panagiotopoulos, G. A. Jeffrey, S. J. LaPlaca, and W. C. Hamilton, *Acta Crystallogr.*, *B30*:1421 (1974).

98. J. Kroon and J. A. Kanters, *Acta Crystallogr.*, *B29*:1278 (1973).

99. E. Bye, *Acta Chem. Scand.*, *A33*:169 (1979).

100. A. M. O'Connell, *Acta Crystallogr.*, *B29*:2320 (1973).

101. A. Meyerhöffer and D. Carlström, *Acta Crystallogr.*, *B25*:1119 (1969).

102. A. Meyerhöffer, *Acta Crystallogr.*, *B26*:341 (1970).

103. R. W. Baker, N. Datta, and P. J. Pauling, *J. Chem. Soc.*, *Perkin II*, 1964 (1973).

104. J. J. Guy and T. A. Hamor, *J. Chem. Soc.*, *Perkin II*, 1875 (1973).

105. J. J. Guy and T. A. Hamor, *J. Chem. Soc.*, *Perkin II*, 1126 (1974).

106. K. N. Coughenour and D. Dennis, *Acta Crystallogr.*, *B31*:1229 (1975).

107. T. T. Petcher, *J. Chem. Soc.*, *Perkin II*, 1151 (1974).

108. D. J. Duchamp, E. C. Olson, and C. G. Chidester, *Amer. Crystallogr. Ass.*, *Ser. 2*, *5*:83 (1977).

109. S. Larsen, *Acta Crystallogr. Suppl.*, *A31*:S168 (1975).

110. M. Simonetta and S. Carra, in *The Chemistry of Carboxylic Acids and Esters* (S. Patai, ed.), Wiley-Interscience, New York, 1969, pp. 3-9.

111. L. Leiserowitz, *Acta Crystallogr.*, *B32*:775 (1976).

112. R. J. Gillespie and R. S. Nyholm, *Quart. Rev. Chem. Soc. London*, *11*:339 (1957).

113. R. J. Gillespie, *J. Amer. Chem. Soc.*, *82*:5978 (1960).

114. R. J. Gillespie, *Can. J. Chem.*, *39*:318 (1961).

115. G. A. Jeffrey and G. S. Parry, *Nature (London)*, *169*:1105 (1952).

116. G. Gurskaya, *The Molecular Structure of Amino Acids*, Consultants Bureau, New York, 1968.

117. J. A. Kanters, J. Kroon, A. F. Peerdeman, and J. C. Schoone, *Tetrahedron*, *23*:4027 (1967).

118. J. D. Dunitz and P. Strickler, in *Structural Chemistry and Molecular Biology* (A. Rich and N. Davidson, eds.), Freeman, San Francisco, 1968, pp. 595-602.

119. L. Leiserowitz and D. vor der Brück, *Crystal Struct. Commun.*, *4*:647 (1975).

120. M. D. Newton and G. A. Jeffrey, *J. Amer. Chem Soc.*, *99*:2413 (1977).

121. N. Mori, Y. Asano, T. Irie, and Y. Tsuzuki, *Bull. Chem. Soc. Jap.*, *42*:482 (1969).

122. B. P. van Eijck, *Rec. Trav. Chem. Pas Bas*, *85*:1129 (1966).

123. G. K. Ambady, Ph.D. dissertation, University of Madras, Madras, India, 1967.

124. W. Klyne and V. Prelog, *Experientia*, *16*:521 (1960).

125. K. Tatsumi, Y. Kishimoto, and C. Hignite, *Arch. Biochem. Biophys.*, *165*:656 (1974).

126. W. Voelter, E. Bayer, G. Barth, E. Bunnenberg, and C. Djerassi, *Chem. Ber.*, *102*:2003 (1969).

127. L. Katzin, *Inorg. Chem.*, *7*:1183 (1968).

128. J. Bolard and G. Chottard, *Inorg. Nucl. Chem. Lett.*, *10*:991 (1974).

129. R. E. Marsh and J. Donohue, in *Advances in Protein Chemistry* (C. B. Anfinsen, Jr., M. L. Anson, J. T. Edsall, and F. M. Richards, eds.), Vol. 22, Academic Press, New York, 1967, pp. 235-256.

130. N. S. Poonia and A. V. Bajaj, *Chem. Rev.*, *79*:389 (1979).

131. J. E. Wertz and O. Jardetzky, *Archiv. Biochem. Biophys.*, *65*:569 (1956).

132. L. E. Erickson and R. A. Alberty, *J. Phys. Chem.*, *66*:1702 (1962).

133. A. Corsini, Q. Fernando, and H. Freiser, *Inorg. Chem.*, *2*:224 (1963).

134. L. E. Bennett, R. H. Lane, M. Gilroy, F. A. Sedor, and J. P. Bennett, Jr., *Inorg. Chem.*, *12*:1200 (1973).

135. D. S. Everhart, M. M. McKown, and R. F. Evilia, *J. Coord. Chem.*, *9*:185 (1979).

136. D. C. Bradley, R. C. Mehrota, and D. P. Gaur, *Metal Alkoxides*, Academic Press, New York, 1978.

137. R. R. Reeder and P. H. Rieger, *Inorg. Chem.*, *10*:1258 (1971).

138. R. D. Gillard and M. G. Price, *J. Chem. Soc.*, *A*, 1813 (1969).

139. A. Tatehata, *Inorg. Chem.*, *15*:2086 (1976).

140. R. D. Gillard and M. G. Price, *J. Chem. Soc.*, *A*, 2274 (1971).

141. G. L. Robbins and R. E. Tapscott, *Inorg. Chem.*, *15*:154 (1976).

142. I. Feldman, C. A. North, and H. B. Hunter, *J. Phys. Chem.*, *64*:1224 (1960).

143. E. Mentasti, *Inorg. Chem.*, *18*:1512 (1979).

144. S. Fronaeus, *Komplexsystem hos Koppar*, Dissertation, Lund, 1948.

145. S. Arhland, *Acta Chem. Scand.*, *7*:485 (1953).

146. R. Portanova, A. Cassol, L. Magon, G. Tomat, *J. Inorg. Nucl. Chem.*, *32*:221 (1970).

147. R. Portanova, G. Tomat, L. Magon, and A. Cassol, *J. Inorg. Nucl. Chem.*, *34*:1768 (1972).

148. A. Sonesson, *Acta Chem. Scand.*, *13*:998 (1959).

149. S. Katz, J. A. Uhrmacher, and R. G. Shinaberry, *J. Coord. Chem.*, *7*:149 (1978).

150. K. B. Dillon and F. J. C. Rossotti, *Chem. Commun.*, 768 (1966).

151. K. B. Dillon and F. J. C. Rossotti, *J. Chem. Soc. Dalton*, 1005 (1973).

152. L. Magon, A. Bismondo, L. Maresca, G. Tomat, and R. Portanova, *J. Inorg. Nucl. Chem.*, *35*:4237 (1973).

153. L. Magon, G. Tomat, A. Bismondo, R. Portanova, and U. Croato, *Gazz. Chim. Ital.*, *104*:967 (1974).

154. I. Feldman, J. R. Hovill, and W. F. Neuman, *J. Amer. Chem. Soc.*, *76*:4726 (1954).

155. T. D. Smith, *J. Chem. Soc.*, 2145 (1965).

156. R. C. Srivastava and T. D. Smith, *J. Chem. Soc.*, A, 2192 (1968).

157. T. W. Gilbert, L. Newman, and P. Klotz, *Anal. Chem.*, *40*:2123 (1968).

158. H. M. N. H. Irving and W. R. Tomlinson, *Chem. Commun.*, 497 (1968).

159. A. Adin, P. Klotz, and L. Newman, *Inorg. Chem.*, *9*:2499 (1970).

160. R. C. Srivastava, T. D. Smith, J. F. Boas, T. Lund, J. H. Price, and J. R. Pilbrow, *J. Chem. Soc.*, A, 2538 (1971).

161. C. G. Ramsay and B. Tamhina, *Talanta*, *22*:437 (1975).

162. W. P. Pijper, *Acta Crystallogr.*, *B27*:344 (1971).

163. E. J. Gabe and M. R. Taylor, *Acta Crystallogr.*, *21*:418 (1966).

164. R. H. Colton and D. E. Henn, *Acta Crystallogr.*, *18*:820 (1965).

165. R. F. Mayers, E. T. Keve, and A. C. Skapski, *J. Chem. Soc.*, A, 2258 (1968).

166. L. Golic and J. C. Speakman, *J. Chem. Soc.*, 2521 (1965).

167. C. K. Prout, R. A. Armstrong, J. R. Carruthers, J. G. Forrest, P. Murray-Rust, and F. J. C. Rossotti, *J. Chem. Soc.*, A, 2791 (1968).

168. A. J. Fischinger and L. E. Webb, *Chem. Commun.*, 407 (1969).

169. A. J. Fischinger, Ph.D. dissertation, Illinois Institute of Technology, 1970.

170. C.-S. Chung, *Inorg. Chem.*, *18*:1318 (1979).

171. Commission on the Nomenclature of Inorganic Chemistry, IUPAC, *Inorg. Chem.*, *9*:1 (1970).

172. I. Grenthe, *Acta Chem. Scand.*, *23*:1253 (1969).

173. I. Grenthe, *Acta Chem. Scand.*, *25*:3721 (1971).

174. L. M. Dikareva, A. S. Antsyshkina, M. A. Porai-Koshits, V. N. Ostrikova, I. V. Arkhangelskii, and A. Z. Zamanov, *Dokl. Akad. Nauk Az. SSR*, *34*:41 (1978).

175. I. Grenthe, *Acta Chem. Scand.*, *23*:1752 (1969).

176. I. Grenthe, *Acta Chem. Scand.*, *25*:3347 (1971).

177. I. Grenthe, *Acta Chem. Scand.*, *26*:1479 (1972).

178. E. L. Muetterties and C. M. Wright, *Quart. Rev. (London)*, *21*: 109 (1967).

179. J. L. Hoard and J. V. Silverton, *Inorg. Chem.*, *2*:235 (1963).

180. D. G. Blight and D. L. Kepert, *Inorg. Chem.*, *11*:1556 (1972).

181. S. J. Lippard and B. J. Russ, *Inorg. Chem.*, *7*:1686 (1968).

182. R. V. Parrish, *Coord. Chem. Rev.*, *1*:439 (1966).

183. B. E. Robertson, *Inorg. Chem.*, *16*:2735 (1977).

184. B. F. Mentzen and H. Sautereau, *Inorg. Chim. Acta*, *35*:L347 (1979).

185. E. J. Corey and J. C. Bailar, Jr., *J. Amer. Chem. Soc.*, *81*:2620 (1959).

186. E. Larsen and I. Olsen, *Acta Chem. Scand.*, *18*:1025 (1964).

187. D. C. Bhatnager and S. Kirschner, *Inorg. Chem.*, *3*:1256 (1964).

188. F. Cariati, F. Morazzoni, G. M. Zanderighi, G. Marcotrigiano, and G. C. Pellacani, *Inorg. Chim. Acta*, *21*:133 (1977).

189. M. Ahlgrén, R. Hämäläinen, and A. Pajunen, *Finn. Chem. Lett.*, 3 (1977).

190. M. Ahlgrén and U. Turpeinen, *Finn. Chem. Lett.*, 129 (1977).

191. M. Ahlgrén and R. Hämäläinen, *Finn. Chem. Lett.*, 239 (1977).

192. K. D. Singh, S. C. Jain, T. D. Sakore, and A. B. Biswas, *Acta Crystallogr.*, *B31*:990 (1975).

193. E. B. Kipp and R. A. Haines, *Inorg. Chem.*, *11*:271 (1972).

194. R. A. Haines and D. W. Bailey, *Inorg. Chem.*, *14*:1310 (1975).

195. M. Krumpolc, B. G. DeBoer, and J. Rocek, *J. Amer. Chem. Soc.*, *100*:145 (1978).

196. N. D. Chasteen, R. L. Belford, and I. C. Paul, *Inorg. Chem.*, *8*: 408 (1969).

197. R. E. Tapscott, *Inorg. Chim. Acta*, *10*:183 (1974).

198. N. D. Chasteen and M. W. Hanna, *J. Phys. Chem.*, *76*:3951 (1972).

199. H. C. Freeman, M. R. Snow, I. Nitta, and K. Tomita, *Acta Crystallogr.*, *17*:1463 (1964).

200. R. D. Gillard, R. Mason, N. C. Payne, and G. B. Robertson, *Chem. Commun.*, 155 (1966).

201. T. S. Cameron and M. Duffin, *Crystal Struct. Commun.*, *6*:453 (1977).

202. C. Sterling, *J. Inorg. Nucl. Chem.*, *29*:1211 (1967).

203. P. B. Chakrawarti and H. N. Sharma, *Sci. Cult.*, *44*:464 (1978).

204. M. Vyas, T. D. Sakore, and A. B. Biswas, *Acta Crystallogr.*, *B34*:1345 (1978).

205. A. Pajunen and E. Näsäkkälä, *Finn. Chem. Lett.*, 189 (1977).

206. J. A. Kanters and J. Kroon, *Science*, *260*:600 (1976).

207. J. A. Kanters, private communication.

208. A. B. Ablov, G. A. Popovich, G. I. Dimitrova, G. A. Kiosse, I. F. Burshtein, T. I. Malinovskii, and B. M. Schedrin, *Dokl. Akad. Nauk SSSR*, *229*:611 (1976).

209. Z. Warnke and E. Kwiatowski, *Roczniki Chem.*, *47*:467 (1973).

210. W. Van Havere and A. T. H. Lenstra, *Acta Crystallogr.*, *B36*:1483 (1980).

211. F. E. Cole, *Amer. Crystallogr. Ass. Abstr. Papers, Summer Meeting*, Paper 009, 188 (1973).

212. J. S. Mariano and V. M. S. Gil, *Mol. Phys.*, *17*:313 (1969).

213. A. Karipedes, *Inorg. Chem.*, *18*:3034 (1979).

214. H. Hoffman and U. Nickel, *Ber. Bunseng. Phys. Chem.*, *72*:1096 (1968).

215. J. E. Berg, S. Brandänge, L. Lindblom, and P.-E. Werner, *Acta Chem. Scand.*, *A31*:325 (1977).

216. Yu. K. Tselinskii, B. V. Pestryakov, A. I. Olivson, and T. M. Pekhtereva, *Zh. Neorg. Khim.*, *24*:1243 (1979).

217. A. Tatehata, *Inorg. Chem.*, *17*:725 (1978).

218. A. Karipedes and A. T. Reed, *Inorg. Chem.*, *15*:44 (1976).

219. W. Van Havere and A. T. H. Lenstra, *Bull. Soc. Chim. Belg.*, *87*: 419 (1978).

220. W. Van Havere and A. T. H. Lenstra, *Bull. Soc. Chim. Belg.*, *89*: 427 (1980).

221. A. T. H. Lenstra and W. Van Havere, *Acta Crystallogr.*, *B36*:156 (1980).

222. R. A. Mariezcurrena and S. E. Rasmussen, *Acta Crystallogr.*, *B29*:1035 (1973).

223. D. H. Busch and J. C. Bailar, Jr., *J. Amer. Chem. Soc.*, *76*:5352 (1954).

224. B. F. Hoskins and C. D. Pannen, *Inorg. Nucl. Chem. Lett.*, *11*: 405 (1975).

225. A. T. H. Lenstra and W. Van de Mieroop, *Bull. Soc. Chim. Belg.*, *85*:721 (1976).

226. A. T. Reed and A. Karipedes, *Acta Crystallogr.*, *B32*:2085 (1976).

227. L. Kryger and S. E. Rasmussen, *Acta Chem. Scand.*, *26*:2349 (1972).

228. T. Doyne and R. Pepinsky, *Acta Crystallogr.*, *10*:438 (1957).

229. C. I. Brändén and B. O. Söderberg, *Acta Chem. Scand.*, *20*:730 (1966).

230. M. A. Porai-Koshits, L. A. Aslanov, G. V. Ivanova, and T. N. Polynova, *Zh. Strukt. Khim.*, *9*:475 (1968).

231. J.-E. Berg and P. E. Werner, *Z. Kristallogr.*, *145*:310 (1977).

232. C. A. Evans, R. Guevremont, and D. L. Rabenstein, in *Metal Ions in Biological Systems* (H. Sigel, ed.), Vol. 9, Dekker, New York, 1979, pp. 41-75.

233. W. L. Alworth, *Stereochemistry and Its Application in Biochemistry*, Wiley-Interscience, New York, 1972, pp. 183-186.

234. J. F. Speyer and S. R. Dickman, *J. Biol. Chem.*, *220*:193 (1956).

235. J. P. Glusker, *J. Mol. Biol.*, *38*:149 (1968).

236. J. Strouse, *J. Amer. Chem. Soc.*, *99*:572 (1977).

237. R. H. Dunhill, J. R. Pilbrow, and T. D. Smith, *J. Chem. Phys.*, *45*:1474 (1966).

238. D. van der Helm, J. P. Glusker, C. K. Johnson, J. A. Minkin, N. E. Burow, and A. L. Patterson, *Acta Crystallogr.*, *B24*:578 (1968).

239. E. J. Gabe, J. P. Glusker, J. A. Minkin, and A. L. Patterson, *Acta Crystallogr.*, *22*:366 (1967).

240. J. P. Glusker, D. van der Helm, W. E. Love, M. L. Dornberg, J. A. Minkin, C. K. Johnson, and A. L. Patterson, *Acta Crystallogr.*, *19*:561 (1965).

241. C. E. Nordman, A. S. Weldon, and A. L. Patterson, *Acta Crystallogr.*, *13*:414 (1960).

242. C. K. Johnson, *Acta Crystallogr.*, *18*:1004 (1965).

243. H. L. Carrell and J. P. Glusker, *Acta Crystallogr.*, *B29*:638 (1973).

244. J. P. Glusker and H. L. Carrell, *J. Mol. Struct.*, *15*:151 (1973).

245. J. Strouse, S. W. Layten, and C. E. Strouse, *J. Amer. Chem. Soc.*, *99*:562 (1977).

246. B. Sheldrick, *Acta Crystallogr.*, *B30*:2056 (1974).

247. (a) J. P. Glusker, *Acc. Chem. Res.*, *13*:345 (1980); W. C. Stallings, C. T. Monti, R. C. Job, and J. P. Glusker, private communication. (b) I. F. Burshtein, G. A. Kiosse, A. B. Ablov, T. I. Malinovskii, B. M. Schedrin, and N. V. Rannev, *Dokl. Akad. Nauk SSSR*, *239*:90 (1978).

248. D. Mastropaolo, D. A. Powers, J. A. Potenza, and H. J. Schugar, *Inorg. Chem.*, *15*:1444 (1976).

249. H. L. Carrell and J. P. Glusker, *Acta Crystallogr.*, *B29*:674 (1973).

250. H. L. Carrell, J. P. Glusker, J. J. Villafranca, A. S. Mildvan, R. J. Dummel, and E. Kun, *Science*, *170*:1412 (1970).

251. G. A. Kiosse, N. I. Golovastikov, and N. V. Belov, *Dokl. Akad. Nauk SSSR*, *155*:545 (1964).

252. Y. Kushi, M. Kuramoto, and H. Yoneda, *Chem. Lett.*, 135 (1976).

253. Y. Kushi, M. Kuramoto, and H. Yoneda, *Chem. Lett.*, 339 (1976).

254. T. Tada, Y. Kushi, and H. Yoneda, *Chem. Lett.*, 379 (1977).

255. A. J. J. Sprenkels, *Koninkl. Ned. Akad. Wetenschap. Proc.*, *Ser. B*, *59*:221 (1956).

256. H. Hinazumi and T. Mitsui, *Acta Crystallogr.*, *B28*:3299 (1972).

257. M. I. Kay, *Ferroelectrics*, *19*:159 (1978).

258. G. K. Ambady and G. Kartha, *Acta Crystallogr.*, *B24*:1540 (1968).

259. F. Mazzi, F. Jona, and R. Pepinsky, *Z. Kristallogr.*, *108*:359 (1957).

260. C. A. Beevers and W. Hughes, *Proc. Roy. Soc.*, *Ser. A*, *177*:251 (1941).

261. S. Mitani, S. Fukui, I. Shibuya, Y. Shiozaki, K. Toyoda, and R. Pepinsky, *Ferroelectrics*, *8*:477 (1974).

262. R. Sadanaga, *Acta Crystallogr.*, *3*:416 (1950).

263. J. Kroon, A. F. Peerdeman, and J. M. Bijvoet, *Acta Crystallogr.*, *19*:293 (1965).

264. J. Kroon and J. A. Kanters, *Acta Crystallogr.*, *B28*:714 (1972).

265. M. Currie and J. C. Speakman, *J. Chem. Soc.*, *Perkin II*, 1549 (1975).

266. A. J. Van Bommel, *Koninkl. Ned. Akad. Wetenschap., Proc., Ser. B.*, *56*:268 (1953).

267. L. K. Templeton and D. H. Templeton, *Acta Crystallogr.*, *A34*: 368 (1978).

268. G. K. Ambady, *Acta Crystallogr.*, *B24*:1548 (1968).

269. M. A. Ivanov and A. L. Kosoy, *Acta Crystallogr.*, *B31*:2843 (1975).

270. L. Johansson and B. Nordén, *Inorg. Chim. Acta*, *29*:189 (1978).

271. R. E. Tapscott, J. D. Mather, and T. F. Them, *Coord. Chem. Rev.*, *29*:87 (1979).

272. B. Nordén, *Inorg. Nucl. Chem. Lett.*, *13*:355 (1977).

273. J. C. Bailar, Jr., H. B. Jonassen, and A. D. Gott, *J. Amer. Chem. Soc.*, *74*:3131 (1952).

274. R. A. Haines, E. B. Kipp, and M. Reimer, *Inorg. Chem.*, *13*:2473 (1974).

275. A. Tatehata, *Chem. Lett.*, 561 (1972).

276. A. Tatehata, *Inorg. Chem.*, *16*:1247 (1977).

277. R. D. Gillard and M. G. Price, *Chem. Commun.*, 67 (1969).

278. R. E. Tapscott and R. L. Belford, *Inorg. Chem.*, *6*:735 (1967).

279. M. E. McIlwain, R. E. Tapscott, and W. F. Coleman, *J. Mag. Reson.*, *26*:35 (1977).

280. R. E. Tapscott, L. D. Hansen, and E. A. Lewis, *J. Inorg. Nucl. Chem.*, *37*:2517 (1975).

281. L. D. Pettit and J. M. Swash, *J. Chem. Soc. Dalton*, 286 (1978).

282. S. K. Hahs and R. E. Tapscott, *Chem. Commun.*, 791 (1974).

283. R. M. Holland and R. E. Tapscott, *J. Coord. Chem.*, *11*:17 (1981).

284. S. Kaizaki, J. Hidaka, and Y. Shimura, *Bull. Chem. Soc. Jap.* *42*:988 (1969).

285. D. Marcovich and R. E. Tapscott, *J. Amer. Chem. Soc.*, *102*:5712 (1980).

286. J. F. Boas, R. H. Dunhill, J. R. Pilbrow, R. C. Srivastava, and T. D. Smith, *J. Chem. Soc.*, *A*, 94 (1969).

287. N. D. Chasteen and R. L. Belford, *Inorg. Chem.*, *9*:169 (1970).

288. L. Johansson and R. Larsson, *Chem. Scripta*, *5*:67 (1975).

289. (a) R. B. Ortega, R. E. Tapscott, and C. F. Campana, *Inorg. Chem.*, *21*, in press. (b) R. B. Ortega, R. E. Tapscott, and C. F. Campana, *Inorg. Chem.*, *21*:672 (1982).

290. R. E. Tapscott, R. L. Belford, and I. C. Paul, *Inorg. Chem.*, *7*:356 (1968).

291. R. B. Ortega, C. F. Campana, and R. E. Tapscott, *Acta Crystallogr.*, *B36*:1786 (1980).

292. S. K. Hahs, R. B. Ortega, R. E. Tapscott, C. F. Campana, and B. Morosin, *Inorg. Chem.*, *21*:664 (1982).

293. J. I. Brauman and L. K. Blair, *J. Amer. Chem. Soc.*, *92*:5986 (1970).

294. J. G. Forrest and C. K. Prout, *J. Chem. Soc.*, *A*, 1312 (1967).

295. G. A. Kiosse, N. I. Golovastikov, A. V. Ablov, and N. V. Belov, *Dokl. Akad. Nauk SSSR*, *177*:329 (1967).

296. B. Kamenar, D. Grdenić, and C. K. Prout, *Acta Crystallogr.*, *B26*:181 (1970).

297. H.-C. Mu, *K´o Hseuh T´ung Pao*, *17*:502 (1966).

298. M. E. Gress and R. A. Jacobson, *Inorg. Chim. Acta*, *8*:209 (1974).

299. K. Matsumoto, S. Ooi, M. Sakuma, and H. Kuroya, *Bull. Chem. Soc. Jap.*, *49*:2129 (1976).

300. K. Yokoho, K. Matsumoto, S. Ooi, and H. Kuroya, *Bull. Chem. Soc. Jap.*, *49*:1864 (1976).

301. Y. Nakayama, K. Matsumoto, S. Ooi, and H. Kuroya, *Bull. Chem. Soc. Jap.*, *50*:2304 (1977).

302. A. Zalkin, D. H. Templeton, and T. Ueki, *Inorg. Chem.*, *12*:1641 (1973).

303. K. Matsumoto, M. Yonezawa, H. Kuroya, H. Kawaguchi, and S. Kawaguchi, *Bull. Chem. Soc. Jap.*, *43*:1269 (1970).

304. K. Matsumoto, private communication of unpublished results.

305. K. Matsumoto, H. Kawaguchi, H. Kuroya, and S. Kawaguchi, *Bull. Chem. Soc. Jap.*, *46*:2424 (1973).

306. R. L. Belford, R. J. Missavage, I. C. Paul, N. D. Chasteen, W. E. Hatfield, and J. F. Villa, *Chem. Commun.*, 508 (1971).

307. R. J. Missavage, R. L. Belford, and I. C. Paul, *J. Coord. Chem.*, *2*:145 (1972).

308. C. K. Prout, J. R. Carruthers, and F. J. C. Rossotti, *J. Chem. Soc.*, *A*, 3336 (1971).

309. G. A. Jeffrey and R. D. Rosenstein, *Advan. Carbohydr. Chem.*, *19*:7 (1964).

310. G. Strahs, *Advan. Carbohydr. Chem. Biochem.*, *25*:53 (1970).

311. G. A. Jeffrey and M. Sundaralingam, *Advan. Carbohydr. Chem. Biochem.*, *30*:446 (1974).

312. G. A. Jeffrey and M. Sundaralingam, *Advan. Carbohydr. Chem. Biochem.*, *31*:347 (1975).

313. G. A. Jeffrey and M. Sundaralingam, *Advan. Carbohydr. Chem. Biochem.*, *32*:353 (1976).

314. G. A. Jeffrey and M. Sundaralingam, *Advan. Carbohydr. Chem. Biochem.*, *34*:345 (1977).

315. T. Taga, Y. Kurada, and M. Ohashi, *Bull. Chem. Soc. Jap.*, *51*: 2278 (1978).

316. J. E. Powell, J. L. Farrell, and S. Kulprathipanja, *Inorg. Chem.*, *14*:786 (1975).

317. E. J. Meehan, H. Einspahr, and C. E. Bugg, *Acta Crystallogr.*, *B35*:828 (1979).

318. T. Taga, M. Ohashi, and K. Osaki, *Bull. Chem. Soc. Jap.*, *51*: 1697 (1978).

319. D. T. Sawyer and J. R. Brannan, *Anal. Chem.*, *38*:192 (1966).

320. A. Mostad and E. Rosenqvist, *Acta Chem. Scand.*, *25*:145 (1971).

321. J. H. Miller, J. E. Powell, R. A. Jacobson, and S. Kulprathi-panja, *Inorg. Chim. Acta*, *18*:25 (1976).

322. C. D. Littleton, *Acta Crystallogr.*, *6*:775 (1953).

323. G. A. Jeffrey and E. J. Fasiska, *Carbohyd. Res.*, *21*:187 (1972).

324. N. C. Panagiotopoulos, G. A. Jeffrey, S. J. LaPlaca, and W. C. Hamilton, *Acta Crystallogr.*, *B30*:1421 (1974).

325. R. Pepinsky, *Phys. Rev.*, *61*:726 (1942).

326. T. Lis, *Acta Crystallogr.*, *B35*:1699 (1979).

327. K. D. Magers, C. G. Smith, D. T. Sawyer, *Inorg. Chem.*, *17*:515 (1978).

328. D. T. Sawyer and M. E. Bodini, *J. Amer. Chem. Soc.*, *97*:6588 (1975).

329. G. D. Smith, A. Fitzgerald, C. N. Caughlan, K. A. Kerr, and J. P. Ashmore, *Acta Crystallogr.*, *B30*:1760 (1974).

330. W. J. Cook and C. E. Bugg, *Acta Crystallogr.*, *B29*:215 (1973).

331. T. Taga, Y. Kuroda, and K. Osaki, *Bull. Chem. Soc. Jap.*, *50*: 3079 (1977).

332. T. Taga and K. Osaki, *Bull. Chem. Soc. Jap.*, *49*:1517 (1976).

333. S. Furberg and S. Helland, *Acta Chem. Scand.*, *16*:2373 (1962).

334. R. Norrestam, P. E. Werner, and M. von Glehn, *Acta Chem. Scand.*, *22*:1395 (1968).

335. P.-E. Werner, R. Norrestam, and O. Rönnquist, *Acta Crystallogr.*, *B25*:714 (1969).

336. C. O. Haagensen and J. Danielson, *Acta Chem. Scand.*, *18*:581 (1964).

337. H. C. Freeman, in *Advances in Protein Chemistry* (C. B. Anfinsen, Jr., M. L. Anson, J. T. Edsall, and F. M. Richards, eds.), Vol. 22, Academic Press, New York, 1967, pp. 257-424.

338. C. J. Hawkins, *Absolute Configuration of Metal Complexes,* Wiley-Interscience, New York, 1971, pp. 94-96.

339. I. L. Karle and J. Karle, *J. Amer. Chem. Soc.*, *88*:24 (1966).

340. G. Smith, J. Whitnall, and C. H. L. Kennard, *Crystal Struct. Commun.*, *5*:749 (1976).

341. G. Smith, C. H. L. Kennard, and A. H. White, *J. Chem. Soc., Perkin II*, 791 (1976).

342. G. Smith, C. H. L. Kennard, A. H. White, and P. G. Hodgson, *Acta Crystallogr.*, *B33*:2922 (1977).

343. (a) G. Smith, C. H. L. Kennard, and A. H. White, *Crystal Struct. Commun.*, *6*:49 (1977). (b) G. Smith, C. H. L. Kennard, and A. H. White, *Acta Crystallogr.*, *B34*:2885 (1978).

344. G. Smith, C. H. L. Kennard, and A. H. White, *Aust. J. Chem.*, *29*:2727 (1976).

345. A. Wagner and G. Malmros, *Acta Crystallogr.*, *B35*:2222 (1979).

346. A. Wagner and G. Malmros, *Acta Crystallogr.*, *B35*:2220 (1979).

347. G. Davey and S. H. Whitlow, *J. Crystallogr. Mol. Struct.*, *3*: 193 (1973).

348. R. A. Moran and G. F. Richards, *Acta Crystallogr.*, *B29*:2770 (1973).

349. M. Senma, T. Taga, and K. Osaki, *Chem. Lett.*, 1415 (1974).

350. T. Srikrishnan and R. Parthasarathy, *Acta Crystallogr.*, *B34*: 1363 (1978).

351. P. Swaminathan, J. McAlister, and M. Sundaralingam, *Acta Crystallogr.*, *B36*:878 (1980).

352. C. M. Weeks, D. C. Rohrer, and W. L. Duax, *J. Steroid Biochem.*, *7*:545 (1976).

353. K. Nimgirawath, V. J. James, and J. D. Stevens, *Crystal Struct. Commun.*, *4*:617 (1975).

354. F. Mo and B. K. Siverston, *Acta Crystallogr.*, *B27*:115 (1971).

355. K. Morikawa, K. Torii, Y. Itaka, and M. Tsuboi, *Acta Crystallogr.*, *B31*:1004 (1975).

356. R. Roques, J. Bellan, J. C. Rossi, J. P. Declercq, and G. Germain, *Acta Crystallogr.*, *B35*:2467 (1979).

357. E. Hogfeldt and S. Nilsson, *Acta Chem. Scand.*, *A33*:559 (1979).

358. M. Ōki and M. Hirota, *Bull. Chem. Soc. Jap.*, *33*:119 (1960).

359. M. Ōki and M. Hirota, *Bull. Chem. Soc. Jap.*, *34*:374 (1961).

360. M. Ōki and M. Hirota, *Bull. Chem. Soc. Jap.*, *34*:378 (1961).

361. C. K. Prout, C. Walker, and F. J. C. Rossotti, *J. Chem. Soc.*, A, 556 (1971).

362. C. K. Prout, M. J. Barrow, and F. J. C. Rossotti, *J. Chem. Soc.*, A, 3326 (1971).

363. C. K. Prout, G. B. Allison, and F. J. C. Rossotti, *J. Chem. Soc.*, A, 3331 (1971).

364. C. K. Prout, J. R. Carruthers, and F. J. C. Rossotti, *J. Chem. Soc.*, A, 554 (1971).

365. C. K. Prout, R. M. Dunn, O. J. R. Hodder, and F. J. C. Rossotti, *J. Chem. Soc.*, A, 1986 (1971).

366. C. K. Prout, P. J. Grove, B. D. Harridine, and F. J. C. Rossotti, *Acta Crystallogr.*, *B31*:2047 (1975).

367. J. R. Carruthers, K. Prout, and F. J. C. Rossotti, *Acta Crystallogr.*, *B31*:2044 (1975).

368. G. Reck and W. Jähnig, *J. Prakt. Chem.*, *321*:549 (1979).

369. C. V. Goebel and R. J. Doedens, *Inorg. Chem.*, *10*:2607 (1971).

370. C.-E. Boman, H. Herbertsson, and Å. Oskarsson, *Acta Crystallogr.*, *B30*:378 (1974).

371. H. Herbertsson and C.-E. Boman, *Acta Chem. Scand.*, *27*:2234 (1973).

372. J. Albertsson, I. Grenthe, and H. Herbertsson, *Acta Crystallogr.*, *B29*:1855 (1973).

373. H. Herbertsson, *Acta Crystallogr.*, *B32*:2381 (1976).

374. J. Albertsson, I. Grenthe, and H. Herbertsson, *Acta Crystallogr.*, *B29*:2839 (1973).

375. J. Albertsson and I. Grenthe, *Acta Crystallogr.*, *B29*:2751 (1973).

376. C.-E. Boman, *Acta Crystallogr.*, *B33*:834 (1977).

377. C.-E. Boman, Ph.D. dissertation, University of Lund, 1977.

378. C.-E. Boman, *Acta Crystallogr.*, *B33*:1529 (1977).

379. C.-E. Boman, *Acta Crystallogr.*, *B33*:838 (1977).

380. V. A. Uchtman and R. P. Oertel, *J. Amer. Chem. Soc.*, *95*:1802 (1973).

381. G. Bombieri, R. Graziani, and E. Forsellini, *Inorg. Nucl. Chem. Lett.*, *9*:551 (1973).

382. G. Bombieri, U. Croatto, R. Graziani, E. Forsellini, and L. Magon, *Acta Crystallogr.*, *B30*:407 (1974).

383. S. H. Whitlow and G. Davey, *J. Chem. Soc.*, 1228 (1975).

384. J. Albertsson, *Acta Chem. Scand.*, *22*:1563 (1968).

385. J. Albertsson, *Acta Chem. Scand.*, *24*:3527 (1970).

386. J. Albertsson and I. Elding, *Acta Chem. Scand.*, *A31*:21 (1977).

387. N. G. Vannerberg and J. Albertsson, *Acta Chem. Scand.*, *19*:1760 (1965).

388. I. Elding, *Acta Chem. Scand.*, *A31*:75 (1977).

389. J. Albertsson and I. Elding, *Acta Crystallogr.*, *B32*:3066 (1976).

390. I. Elding, *Acta Chem. Scand.*, *A30*:649 (1976).

391. S. Furberg, H. Hammer, and A. Mostad, *Acta Chem. Scand.*, *17*: 2444 (1963).

392. G. E. Gurr, *Acta Crystallogr.*, *16*:690 (1963).

393. L. DeLucas, C. E. Bugg, A. Terzis, and R. Rivest, *Carbohyd. Res.*, *41*:19 (1975).

394. S. Thanomkul, J. A. Hjortas, and H. Sorum, *Acta Crystallogr.*, *B32*:920 (1976).

395. S. E. B. Gould, R. O. Gould, D. A. Ress, and W. E. Scott, *J. Chem. Soc.*, *Perkin II*, 237 (1975).

396. S. E. B. Gould, R. O. Gould, D. A. Rees, and A. W. Wight, *J. Chem. Soc.*, *Perkin II*, 392 (1976).

397. T. Anthonsen, B. Larsen, and O. Smidsrød, *Acta Chem. Scand.*, *26*:2988 (1972).

398. T. Anthonsen, B. Larsen, and O. Smidsrød, *Acta Chem. Scand.*, *27*:2670 (1973).

399. J. P. Glusker, J. A. Minkin, and C. A. Casciato, *Acta Crystallogr.*, *B27*:1284 (1971).

400. J. P. Glusker, J. A. Minkin, and F. B. Soule, *Acta Crystallogr.*, *B27*:2499 (1972).

401. H. M. Berman, H. L. Carrell, and J. P. Glusker, *Acta Crystallogr.*, *B29*:1163 (1973).

402. A. J. Morris, A. J. Geddes, B. Sheldrick, and D. Akrigg, *Crystal Struct. Commun.*, *8*:237 (1979).

403. A. J. Morris, A. J. Geddes, B. Sheldrick, and D. Akrigg, *Crystal Struct. Commun.*, *8*:231 (1979).

404. A. A. Balchin and C. H. Carlisle, *Acta Crystallogr.*, *19*:103 (1965).

405. R. C. Elder, L. R. Florian, R. E. Lake, and A. M. Yacynych, *Inorg. Chem.*, *12*:2690 (1973).

406. S. Jeannin, Y. Jeannin, and D. Lavigne, *J. Organometal. Chem.*, *40*:187 (1972).

407. I. Hansson, *Acta Chem. Scand.*, *22*:509 (1968).

408. B. V. DePamphilis, A. G. Jones, M. A. Davis, and A. Davison, *J. Amer. Chem. Soc.*, *100*:5570 (1978).

409. P. J. M. W. L. Birker, *Inorg. Chem.*, *18*:3502 (1979).

410. S. Paul, *Acta Crystallogr.*, *23*:490 (1967).

411. E. Martuscelli, L. Mazzarella, and A. Zagari, *Gazz. Chim. Ital.*, *103*:563 (1973).

412. M. G. Drew, D. A. Rice, and C. W. Timewell, *J. Chem. Soc. Dalton*, 144 (1975).

413. S. H. Whitlow, *Acta Crystallogr.*, *B31*:2531 (1975).

414. T. Malmborg and A. Oskarsson, *Acta Chem. Scand.*, *27*:2923 (1973).

415. J. Loub and J. Podlahová, *Inorg. Nucl. Chem. Lett.*, *7*:409 (1971).

416. V. M. S. Gil, *J. Inorg. Nucl. Chem.*, *42*:389 (1980).

417. Yu. K. Tselinskii, B. V. Pestryakov, and T. M. Pekhtereva, *Zh. Obshch. Khim.*, *50*:347 (1978).

418. B. F. Mentzen and H. Sautereau, *Acta Crystallogr.*, *B36*:2051 (1980).

419. J. P. Behr, J. M. Lehn, D. Moras, and J. C. Thierry, *J. Amer. Chem. Soc.*, *103*:701 (1981).

420. G. Smith, E. J. O'Reilly, C. H. L. Kennard, K. Stadnicka, and B. Oleksyn, *Inorg. Chim. Acta*, *47*:111 (1981).

421. T. Lis, *Acta Crystallogr.*, *B36*:701 (1980).

422. P. B. Moore, J. J. Pluth, J. A. Molin-Norris, D. A. Weinstein, and E. L. Compere, Jr., *Acta Crystallogr.*, *B36*:47 (1980).

423. H. Stoeckli-Evans, *Acta Crystallogr.*, *B36*:3150 (1980).

424. K. Kamiya, M. Takamoto, Y. Wada, M. Fujino, and M. Nishikawa, *J. Chem. Soc., Chem. Commun.*, 438 (1980).

425. W. C. Stallings, C. T. Monti, J. F. Belvedere, R. K. Preston, and J. P. Glusker, *Arch. Biochem. Biophys.*, *203*:65 (1980).

426. S. V. Kumar and L. M. Rao, *Acta Crystallogr.*, *B36*:1218 (1980).

427. G. Smith, C. H. L. Kennard, A. H. White, and P. G. Hodgson, *Acta Crystallogr.*, *B36*:992 (1980).

428. G. Smith, C. H. L. Kennard, and A. H. White, *Acta Crystallogr.*, *B37*:275 (1981).

429. P. Swaminathan, J. McAlister, and M. Sundaralingam, *Acta Crystallogr.*, *B36*:878 (1980).

Author Index

Numbers in parentheses are reference numbers and indicate that an author's work is referred to although no name is cited in the text. Italic numbers give the page on which the complete reference is listed.

Subject Index